PALGRAVE MACMILLAN TRANSNATIONAL HISTORY SERIES

Series Editors: Akira Iriye (Harvard University) and **Rana Mitter** (University of Oxford)

This distinguished series seeks to develop scholarship on the transnational connections of societies and peoples in the nineteenth and twentieth centuries; provide a forum in which work on transnational history from different periods, subjects, and regions of the world can be brought together in fruitful connection; and explore the theoretical and methodological links between transnational and other related approaches such as comparative history and world history.

Published by Palgrave Macmillan:

THE NATION, PSYCHOLOGY, AND INTERNATIONAL POLITICS, 1870–1919
By Glenda Sluga

COMPETING VISIONS OF WORLD ORDER: GLOBAL MOMENTS AND MOVEMENTS, 1880s–1930s
Edited by Sebastian Conrad and Dominic Sachsenmaier

PAN-ASIANISM AND JAPAN'S WAR, 1931–1945
By Eri Hotta

CHINESE IN BRITAIN, 1800–PRESENT: ECONOMY, TRANSNATIONALISM, IDENTITY
By Gregor Benton and Edmund Terence Gomez

1968 IN EUROPE: A HISTORY OF PROTEST AND ACTIVISM, 1956–1977
Edited by Martin Klimke and Joachim Scharloth

RECONSTRUCTING PATRIARCHY AFTER THE GREAT WAR: WOMEN, CHILDREN, AND POSTWAR RECONCILIATION BETWEEN NATIONS
By Erika Kuhlman

THE IDEA OF HUMANITY IN A GLOBAL ERA
By Bruce Mazlish

THE TRANSNATIONAL UNCONSCIOUS: ESSAYS IN PSYCHOANALYSIS AND TRANSNATIONALISM
Edited by Joy Damousi and Mariano Ben Plotkin

THE PALGRAVE DICTIONARY OF TRANSNATIONAL HISTORY
Edited by Akira Iriye and Pierre-Yves Saunier

TRANSNATIONAL LIVES: BIOGRAPHIES OF GLOBAL MODERNITY, 1700–PRESENT
Edited by Angela Woollacott, Desley Deacon, and Penny Russell

TRANSATLANTIC ANTI-CATHOLICISM: FRANCE AND THE UNITED STATES IN THE NINETEENTH CENTURY
By Timothy Verhoeven

COSMOPOLITAN THOUGHT ZONES: SOUTH ASIA AND THE GLOBAL CIRCULATION OF IDEAS
Edited by Kris Manjapra and Sugata Bose

IRISH TERRORISM IN THE ATLANTIC COMMUNITY, 1865–1922
By Jonathan Gantt

EUROPEANIZATION IN THE TWENTIETH CENTURY: HISTORICAL APPROACHES
Edited by Martin Conway and Klaus Kiran Patel

NEW PERSPECTIVES ON THE TRANSNATIONAL RIGHT
Edited by Martin Durham and Margaret Power

TELEGRAPHIC IMPERIALISM: CRISIS AND PANIC IN THE INDIAN EMPIRE, C. 1830–1920
By D. K. Lahiri-Choudhury

THE ESTABLISHMENT RESPONDS: POWER, POLITICS, AND PROTEST SINCE 1945
Edited by Kathrin Fahlenbrach, Martin Klimke, Joachim Scharloth, and Laura Wong

EXPLORING THE DECOLONIAL IMAGINARY: FOUR TRANSNATIONAL LIVES
By Patricia A. Schechter

RED GAS: RUSSIA AND THE ORIGINS OF EUROPEAN ENERGY DEPENDENCE
By Per Högselius

CHALLENGING GLOBAL CAPITALISM: LABOR MIGRATION, RADICAL STRUGGLE, AND URBAN CHANGE IN DETROIT AND TURIN
By Nicola Pizzolato

OTTOMANS IMAGINING JAPAN: EAST, MIDDLE EAST, AND NON-WESTERN MODERNITY AT THE TURN OF THE TWENTIETH CENTURY
By Renée Worringer

SCIENCE, GENDER, AND INTERNATIONALISM: WOMEN'S ACADEMIC NETWORKS, 1917–1955
By Christine von Oertzen; Translated by Kate Sturge

Science, Gender, and Internationalism

Women's Academic Networks, 1917–1955

Christine von Oertzen

Translated by Kate Sturge

palgrave
macmillan

SCIENCE, GENDER, AND INTERNATIONALISM

The translation of this work was funded by Geisteswissenschaften International—Translation Funding for Humanities and Social Sciences from Germany, a joint initiative of the Fritz Thyssen Foundation, the German Federal Foreign Office, the collecting society VG Wort, and the Börsenverein des Deutschen Buchhandels (German Publishers & Booksellers Association).

First published in Germany as *Strategie Verständigung: Zur transnationalen Vernetzung von Akademikerinnen, 1917–1955* by Wallstein Publishing Ltd.

First published in English in 2014 by
PALGRAVE MACMILLAN®
in the United States—a division of St. Martin's Press LLC,
175 Fifth Avenue, New York, NY 10010.

Where this book is distributed in the UK, Europe and the rest of the world, this is by Palgrave Macmillan, a division of Macmillan Publishers Limited, registered in England, company number 785998, of Houndmills, Basingstoke, Hampshire RG21 6XS.

Palgrave Macmillan is the global academic imprint of the above companies and has companies and representatives throughout the world.

Palgrave® and Macmillan® are registered trademarks in the United States, the United Kingdom, Europe and other countries.

ISBN: 978–1–137–43888–1

Library of Congress Cataloging-in-Publication Data

Oertzen, Christine von.
Science, gender, and internationalism : women's academic networks, 1917–1955 / Christine von Oertzen.
pages cm
Includes bibliographical references and index.
ISBN 978–1–137–43888–1 (hardback)
1. International Federation of University Women—History. 2. Women—Education (Higher)—History. 3. Women in education—History. I. Title.

LC1551.I53 O47 2014
371.822—dc23 2014004682

A catalogue record of the book is available from the British Library.

Design by Newgen Knowledge Works (P) Ltd., Chennai, India.

First edition: July 2014

10 9 8 7 6 5 4 3 2 1

Contents

Series Editors' Foreword

If you do not have enough time now to read this book from cover to cover, we would suggest that you first go to chapter 6. You will find there a fascinating account, never before told in such detail, of the activities by the International Federation of University Women (IFUW) during the late 1930s into the war years to help female scholars in Germany, Austria, and elsewhere who were victimized by Nazi Germany's race policy. Many of them were eminent Jewish academics who had been deprived of their teaching positions and research appointments and had lost all means of support. Others were younger scholars with no prospect for a regular position because they were identified as Jewish. The IFUW, through its headquarters in London, offered help via its members' initiatives and also in cooperation with other like-minded organizations that had established close ties among European and American academic women. There existed "a transnational network to assist academic women in their flight from persecution," the author writes.

"A transnational network to assist academic women"—these words sum up the main subject of the book. It is about female academics who, in the aftermath of the Great War, sought to establish a "transnational network" of organizations and individuals, all sharing a commitment to international understanding, which they considered the key to a durable peace. In times of crisis, such as the Nazi era, this network played a crucial role in giving assistance to refugee scholars, many of whom would go on to pursue distinguished careers after the war.

Networks, associations, connections—these are among the key terms that inform transnational history, namely the study of history in a transnational perspective. The Palgrave Macmillan Transnational History series has already published a number of volumes on "transnational women," and the present book sheds additional light by focusing on scholarly and academic woman intellectuals.

It may well be that women find it easier than men to establish transnational networks, both personal and organizational. Traditionally, while men wielded power and developed their own political, business, and professional ties from which women tended to be excluded, women willingly and successfully developed their own networks across borders. Moreover, with their involvement in a country's governmental and military institutions, men's networks were more often intra-national rather than cross-national. Women may have been less inhibited in establishing their own connections with one another all over the world. And, as this book shows, even when nations go to war, women's associations have, historically, been more tolerant than men's in reestablishing contact with one another after the hostilities have ended, as demonstrated when the IFUW admitted German and Austrian women scholars' organizations almost immediately after the Great War.

Of course, such ties among academic women did not prevent the rise of Nazism or the coming of another war. As the book shows, during the 1930s, a number of Germany's academic women embraced race politics and were willing to cut off connections with their counterparts in western Europe. Personal connections, in other words, are not always sufficient to prevent or mitigate international tensions. Even today, when there are infinitely more personal and private organizational networks in the world than ever before, interstate relations go on as if with their own momentum, creating, in the process, tragic consequences for all people. This book will be of invaluable help when we try to understand what is certainly one of the most fascinating questions in transnational history namely, the relationship between interpersonal nexus and interstate affairs.

Akira Iriye
Rana Mitter

Acknowledgments

Many individuals generously contributed to the timely completion of this work. I would like to thank especially Karin Hausen, the former director of the Center for Women's and Gender Research at the Technical University of Berlin. Karin's center provided optimal conditions to launch my study, and her critical advice and active encouragement were indispensable from start to finish. While director at the German Historical Institute in Washington, DC, Christof Mauch provided me with ample opportunities to discuss ideas with him and many other colleagues; my time as a research associate at Christof's institute shaped my thinking about transnational history. My debt to Lorraine Daston, who brought me to the exceptionally vibrant interdisciplinary environment of the Max Planck Institute for the History of Science in Berlin, is abiding. Without Lorraine and my colleagues at the MPI, including librarians Ellen Garske, Ruth Kessentini, and Matthias Schwerdt, my study would not have been possible. Three colleagues from the University of Braunschweig, Ute Daniel, Simone Lässig, and Herbert Mehrtens, deserve special mention; I am also grateful to Mineke Bosch, Ulrike Gleixner, Susanne zur Nieden, Carola Sachse, and Angelika Schaser for insightful comments on earlier versions of the manuscript.

I had the good fortune to spend many hours in archives and libraries in France, Germany, the Netherlands, Switzerland, the United Kingdom, and the United States. The present-day members of the women's associations discussed in this book made available to me both sources and valuable suggestions; among them, the American Association of University Women in Washington, DC, warrants mention for providing me, free of charge, with hundreds of copies of documents. Gillian E. Murphy and Anna Towlson aided me in efforts to obtain access to the records of the British Federation of University Women. I am pleased to report that this body of documents, long inaccessible to researchers, is now housed at the London School of Economics.

The biographical appendix to this work was compiled with assistance from Carsten Ahrent, Samuel Coghe, Sandra Hoehn, and Mirjam Thulin. Ulla Drenckhan designed the electronic version of this annotated appendix. Her creative work may be viewed at http://uwind.mpiwg-berlin.mpg.de/en/home. Kiran Patel put me in touch with Palgrave. Akira Iriye and Rana Mitter graciously agreed to include my study in Palgrave's Transnational History Series. Chris Chappell, Mike Aperauch, and Deepa John guided the manuscript through the production process. Translation costs were borne by Geisteswissenschaften International. Kate Sturge shouldered the burden of translating the text from German to English; her meticulous attention to detail made it a true pleasure to work with her. For his support in putting together the manuscript, not to mention the day to day of our shared life, I want to thank my husband, Keith R. Allen, as well as our wonderful son, Anouk. This book is dedicated to Anouk and Keith.

Abbreviations

AAC	Academic Assistance Council
AAUW	American Association of University Women
ACA	Association of Collegiate Alumnae
ADF	Allgemeiner Deutscher Frauenverein (General Association of German Women, founded 1865)
AFFDU	Association française des femmes diplômées des universités (French Association of University Women)
AUUE	American University Union in Europe
BArch	Bundesarchiv (German Federal Archives)
BBC	British Broadcasting Corporation
BDÄ	Bund Deutscher Ärztinnen (Federation of German Women Physicians)
BDC	Berlin Document Center
BDF	Bund Deutscher Frauenvereine (Federation of German Women's Associations, founded 1884)
BDM	Bund Deutscher Mädel (League of German Girls)
BFUW	British Federation of University Women
CPI	Committee on Public Information
DAAD	Deutscher Akademischer Austauschdienst (German Academic Exchange Service)
DAB	Deutscher Akademikerinnenbund (German Federation of University Women)
DDP	Deutsche Demokratische Partei (German Democratic Party)
DNVP	Deutschnationale Volkspartei (German National People's Party)
DVAF	Deutscher Verband Akademischer Frauenvereine (German League of University Women's Associations)
HICOG	High Commission for Occupied Germany
IAW	International Alliance of Women
IAWSEC	International Alliance of Women for Suffrage and Equal Citizenship
ICW	International Congress of Women
IFUW	International Federation of University Women
IRC	International Research Council
IWSA	International Woman Suffrage Alliance
KWI	Kaiser Wilhelm Institute
LAB	Landesarchiv Berlin (Berlin State Archive)
MWIA	Medical Women's International Association
NSDAP	Nationalsozialistische Deutsche Arbeiterpartei (Nazi Party)

NSLB	Nationalsozialistischer Lehrerbund (National Socialist Teachers' Association)
RDA	Reichsbund Deutscher Akademikerinnen (Reich Federation of German University Women)
RHUL	Royal Holloway, University of London
SATC	Student Army Training Corps
SPD	Sozialdemokratische Partei Deutschlands (Social Democratic Party)
SPSL	Society for the Protection of Science and Learning
StA	Stadtarchiv (municipal archives)
SVA	Schweizerischer Verband der Akademikerinnen (Swiss Association of University Women)
VAÖ	Verband der Akademikerinnen Österreichs (Association of Austrian University Women)
VdN	Vereinigung der Nationalökonominnen (Women Economists' Association)
WILPF	Women's International League for Peace and Freedom
WL@LSE	The Women's Library @ London School of Economics

1
Introduction

When Caroline Spurgeon, professor of English literature at Bedford College, London, stepped off her ocean liner in New York, the end of World War I was imminent. It was October 12, 1918, and armistice negotiations with Austria-Hungary and the Ottoman Empire would begin two days later; Germany's capitulation was only a matter of time. Caroline Spurgeon had traveled from Britain to the United States to further the Allied war effort against the Central Powers. Along with her younger colleague Rose Sidgwick, a lecturer in ancient history at Birmingham University, Spurgeon was part of the official British Educational Mission: a committee of seven respected British university lecturers that had been appointed by the Foreign Office in summer 1918 and was in the United States at the invitation of the US government and the American Emergency Council on Education.[1] The committee's task was to visit 46 American colleges and universities over the subsequent six weeks and, based on their observations, to draw up proposals for enhancing exchange between British and American students, teachers, and scholars. The initiative ultimately sought to disengage the United States from its close academic ties with the German Reich.

The arrival of the British professor and her young colleague in New York in October 1918 marks the beginning of women academics' transnational networking. Prompted by wartime educational policy requirements, Spurgeon's tour introduced the British women to leading American colleagues. Subsequent discussions on what the inter-Allied work on higher education would mean for female students and teachers, and how their concerns could best be given a voice, culminated in the idea of establishing a new network of academic women, initially within the Allied sphere.

The Versailles peace negotiations, the founding of the League of Nations, and the introduction of women's suffrage after hard years of campaigning fueled ambitions to achieve more than an inter-Allied female educational alliance. Instead, American and British initiators envisaged the formation of a multinational female educational elite that would lay claim to a role in global politics. The new international affiliation of university- and college-trained women would call on its members to commit to the values of a "world community" then forming around the League of Nations, to act for world peace, and, at the same time, to ensure women's access to science and higher education worldwide. To this end, the international association was to establish a dense web of personal friendships among female academics across national and disciplinary boundaries; promote international exchange between women students, teachers, and researchers; and support women's advancement in the academic sphere.

The International Federation of University Women (IFUW) was founded in London in spring 1919. By 1922, its coverage had grown from 8 to 22 national member associations; by 1930, the IFUW united twenty-four thousand academic women from 30 countries. The organization joined the spectrum of non-state actors arising around the League of Nations, a landscape that has aptly been described as a "transnational civil society."[2] A German organization of university women, the Deutscher Akademikerinnenbund (DAB), was formed in 1926 and joined the IFUW the same year.

My study takes as its starting point the birth of the IFUW and the 30 member associations it rapidly acquired. I explore how the idea of the IFUW gained focus and substance, and reconstruct the growth and workings of the new organization, which for the first time brought together women academics from many different, mainly European, countries. I also investigate the degree to which the organization succeeded in realizing its goals over the subsequent four decades, in the face of turbulent global economic and political conditions. Which actors, models, and visions carried the organization forward, and how should we locate them within the international context of the politics of gender and scholarship during the twentieth century?

Tracing the history of this international umbrella organization is, then, one key focus of the present study. Equally important is the question of what the IFUW, dominated as it was by Britain and America, meant for those members who had been socialized within the academic systems of Continental Europe. The interface of international objectives and principles with national interests, needs, and convictions proved remarkably fraught, as the case of the German organization illustrates. For the entire period under study, the relationship between the IFUW and German academic women was one of particular tension—a tension that offers vivid insights into both the potential and the limitations of transnational networking. This study focuses especially on the degree to which German women scholars' entry into the new, international female academic community enabled them to forge new professional or political opportunities and personal bonds—before, during, and after the Nazi dictatorship.

The interest in transnational relationships has expanded markedly in recent years, a trend common in German and English-language historiography.[3] This transnational turn has encouraged renewed interest in women's international activities.[4] And yet, little attention has so far been devoted to the IFUW and its national member organizations.[5] The IFUW lies at the intersection of several fields of research rarely addressed in common: the cultural and gender history of science and the history of higher education, the history of international relations, and the history of national and international women's movements. Historians of science investigating women's access to higher education and their academic contributions have explored specific institutional or disciplinary contexts within national frameworks.[6] Similarly, historical studies of the women's movement in Britain, the United States, Germany, and elsewhere in Europe have tended to focus on national contexts.[7] Research on the international women's movement, in turn, has concentrated on women's politics more generally.[8] Finally, the history of international relations has accorded only limited attention to gender history; the same can be said for the history of international education and scientific networking.[9] In terms of specific national, disciplinary, and biographical literature, my study builds on excellent work in all domains; it aims to draw those fields together to tease out relationships between national and disciplinary, as well academic and non-academic contexts. My study highlights the way the IFUW functioned and intervened as an international women's organization and as a transnational, gender-specific academic network. Tracing overlapping national and international commitments, the study presents a new approach to the cultural history of international relations.[10]

In this book, I approach the IFUW—in the spirit of its founders—as an academic network. This means that the biographies of historical actors are of critical importance, as they cast light on the ways that personal connections and informal links arose and flourished via the IFUW's networks. To a considerable extent, the female academic network I explore here was built on personal ties and traditions, social practices whose origins took shape in the last decades of the nineteenth century with the emergence of national and international women's movements. Since the publication of Carrol Smith-Rosenberg's groundbreaking paper on "The Female World of Love and Ritual," these emotionally intense, long-lived friendships have offered gender historians a fruitful means to analyze women's networks and organizations.[11] "Friendship" in this sense was, as Edith Saurer has argued, both a personal and a public commitment.[12]

Especially in its foundation phase, the IFUW was shaped by personal attachments and emotional synergies of this kind. The combination of personal affection and public obligation is exemplified by the relationship of the IFUW's Anglo-American founding couple, the New York college dean Virginia Gildersleeve and the British professor Caroline Spurgeon. The women met in fall 1918 during the British Educational Mission's tour through the United States, and entered into a lifelong transatlantic companionship that proved highly productive in terms of both scholarship and the politics of science. Their bond persisted until Caroline Spurgeon's death in 1942.[13] The founders of the IFUW also declared friendship in general to be an essential pillar of their international organization. One of the central concerns of this book is the extent to which friendships and other personal encounters and bonds, as well as professional and academic contacts within and outside the organization, contributed to the structure, stability, and continuity of the IFUW and its member associations.

The IFUW was a product of the immediate post–World War I era, and its founding statement in 1919 established academic internationalism as a binding ethical maxim.[14] According to the federation's leading representatives, it was their sex that made them particularly well qualified to stand up for general international understanding in the name of academic objectivity. With this claim, they positioned themselves within a colorful array of intellectuals, writers, scholars, and politicians—women and men alike—who supported the League of Nations and successfully argued for the League to be granted responsibility not only for political and economic concerns, but for intellectual and scientific matters as well.[15] In 1922, the efforts of this internationalist circle bore fruit with the establishment of the League of Nations Committee on Intellectual Cooperation, and it emphatically welcomed the International Institute of Intellectual Cooperation that was formed shortly afterward in Paris. The IFUW managed to ensure that women were included among those appointed to the institute's new academic posts. In line with David Livingstone's advice to see internationalism in science as "a social achievement, not the inevitable consequence of some inherent scientific essence,"[16] this book explores the form of internationalism pursued by the IFUW in the politically turbulent period following World War I. How far did the organization succeed in convincing its own multinational academic membership to commit to the conceptual triad of science, womanhood, and international community, and how far did it manage to mediate between internationalism and the various nationalisms of the member associations? Did the member associations, especially the DAB, accept the IFUW's principles in this respect, or were some academic women inclined to develop models that altered the balance of priorities between internationalism and nationalism?

In 1922, Elise Richter, a 54-year-old Viennese teacher of Romance languages, was asked by a British acquaintance whether she would be prepared to found an Austrian association of women academics and enter the international federation. Richter confessed that she did not personally know a single woman who had pursued a path similar to her own. Richter, who was appointed Austria-Hungary's first female lecturer in 1907 and in 1922 became the new Austrian republic's first female associate professor, may have been exaggerating the point slightly in her memoirs for effect. Her observation is all the more surprising in view of the liberal attitude toward women that prevailed in Viennese academia around 1900, a climate portrayed so convincingly by Maria Rentetzi's study.[17] But Richter's comment does point to gender-specific differences between the academic cultures of Continental Europe on the one hand and Britain and the United States on the other. Women scientists in Central Europe barely knew each other; they did not nurture intensive personal ties. This is not to say that the phenomenon of female friendship was necessarily unknown to them; quite the contrary. Elise Richter, for example, lived with her elder sister Helene, a self-taught expert on Shakespeare and a respected theater critic. Nor were Richter's views far removed from the objectives of the women's movement. As a scholar, however, she cultivated intense relationships solely with men; indeed, she actively avoided contact with women, especially those reputed to be involved in the women's movement. Drawing additional attention to her gender seemed to her to carry the risk of damaging her prospects within the university.[18] It is remarkable that before the end of World War I, collegial friendship among women academics was virtually unknown within the German-speaking universities. Rather, women focused on proving their worth as individuals in a masculine world: almost without exception, women in the German academic system struggled in isolation.

The reasons why Elise Richter nonetheless decided in 1922 to found the Austrian association of academic women and thus enter the public limelight as a woman and as an academic will be explored in subsequent chapters. In the case of Germany, I will outline why and how academic women sought to convince their female colleagues inside and outside the universities to join together as educated women under the umbrella of the DAB and the IFUW. At issue was the decision to participate in an international initiative at all—a politically explosive choice in Germany at this time. An equally important facet of any analysis of German academics' attitude to the international community is the extent and manner in which the Anglo-American model of female academic networking was adopted and anchored in German academic life from the 1920s onward.

This account draws on a range of archival sources, most of them held in American and British archives. In the German case, unpublished source material is more difficult to find. The very sparseness of the DAB's business papers for the Weimar period, which form part of the Helene Lange Archives in the Berlin state archives (Landesarchiv Berlin), illustrates the precarious financial situation of the young DAB in the 1920s: at the time, a lack of money and of paper meant it was virtually impossible even to send out newsletters. The DAB archive material also shows marks of having been moved from place to place each time the association's presidency changed. For the Nazi period, we must rely exclusively on sources published by the organs of the Deutsches Frauenwerk, the National Socialist "German Women's Agency." And for the period after 1945, the holdings of the German Federal Archives in Koblenz do not, at least with regard to the initial postwar years, suffice to piece together the networks that were then being rebuilt between women in West Germany. It is rare that a woman scholar's personal papers survive in a German public archive; the most fruitful sources in this respect have proved to be the papers of certain DAB presidents, members of the Reichstag, and later members of the West German Bundestag, notably those of Marie-Elisabeth Lüders, the cofounder of the DAB and later honorary president of the Bundestag.

In contrast to the dearth of German sources, the richness of sources in the United States indicates how differently women's academic networks in America were structured and anchored in society. The archives of the American Association of University Women (AAUW) have been housed in the organization's Washington offices since 1917. They contain not only extensive material on the AAUW's history, but also important documents from the early years of the IFUW. The correspondence of the AAUW's International Relations Committee offers insight into the network of personal friendships and mutual assistance linking many different parts of Europe and America, especially during the period of National Socialism and the persecution and emigration of European women scholars. Important documents can also be found in the extensive and well-ordered papers of former college deans, many of whom were the most committed protagonists of the new, transnational female educational elite. These sources have enabled me to reconstruct in detail the organization and policies of the IFUW. Of special value here are the personal papers of Dean Virginia Gildersleeve at Barnard College and Columbia University, New York, and of M. Carey Thomas, the longtime president of Bryn Mawr College in Philadelphia.

Using German-language sources alone, the Nazi period, in particular, would have been impossible to investigate in any detail, in terms either of the train of events within Germany or of the subsequent careers of the DAB's Jewish members, who, in 1933, were excluded by their colleagues, dismissed from their employment, and forced into emigration. Fortunately, the British Federation of University Women's archives contain documents on the BFUW's assistance for emigration and for the refugees, enabling a detailed understanding of the organization of rescue operations and the personal decisions, and later professional lives, of academic women in exile. Until the mid-1990s, the BFUW archives were housed in the association's international hall of residence, Crosby Hall; when the hall was closed, the papers were moved to the Portsmouth University library. The BFUW subsequently transferred the entire holdings to the Women's Library in London, with the aim of keeping them in appropriate archival conditions and making them more easily accessible. Unfortunately, however, the collection remained closed for more than a decade after the move.[19] It was reopened in early 2014, when the Women's Library found a new home in the library of the London School of Economics. As this book goes to press, the BFUW archives are in the process of being recatalogued. In the following, my citations indicate both the new references and the old Portsmouth filing.

This study is divided into seven main chapters, arranged chronologically to address different aspects of the international network and its interface with the German member association. Chapter 2 follows Caroline Spurgeon and Rose Sidgwick on their official tour through the United States at the end of World War I, reconstructing the motivation and context of the IFUW's formation. This chapter examines in detail how the IFUW's founders achieved such rapid success in their organization's networking and growth. It applies a gender-historical perspective to the emergence of the Anglo-American predominance within international science and education policy after World War I: in the initial stages of the IFUW's development, women academics from the former Central Powers were excluded from the organization.

In chapter 3, I turn to the IFUW's agenda and policies during the 1920s, with a particular focus on the federation's two most important initiatives. The first of these is the IFUW's establishment of three international guesthouses—in Washington, Paris, and London. Each offered accommodation for around 50 traveling women academics and provided good, reasonably priced meals, a well-furnished library, and spacious clubrooms. The second is the federation's creation of an international fellowship program for women scholars, which during the 1920s was already lending the IFUW a high degree of credibility as an institution of nonpartisan academic internationalism. The IFUW's policy of promoting scholarship, as practiced within this program, laid the foundations for its later rapprochement with former wartime enemies.

In chapter 4, the book's perspective shifts from an international to a national context. I explore the protracted disputes within Germany on the question of whether, and when, German women academics should take part in the new female network. These debates reflected the bitter feuds around science policy and national academic sensibilities that characterized the international situation in the years following World War I. This chapter also reveals the deep crisis in which female academics found themselves at the beginning of the Weimar Republic. This is the context for my discussion of the extent to which women academics in Germany managed, more quickly than their male colleagues, to grasp the academic internationalism proposed after World War I by new international organizations as an opportunity—on the one hand, to benefit from the resources of the IFUW and promote Germany's political interests internationally, and on the other, to tap into the momentum of the IFUW's energy to create a new form of female networking within Germany. I examine the founding of the DAB as a transnational project and ask how far this female academic umbrella organization should be regarded as an attempt to transfer to the German context an essentially Anglo-American model of nurturing female academic traditions.

Chapter 5 addresses the extraordinary political challenges that faced the DAB, its members, and the IFUW as a whole in the wake of the National Socialist "seizure of power" in January 1933. I show first of all that the DAB's political survival in Germany was inextricably tied to its membership in the IFUW. I outline the process by which the association underwent *Gleichschaltung*, or alignment with the regime's policy and ideology, describing this process in terms of both the IFUW's stance and German women's personal connections with their colleagues abroad. I am interested here in the form in which the transnational networking of female academics survived under the conditions of the Nazi dictatorship. The DAB remained in the IFUW until 1936. Shortly thereafter, it was dissolved into a larger Nazi women's organization, the German Women's Agency. I also pursue the question of whether, and how, female academic networks within Germany continued to exist in isolation from the international community, and show how women once associated with the DAB protected their own interests by distancing themselves energetically from the values of the bourgeois, civic world community, by allying themselves to female forms of the racist science of the day, or by seeking out new networks within Nazi structures.

Chapter 6 sets these developments against the IFUW's reactions to the National Socialist revolution within Germany. I show that the practical continuation of an international federation functioning outside Germany, and its transnational network, was vitally important for those women academics in Germany (and in Germany's growing sphere of influence) who had been dismissed from public service positions and barred from the DAB. This chapter highlights a development long overlooked in science studies and exile studies: between 1933 and 1945, the academic networks of the female international community functioned efficiently to assist persecuted members in escaping Nazism. I indicate the specific areas where the aid of the IFUW and its member associations was concentrated, identify the individuals who carried out and funded that aid, and show which academic women benefited from it.

Chapter 7 asks how Jewish women academics within Germany, and later German-dominated Europe, reacted to the Nazi persecution. Through the correspondence housed in the BFUW and AAUW archives, I retrace individual women's dilemmas, their options—or lack of options—and the choices they pursued.

Chapter 8 shifts attention back to Germany. In this chapter, I focus on the ways in which academic women there sought to recast national and international networks. By exploring German developments in larger international contexts, I go beyond myths of a "completely new beginning" after May 1945 to confront questions of connections and continuity, of national and transnational memory—areas of scholarly investigation that deserve further attention in the years to come.

2
Global War, Global Citizens, Global Mission: The Anglo-American Project of an International Federation of University Women

Academic Mobilization and Educational Diplomacy during World War I

The origins of a world alliance of women academics date back to the United States' entry into World War I. Three days after President Wilson's declaration of war on Germany on April 6, 1917, the general meeting of the supraregional federation of American women college graduates, the ACA or Association of Collegiate Alumnae, was held as planned in Washington, DC. Under the impact of events, the delegates decided—in common with their peers all over the country—to place their organization at the service of the nation.[1] A War Service Committee was appointed to draw up and implement practical measures. The new Committee's eight members included the presidents of Bryn Mawr, Wellesley, and Mount Holyoke colleges (Carey Thomas, Ellen Pendleton, and Mary Woolley), along with the ACA's president, Lois Kimball Rosenberry, and its general secretary, Gertrude S. Martin.[2]

Some women's colleges had already been preparing for the worst case well before the United States entered the European war. Following a call by Columbia University's president Nicholas Murray Butler, made five days after diplomatic relations with Germany were severed in February 1917, the Columbia University Committee on Women's War Work had been set up in New York under the leadership of Dean Virginia Gildersleeve. As early as October 1916, Gildersleeve had joined the executive board of the Women's Advisory Committee of the Council of National Defense, the central civilian body coordinating the country's economic and social mobilization. Her office at Barnard College became the focal point for the most important initiatives in and around New York City regarding the women's colleges' preparations for the anticipated declaration of war by the United States. For the Committee on Women's War Work, the main task was to find out where (if and when the declaration came) "women's services would probably be needed and what training they would require." The Committee had also set up an information bureau within the University, where women students could register for volunteer war work. In spring 1917, this bureau's card index held the names of some ten thousand students and alumnae.[3]

The ACA's War Service Committee, too, initially concentrated its energies on choosing the best place to invest the association's resources. President Wilson's administration helped the Committee find that domain when the "Committee on Public Information" (CPI) was formed in April 1917. The 19 sections of the CPI, a powerful propaganda machine directly answerable to the president, were tasked with enlisting support for the United States' military intervention in the Old World from America's heterogeneous, immigration-based society with its high proportion of recent arrivals from central and eastern Europe.[4] In spring 1917, the CPI put its considerable media resources into a nationwide search for influential people working in the arts, business, journalism, and higher education who would agree to offer their public support for the war and the government's war aims. This task, noted the director of the CPI's "Speaking Division," Arthur Eugene Bestor, was second in importance only to the soldiers' military service. The war, he added, would not be decided on the battlefield alone, but also in the United States itself: everything depended on how far it proved possible to attract public consent for the cause on the home front. Indeed, it was

> only through the united efforts of enlightened and enthusiastic Americans that the full strength of the Nation can accomplish those things for which she entered this conflict. Upon the leaders of our public opinion therefore, rests a responsibility heavier than perhaps ever rested upon any group of people in our entire history. It is a task which can be performed only by men and women who themselves are well informed and who cooperate with all patriotic organizations, governmental and private, which are unifying public opinion in support of the national purpose.[5]

According to a report published by the ACA War Service Committee, this appeal provided the initial momentum for the ACA's commitment to the dissemination of government propaganda. The call, circulated by the Committee chair, Bryn Mawr's president Carey Thomas, had inspired the committee: "Here was our opportunity and here was our responsibility.... It seemed that there was no other form of service for which the Association was so well fitted, and that there was no other body of women so well prepared by training and experience to undertake the task."[6] Referring to the CPI's objectives, the ACA's War Service Committee resolved to make its most important objective the guidance of public opinion toward the government's line. At the same time, the members regarded this commitment as an opportunity to prove that the education of female college graduates permitted them, like their male counterparts, to take on leadership roles within society during this period of social and political crisis. In April 1917, the War Service Committee's first public statement found that "college women in general" were "particularly well fitted" to join in one of the most urgent tasks of US mobilization, "namely, the task of effectively informing the millions of persons who make up this greatest of democracies as to the significance and the necessity for this war and the peril of a premature peace."[7]

The plan for participation in the "patriotic education campaign," drawn up by the Committee under the leadership of Carey Thomas, provided firstly for the creation of a list of experienced speakers who were capable of addressing a large audience. Secondly, it proposed strategies that had been proving successful within the ACA for many decades: like every ACA campaign, this one, too, would aim for the self-improvement of ACA members. The call for patriotic work in rural areas and small towns was directed particularly at married college graduates, who would, it was hoped, take up a public role as representatives of educated America. The ACA campaign thus saw its focus as "intensive work in the home community, with a special effort to reach the foreign element, the rural districts, and the ignorant and misinformed."[8] To equip its local branch members for their national task, the ACA put together an extensive collection of texts for them to use. Precise templates for short and longer speeches were provided, explaining the American war objectives and asking for donations. These activities were to be carried out in close cooperation with the CPI and other organizations that served the propagandistic mobilization of the country.[9]

The great importance accorded the campaign by the ACA council demonstrates how rapidly the American educational elite adopted a perspective that was historically new in the United States in addressing the nation as a whole. Entry into the war pushed large parts of America into a patriotic rapture equal to the "war euphoria" that broke out in Europe in August 1914.[10] However, if mobilization efforts aroused noble feelings for the American nation and the European allies, speculation on the strategic benefit of the war for women's position within America was not without its relevance as well. Already in peacetime, the ACA's goal had been to make women visible within the American educational landscape. The Association's feverish search for as lofty a patriotic task as possible indicates how enthusiastically US colleges and universities responded to the call to arms. Most scientists were putting their work at the service of the government—whether by directly taking up posts in Washington or by redirecting their research resources to cater to the needs of the military.[11] In 1917, too, with generous state support, hundreds of American colleges and universities were eagerly transforming themselves into training camps, in many cases at the expense of the institutions' educational role. In the course of the war, the Student Army Training Corps (SATC), formed in July 1917, trained 125 thousand students as reserve officers on various campuses, including Harvard, Yale, and Princeton. David Levine has pointed out that for these colleges and universities, such efforts not only secured a financial base, but also allowed them to position themselves differently in society. By demonstrating their readiness to take up arms, argues Levine, they succeeded in convincing both politicians and the public that the male educational elite was well qualified to take on leadership tasks in both the military and civilian spheres.[12]

In spring 1917, organized female academics in the United States held similar goals. However, they had to seek different fields of activity as showcases for their patriotic leadership qualities. "Our Association must rouse itself sufficiently to see its own importance among organizations and its duty to itself and to the Government in the present crisis," insisted the council in its call to the branch presidents. "What is needed is a full realization by members of the potential strength of this great body of trained women to which they belong and a willingness to cooperate toward any desired end."[13] Not unlike their male colleagues, the ACA functionaries regarded the war as a chance to show that the female educational elite of the United States was ready to take up a leadership position in society, outside the college walls. Proving themselves to be loyal educators of the nation seemed the ideal form for this undertaking. The educated woman of America was to fight like a soldier, "accepting her responsibility as a leader of public opinion. So she can make her training count in service to her country."[14]

The ACA's reports on the structure and successes of the "patriotic education campaign" show that members all over the country were following the call to the lectern, with particularly large numbers reinforcing the volunteer army of amateur speakers—the CPI's "Four-Minute Men"—in those cases where members active at state level created the necessary conditions and tirelessly recapitulated their calls for cooperation.[15] Pennsylvania, in particular, did an exemplary job. This state was home to most of the country's long-established, six-million-strong German-American community. The local political climate was tense. From her Bryn Mawr office, Carey Thomas drove on and supervised the efforts for public education with almost military efficiency. In other areas with significant German and central European populations, such as Albany, Washington, Chicago, Detroit, Columbus, Boston, New York, New Haven, Buffalo, St. Louis, Cleveland, or Iowa City, many hundreds of women signed up as volunteer speakers. They attended training courses in rhetoric and took the oath of loyalty that was required to qualify as a public speaker in the campaign.[16] The ACA's Minneapolis branch alone organized more than 6,100 speeches, "long and short, most of which were accompanied with lantern slides on American history and government, for the newly arrived immigrants."[17] Once established, the networks of the propaganda-led Americanization campaign were not only deployed directly to garner support for Wilson's war policy in Europe, but also to appeal for the third Liberty Bond in April 1918, which Americans of German origin were specifically pressured to buy in order to demonstrate their loyalty.[18]

For the ACA, this reorientation toward a national endeavor—outside its previous, more regionally oriented policy interests—brought with it an unprecedented degree of politicization. The Association was not alone in experiencing this development. For American science and scholarship as a whole, the

United States' involvement in the events of the European war resulted in a new, strongly patriotic focus and a structural centralization with implications that reached far beyond both American borders and the duration of the war.[19] A similar process occurred throughout the entire field of higher education, albeit at a slower pace. March 1918 saw the first meeting of the Emergency Council on Education, a body made up of 15 of the large organizations of higher education.[20] Similarly to the National Research Council, founded in 1916 and charged with placing US science systematically and efficiently in the service of the war, the Emergency Council aimed to pool the nation's higher education resources to enable a fast, coordinated response to all the measures the war would continue to require.

The effect of the call to arms in the United States was not merely to politicize life within North America. The summons also directed attention beyond the continent's borders. In July 1917, colleges and universities had already joined to form the American University Union in Europe (AUUE), as a way of taking care of their male students in arms in Europe. A large hotel was requisitioned for the Paris headquarters of the AUUE, and more than thirty thousand registrations were accepted. In Rome and London, registrations ran to 5,000 in each city. During the war years, the AUUE developed close contacts with French and British universities, ensuring, among other things, that after the armistice more than 7,000 college students were able to study for three months in France, and another 500 at British institutions.[21]

Interest in networking with the Allies in Europe was not restricted to practical considerations and the social assistance that the AUUE offered so extensively to the colleges' own troops in 1917. From the start, the Emergency Council on Education saw its most important role as being to stimulate international cooperation around issues of education, not unlike the National Research Council's view of its role in inter-Allied scientific exchange.[22] It is important to note that, up to the end of the war, US interest in the "international relations" aspect of educational and science policy was directed exclusively toward its allies in Europe, especially Britain and France; Germany and the Central Powers were explicitly excluded. In spring 1918, the Emergency Council on Education's first official act was, jointly with the Department of State and the Council of National Defense, to invite 135 female graduate students from France to visit the United States. The French women were to be given the opportunity to study at American universities for a year, and in return would teach French at schools, colleges, and universities.[23] This program was intended to make the American college and university system better known in France; at the same time, American school and college students were to enjoy better language instruction so that more of them would be able to study in France. France's suitability for the exchange program was enhanced by the fact that the French universities—and the French government—had shown greater interest in intensifying academic relations with the United States since 1915.[24] By 1917, a series of awkward bureaucratic obstacles had been eliminated, and after the end of the war most French universities established summer schools teaching the French language. The Sorbonne in Paris also designed special programs for college freshmen and certificates for advanced students, both carefully tailored to the requirements of their new American clientele.[25]

The efforts of the Emergency Council on Education to make France the main European port of call for a new generation of US academics were based on broad assent among American professors. It was from these circles that attempts came to underpin the Council's policy from a history-of-science perspective. Writing in the *Journal of the Association of American Colleges*, John H. Wigmore, the president of the American Association of University Professors (founded in 1915), argued that French services to science had, wrongly, long been neglected in favor of Germany; truthfulness and science itself now demanded that this state of affairs be remedied. Additionally, wrote Wigmore, France was

> a democracy like ourselves, both politically and socially. It is therefore highly probable that the conditions and methods, the aims and the spirit of her universities and her savants, and the inspiration of her learning, will be in harmony with our own, and a helpful and healthy influence for our young men as the future teachers of our youth and the moulders of public opinion.[26]

Strengthening academic relationships with France was the first important initiative of the Emergency Council on Education. A second was undertaken soon after the Council's formation, when a group of high-ranking British professors was invited to make a six-week tour through the East, South, and Midwest of the United States. This initiative, too, included an important contribution by the National Council of Defense and the US State Department. The tour members' goal was to draft guidelines on intensifying exchange between American and British students.

The members of the resulting British Educational Mission included two female academics. English literature specialist Caroline Spurgeon of Bedford College, University of London, was England's first woman professor and the president of the British Federation of University Women (BFUW); Rose Sidgwick was a young teacher of ancient history at the University of Birmingham. The British Educational Mission arrived in New York in October 1918 to begin its round of visits to colleges and universities, and was received just a few days later by President Wilson in the White House.[27]

By that time, an ACA Committee on International Relations had been established. It was set up at the ACA's general meeting in April 1917, at the same time as the War Service Committee. As for its membership, this was to be decided at the ACA Council meeting in Chicago in April 1918.[28] The selection of members for the Committee on International Relations indicates how closely the Association's budding internationalization was tied to the United States' entry into the war: the Committee included the stalwarts of America's female academic mobilization, with President Carey Thomas of Bryn Mawr College, who chaired the ACA's War Service Committee, and as its chair Dean Virginia Gildersleeve.[29] Gildersleeve and Thomas especially, but also Mary

Woolley and several other college presidents who were particularly committed to the cause of national mobilization, were the ACA functionaries who considered it necessary for the Association's activities to be extended beyond the borders of the United States, and who were prepared to make this a practical reality. Their geographical focus, again, was initially limited to fostering loyalties among America's wartime alliances.

When the ACA's Committee on International Relations met for the first time, in June 1918, the thrust of its discussions was determined by the activities of the Emergency Council on Education and the AUUE. First, the committee debated a proposal by Carey Thomas to open a "center for college women" in Paris—an idea clearly inspired by the AUUE's activities in Paris. The committee agreed to inquire as to whether an additional American institution, specifically oriented toward the needs of female students, was genuinely necessary and affordable. After this point in the meeting, by far the most time was spent on working out how the Committee could ensure the ACA was appropriately represented in the planned American educational commissions on Italy, France, and Britain. In the end it was resolved that the ACA must manage to send at least two female professors, to be nominated by the Association itself.[30]

This first session of the ACA's Committee on International Relations makes clear that the newly emerging field of international educational diplomacy was going to demand battles that differed little from those being fought on a national, regional, or local level. Here, too, it would be important to insist on women's participation and to be represented in person, as women, in order to advocate women's interests and ensure their success—especially as the conditions for men and women students differed greatly in Europe as well as at home. It is evident what issues were at stake, for students in general and women students in particular, when the possible "reorganization of some of the courses of graduate study" in French, Italian, and British universities was debated: efforts to enhance interchange between the Allies in the domain of higher education were, as the ACA Committee on International Relations pointed out in June 1918, strongly oriented on opening up access to higher degrees in Europe for those college graduates "who no longer wish to study for the degree of Ph.D. in Germany."[31]

It is well known that America's entry to the European war triggered a widespread hostility to everything German throughout US society.[32] Considerable research has also been carried out on the ways that this stance was reflected in American educational institutions. Concerted efforts were made to minimize the German influence on education and severely prune German-related components in the curriculum.[33] However, little attention has been devoted to the contemporaneous efforts to cut back personal academic connections with Germany itself and to encourage instead both teachers and the new generation of scholars to shift their focus of interest to other European nations, specifically France and Britain.[34]

This reorientation within higher education outlived the armistice with Germany in November 1918. A conference in December 1918 on the topic of "After-War Problems in the Higher Education of Women" indicates as much. The ACA's International Relations Committee, together with the Emergency Council on Education, had invited participants to a small but influential meeting at Radcliffe, the women's college of Harvard University in Cambridge, Massachusetts, for a detailed discussion of the opportunities and obstacles associated with studies abroad in Britain and the United States. Among those present were the presidents of the respected women's colleges of the East Coast, the "Seven Sisters," and the two women members of the British Educational Mission, Caroline Spurgeon and Rose Sidgwick. Just a few days after the armistice, the two British women had separated from the rest of the mission in order to concentrate on the situation of the women's colleges and universities in the United States and to spend more time exchanging views and ideas with their American women colleagues on the possibility of creating transatlantic networks.[35]

The participants at the Radcliffe College conference agreed that student exchange should focus on postgraduates. On both sides of the Atlantic, they decided, students' education at college or university was not amenable to interruption prior to the bachelor's examinations. Their next topic was more delicate: the difficulty in intensifying Anglo-American exchange, especially as regards women, seemed to lie with the British universities. For American women, studying in Germany, Austria, or Switzerland was made much more attractive by the fact that such studies could result in a doctoral degree, the Dr. phil. that was so widely respected in the United States. In Germany, there had been an increasing tendency to award a Dr. phil. to foreigners on less rigorous criteria than to German students, leading to a corresponding decline in the German degree's standing within the United States.[36] Even so, feared the British professor Caroline Spurgeon, it would be very difficult to persuade American women graduates to choose Britain in preference to Germany as a destination for their study or research visits unless they were offered the prospect of gaining a standardized academic degree that was recognized and respected in the United States.[37]

The British Foreign Office had been trying since 1916 to push its British universities to introduce postgraduate courses for American students leading to a PhD. The aim was to weaken German influence within the American educational system.[38] A memorandum by Lord Bryce, commissioned by the Foreign Office and titled "As to American Students at British Universities," of October 31, 1916, put the case for such measures in no uncertain terms. In Lord Bryce's view, it was self-evident that nothing could be better for the political and cultural relations of the two countries than a process by which "the number of Americans of the most educated class who are attached to England by those ties of affection which a man forms at the most susceptible period of his life, should be greatly enlarged, and a wider basis for a mutual good understanding of one another's aims and sentiments created."[39] Oxford introduced the new degree of "D.Phil." on June 12, 1917, as the first British university to do so; the last to follow Oxford's lead—under increasing pressure from the Foreign Office—were Cambridge, on February 22, 1919, and the University of London, on May 28, 1919.[40]

At the time of the Radcliffe conference in December 1918, it was already becoming clear that in the near future students would be able to attain a PhD anywhere in Britain. In the case of Oxford and Cambridge, however, this applied only to male candidates. Spurgeon stressed the fact that women at these universities continued to be denied any degree title whatsoever. The men and women attending the conference reacted to her comment in different ways. The president of Vassar College, Henry Noble MacCracken, argued that the status quo should not be attacked and that, instead, the objective should be for American colleges to give due recognition to time spent studying abroad even without the award of a doctoral degree. Carey Thomas dismissed this notion. It was of the utmost urgency, Thomas stressed, that the ties between British, French, and American universities be strengthened, and in order for this to happen, the newly created degrees must be accessible to women in the same way as to men.[41] The conference finally passed a resolution to that effect: "as far as degrees in British universities are open to American students, they should be open to women as well as to men."[42]

Following her official visit, Caroline Spurgeon took the Radcliffe conference as an occasion to add a supplementary evaluation to the British Educational Mission's report, in which she outlined the special situation of transatlantic exchange with respect to women. Spurgeon concurred with her male colleagues' judgment on many of the report's points, but considered it necessary to offer her own observations. After expressing her admiration of the facilities, beauty, eminence, and impressive wealth of the American colleges and universities, she added:

> The vast sums of money freely spent, both by state and private individuals, on university education in America creates envy in the mind of any English man or woman who cares for and believes in education... Women have specially benefited from this liberality for educational purposes, and it can scarcely be realised by those who have not recently visited America what magnificent buildings and equipment have been provided... in the last few years, in many cases within the last five or ten years.[43]

This impressive blossoming of American higher education did not, she noted, benefit merely a small number of young women whose goal was to pursue a professional vocation; the institutional division between general and vocational education practiced by American colleges and universities, unlike those in Britain, had opened up access for a broad stratum of women to receive a sound education not directed exclusively at academic careers. Spurgeon emphasized the generous facilities provided for both male and female students for their own activities: clubhouses with spacious lobbies and rooms for reading, working, and meetings; kitchenettes; lecture theaters and auditoria; gymnasia and swimming pools; bowling alleys and riding stables. These amenities, she argued, made an inestimable contribution to the development of social life on campuses, and equipped American middle-class women far more effectively than their British sisters "to take part in public work." She had also been able to sense quite palpably a spirit of freedom, especially in the coeducational universities of the Midwest: "The free and natural intercourse between men and women is very pleasant to see.... The system of student self-government adds to this sense of freedom."[44] In comparison with the United States, wrote Spurgeon, Britain made a particularly poor showing with respect to physical education. In American colleges and universities, sport was taken for granted as a part of women's education—as had been laid down in the classical model of education among the ancients. In the United States, a liberality in the educational system had been achieved that vouchsafed women, by right and tradition, the enjoyment of a higher education in a worthy setting. It was this right that gave American women the "independence of outlook and confidence without need for self-assertion which is characteristic of the American college woman."[45]

Spurgeon concluded that a study or research visit to the United States was extremely desirable and beneficial for British graduate students. It was also

> on the whole more important for women than for men, and especially for English women of the professional and teaching class. Men, owing to their work as soldiers, sailors, engineers, administrators and so on, have naturally more opportunity to travel than women. Yet women, owing to the very fact that they are perhaps less adventurous in spirit, more restricted to the home atmosphere and more absorbed in detail, have particular need of the broadening and widening experience of travel and of life in countries other than their own.

It was important to take in life from a different perspective, to learn of different social problems and solutions, to become acquainted with people of different nationalities. Spurgeon found it

> difficult to imagine many experiences more stimulating or educative for a woman graduate of one of our provincial universities, who is going to make teaching her profession, than to go out for a year to one of the great American co-educational Universities of the West or Middle-West, either to do advanced work under a selected teacher, or as a junior teacher in her own subject.... The free discipline, glorious surroundings, and opportunities for physical development, as well as the enlarged experience, would be peculiarly valuable to these students, and would tend to raise their standard as to the conditions under which educational work should be carried on.[46]

Her own impression was that American women college graduates showed "a very general desire" to pursue studies or research in Britain. Their interest was primarily in the ancient, aristocratic institutions with a global reputation, something that the American system could not yet offer to the same extent. Spurgeon considered it an urgent matter for the British to respond with alacrity and flexibility to this interest: "From an international point of view it is important that we on our side should as speedily as possible do all we can do to facilitate this."[47] She recorded word for word the resolutions of the December 1918 conference at Radcliffe College, concluding from these, as her final recommendation in the report, that pressure must "be put on the Universities of Oxford and Cambridge to open their degrees to women, as this has now become a matter of international as well as national importance."[48]

Spurgeon's report makes clear that reflections on academic networking within the Entente, begun during the war, had lost none of their relevance in early 1919. In her insistence on women finally being permitted to obtain degrees at Oxford and Cambridge, one discerns British academic women's hopes to elevate women's educational issues to matters of international politics. But a further point in the Mission's report indicates that, above and beyond this aim, the female academic elite in Britain and America was concerned to assert itself within the newly emerging structures of international educational policy. Caroline Spurgeon's supplement made special note of another of the problems discussed at the Radcliffe conference: the well-informed Virginia Gildersleeve had told the assembled participants about proposals to establish an Institute of International Education in New York. According to the plans, this institute would take responsibility for student concerns both inside and outside the United States, and function as a clearinghouse for international academic exchanges. In response, a resolution by the conference participants urged that "all steps taken in the establishment and maintenance of an Institute of International Education should contemplate representation of Women's Colleges in the committee of control."[49] Spurgeon concluded her supplementary report by repeating this demand verbatim, adding that "among the staff of any such Committee or Institute there should be a woman official as assistant to the Director."[50]

One of the strategies pursued by both American and British women was to work for targeted, individual appointments to the key executive committees, as a means of ensuring that women were represented in the emerging institutions and that their interests would be taken adequately into consideration. A second strategy also took shape in the course of fall 1918, during Spurgeon and Sidgwick's stay in the United States: the creation of international structures specifically addressing female college graduates. It was not initially clear what form such structures should take, but the issue was already on the horizon at the Radcliffe conference, where point six of the agenda noted "the desirability of an international association of college women, or the establishment of relations between the ACA and the Federation of University Women in Britain,"[51] although the point was not discussed in any more detail than that.

To a large extent, the idea of founding an international league of academic women can be traced back to Caroline Spurgeon. In July 1918, even before being appointed to the British Educational Mission, Spurgeon had tried, with the Duchess of Marlborough and in cooperation with the AUUE, to set up a trust charged with establishing "scholarships for women of the Empire and the United States of America." She hoped that creating a program of this kind would ensure that

> a continual succession of chosen women from all parts of the world where English is spoken will come into close touch with life and ideas of English-speaking peoples other than their own; and they will, on their part, be able to diffuse knowledge and understanding of the countries from which they have come.[52]

The idea of a scholarship program did not go beyond these preliminary stages. However, Spurgeon evidently saw the cooperation between the BFUW and the ACA as offering her a chance to realize, through a different route, her vision of an international exchange organization for women college graduates. During the tour through the United States that she undertook with her colleague Rose Sidgwick, Spurgeon met the president of the ACA, historian Lois Kimball Rosenberry, in Madison, Wisconsin. Rosenberry gave her a semiofficial undertaking that the ACA would be ready to act as a partner organization with the BFUW in any future international cooperation.[53] The fact that this cooperation actually took shape, and the way in which it did so, was a product of the extraordinary circumstances of the British Educational Mission's six-week tour. The British women's trip across the United States is portrayed in the sources as a great journey of mutual discovery. Spurgeon and Sidgwick were visibly overwhelmed by the quantity and quality of the progress made by American higher education in the recent past, and in particular by what it offered to women. The American women reveled in the evident esteem in which they were held by their distinguished European colleagues. Much shared ground was also identified, relating to ideas about how the postwar world should be shaped and rebuilt, and about the women's own future role in this process:

> We realised...that their conception of the sort of world for which they wanted to work was the same as ours in the universities of America, and that there was no barrier, no difference in ultimate purpose, which need keep the women of Great Britain and the United States from working side by side for common needs.[54]

The news of the end of the war arrived during the British Educational Mission's tour. The women's feeling that the greatest and most cruel of all armed conflicts had now ended with victory only reinforced their sense of affinity and of shared responsibility for the future. From now on, the debate on shared educational concerns would allocate an increasingly central role to the maintenance of international peace, something that had previously played no part at all.

An additional factor intensified the women's enthusiasm for their common cause: the sudden death of Rose Sidgwick. Shortly before the British Educational Mission was due to return home, both Spurgeon and Sidgwick fell ill with the vicious influenza that, in the winter of 1918–19, would claim more victims worldwide than four years of armed conflict. Rose Sidgwick succumbed to the virus just as the year 1918 drew to a close. This eloquent historian had not only been by far the youngest member of the British Educational Mission, but had, in the course of the trip, become a kind of "star" of the Mission. Looking back on her life, Rose Sidgwick's companions saw her as an embodiment of the ideal—the youthful, energetic, educated, cosmopolitan, and open-minded personality—toward which the project of the Anglo-American educational alliance should direct its efforts.

The later report of the British Educational Mission recommended that the memory of fallen members of colleges and universities be kept alive and that Memorial Fellowships be instituted to this end; Rose Sidgwick was probably one of the first college graduates to receive this honor in the name of international academic understanding.[55] At the large and carefully orchestrated funeral service in Columbia University's St. Paul's Chapel, leading lights from education and politics gathered around Sidgwick's casket. On the political side, Senator Elihu Root, the Nobel Peace Prize winner of 1912 and a Republican elder statesman in matters of American foreign policy, was present, as was Commander A. T. Blackwood, the naval attaché of the British Embassy in Washington, DC, and a representative of the British War Mission. Among the important figures from academic life paying their respects were a representative from Harvard University; the president of Yale University; the provost of Columbia University; the chancellor of New York University; the nearly 80-year-old classical philologist and liberal university reformer Henry Jackson of Trinity College, Cambridge;[56] and Professor Stephen Duggan of the College of the City of New York, who would soon afterward be appointed director of the Institute of International Education.[57]

Alongside all the undoubted personal sympathy and grief, the rhetorical handling of Sidgwick's death shows how perfectly the event fulfilled the task of representing female college graduates as—no less than men—indispensable combatants in the past war, and how successfully the young British woman's destiny could be made to yield capital for the new transatlantic educational movement in general and its female representatives in particular. Virginia Gildersleeve noted in her memoirs, 40 years later, that Sidgwick had died like a soldier for her country: "I felt she had died as truly in the service for her country as had the thousands of her young countrymen who had fallen on the fields of Flanders and of France."[58] In a memorial address for Sidgwick given to British students in January 1919, the philosopher and political economist John Henry Muirhead, Sidgwick's former colleague at the University of Birmingham, proclaimed that "already over the grave of this brilliant young lecturer the union of hearts of the educational world in England and America has set a seal on the work of the Mission in which she played so distinguished a part, and dedicated it to the high heavens."[59]

Virginia Gildersleeve immediately set about organizing collections for a "Rose Sidgwick Memorial Fund." The fund's illustrious committee included not only all the presidents, male and female, of the women's colleges and major state universities across the American Midwest, but also Stephen Duggan, the future head of the Institute of International Education. The daughter of the American Ambassador in London and trustee of Barnard College, heiress Mabel Choate,[60] became the fund's treasurer, while the equally wealthy writer Rebecca Hooper Eastman was its secretary.[61] Within a few months, the body had accumulated almost $6,000; the largest sums were donated by Mabel Choate herself and by the steel magnate and philanthropist Andrew Carnegie, the initiator of the Carnegie Endowment for International Peace (founded in 1910). As the president and director of the philanthropic Carnegie Corporation of New York, he, like Choate, contributed $1,000;[62] the heiress and influential widow Dorothy Whitney Straight donated $250 in memory of her late husband Willard, who himself had died of influenza in Paris shortly before.[63] Many small gifts from the circles associated with the women's colleges, furthermore, show that large numbers of the ACA's members followed the call to donate in memory of their late colleague. Within the ACA, the news of Sidgwick's death had given rise to a groundswell of interest in the organization's new, international commitments.[64] By 1921, a total of $10,000 in private donations had been collected.

The Rose Sidgwick Memorial Fund was designed to help British university women spend one year studying in the United States. In 1921 the first grant recipient, a British economist, arrived for her year-long visit to Columbia University.[65] Administered and financially supported by the ACA, the Fund guaranteed the award of a twelve-month grant for a British woman scholar every subsequent year, and indeed is still doing so today.[66] The Rose Sidgwick Memorial Fellowship came to symbolize the ACA's success in linking its wartime commitment with its postwar ambitions, and provided those ambitions with a respectable anchor in society. The grant was the foundation of the activities to which the IFUW would sign up on a larger scale shortly afterwards.

Transnational Networking: The Founding Years, 1919–25

Above and beyond all the public declarations and rhetorical flourishes, Rose Sidgwick's death set the seal on a transatlantic "union of hearts" of a very personal kind. While the lecturer was carried to her grave amid full honors, her weakened colleague Caroline Spurgeon struggled to overcome the influenza. Unfit to travel, Spurgeon remained in Gildersleeve's care in New York for several weeks, until mid-January 1919. It was during this period of convalescence that the spontaneous rapport between Virginia Gildersleeve and Caroline Spurgeon deepened, blossoming into a transatlantic companionship that would flourish until Spurgeon's death in 1942.[67]

The connection between Spurgeon and Gildersleeve formed the nucleus of academic networking across the Atlantic. Regarding the American women, the most important figures have already been named: heading the list was the president of the women's college Bryn Mawr, Carey Thomas; also important were Mary Woolley, president of Mount Holyoke College[68] and chair of the AAUW, and Lois Kimball Rosenberry, historian and Dean of Women at the University of Madison, Wisconsin. Responsibility for the Association's policies and activities lay with the AAUW's leadership, which in turn recruited largely from the presidents of the women's colleges affiliated with the Association.[69] In the 1920s, at a time when US foreign policy was guided by political abstention after the failure of Wilson's call for the United States to join the League of Nations and during the presidencies of Harding and Coolidge, the officers of the AAUW were part of a small but influential class of "internationalists" that represented America in Europe and the world and tried to bring about a liberal, international global order through the force of their personal commitment.[70]

On the British side, the inner circle of founders included, alongside Spurgeon, Winifred Cullis, England's first woman professor of medicine, and the biochemist Ida Smedley MacLean, who worked at the Lister Institute of Preventive Medicine in London. All the women I have named belonged to the first generation who had managed to make the leap into a university career. Carey Thomas (born 1857) and Caroline Spurgeon (born 1869) were the oldest in the group—pioneers and dominating personalities. Gildersleeve, Cullis, MacLean, Rosenberry, and numerous other activists (all born in the mid-1870s) could also draw on many years of experience in research, teaching, higher education policy, and administration. Rosenberry and Gildersleeve aside, each had pursued studies or research abroad, mostly as a way of circumventing barriers that were placed in the path of their academic careers at home. Thomas, who had been rejected for doctoral research by Johns Hopkins, studied instead at Leipzig University and obtained her doctorate in Zurich; Spurgeon gained a doctorate at the Sorbonne that was recognized in Britain,[71] as did Cullis for Canada and MacLean for the United States.

The initial foundations of the IFUW took little more than a year to build. It was in March 1919 that the British university women made their official call to the ACA to establish a joint organization to promote the international exchange of women students, teachers, and researchers, and July 1920 already saw the constitutive convention of the IFUW, with representatives from 15 countries. Besides America and Britain, women attended from Canada, India, South Africa, Australia, Italy, France, Czechoslovakia, the Netherlands, Spain, Belgium, Denmark, Norway, and Sweden.[72] In none of these countries was there an organized community of women academics at that time, a fact that highlights the tenacious and professional efficiency required to set up the new organization so fast. It also illustrates the close connection between the development of the international network and the establishment of individual, national organizations of academic women: the IFUW only accepted one representative from each nation.[73] For Europe, this rule meant that national associations had to be formed from scratch, but even the United States' own organizations did not, in 1919, fulfill the required criteria. Because the ACA only represented the interests of the colleges of the northern states and their graduates, the establishment of the IFUW prompted it to merge with the much smaller association of the southern states and form a single, national organization.[74] This fusion also opened up the opportunity for a change of name, something for which Gildersleeve had long advocated: "We cannot expand and become influential under our absurd name," she reasoned before the general meeting in 1919, and succeeded in having the previous name "Association of Collegiate Alumnae" (ACA) replaced by "American Association of University Women" (AAUW). Under this title, the Association would, Gildersleeve asserted, emerge as "an international institution with affiliations with all other university women abroad."[75]

During the first two years, the practical organizational work of the IFUW fell to the British and American associations and their respective Committees for International Relations. Both these bodies were able to secure external support in the period. With Spurgeon's assistance, the British women obtained the use of an office and a secretary's salary from the British government's Universities' Bureau,[76] while Gildersleeve helped the American women gain a similar degree of support from the Carnegie Institute of International Education in New York.[77]

The rapid growth of the international association was made possible by the availability of sound resources in the administrative domain; it was driven on by the carefully targeted letter-writing and traveling activities of the American and British associations' founder members. Carey Thomas, for example, requested a year's leave of absence from her college in 1919 in order to travel around Europe, Africa, and Asia. Armed with an official mandate from the AAUW, she used this journey to establish the contacts that would form the basis of the IFUW's development.[78] In January 1920, Thomas was able to report that she had found just the right women to organize Spanish and French academics on a national level. These included the educationist Maria de Maeztu, who had studied in Spain and in Marburg, Germany, and received an honorary doctorate at Smith College, Massachusetts, in 1919. Maeztu had recently established a residence for women students in Madrid, where she worked as a university professor, and was an authoritative figure in the Spanish women's and educational reform movement.[79] For France, Thomas was equally enthusiastic about the mathematician Anne Amieux. Amieux had gained her doctorate at the Sorbonne in 1889 as one of the first women in France to do so. She was awarded the prestigious Albert Kahn "Around the World" fellowship by the Sorbonne in 1905, and in 1918, when women's higher education in France was in a state of upheaval after World War I, she became the director of the École normale supérieure des jeunes filles in Sèvres, near Paris.[80] Amieux was one of those working toward the "assimilation" of girls' schooling into the boys' schooling system. As the new director of the École normale supérieure, she did not wait for the final political decision to be taken, but set to work immediately reorienting her school toward the standards of the *baccalauréat* and making the school into a female elite institution that would prepare girls equally for university, teaching, and public service.[81] This was probably one reason why Carey Thomas considered her "a perfect darling with all our views."[82]

Other contacts with France were forged through Marguerite Mespoulet, a teacher at the Lycée Victor Hugo in Paris. The first scholar of English literature to gain a professorial qualification in France, Mespoulet—like Amieux before her—had been awarded the Albert Kahn "Around the World" traveling fellowship in 1911. Again like Amieux, she belonged to the Club Autour du Monde of former Kahn fellowship winners, who were personally committed to the philanthropist's utopian vision of world peace.[83] Mespoulet first visited Barnard College in New York just after the end of World War I, and later found a long-term livelihood at the college. Through Mespoulet, the IFUW established contact with Marie Bonnet, who in 1901 had founded a "settlement" for women students in Paris, based on an American model. Amieux, Mespoulet, Bonnet, the historian Marie Monod (who also taught at the École normale supérieure in Sèvres), and the German literature specialist Marie-Louise Puech, who had taught at the francophone McGill University in Montreal for eight years, together became the nucleus of a national French association of female academics that included around a hundred women. It was established in 1919 and quickly became a member of the IFUW.[84]

In Italy, Czechoslovakia, and the Scandinavian countries, too, national organizations sprang from personal contacts. During her stay in Rome, Florence, and Milan, the Englishwoman Christine Burrows, principal of a women's college in Oxford, met the Italian academics Isabella Grassi and Luisa Ancona. A. C. Paues, who had studied at Newnham College, Cambridge, and gained her doctorate at the University of Uppsala, spread the word among her colleagues in Sweden, Norway, and Denmark.[85] Ties to academic women in Czechoslovakia were created with the help of Alice Masaryk, the daughter of the new Republic's first president and his American wife, Charlotte Garrigue. The economist Masaryk, who held a doctorate in history, was the founder of a school for women social workers. In 1904, upon completing her studies, Masaryk had worked in the settlements of Chicago, and since that time had been a close friend of Julia Lathrop, a prominent member of the AAUW from the circle around Jane Addams.[86]

The creation of new associations in Europe was not the only objective of these international networking processes. When the British women Spurgeon, MacLean, and Cullis, on the invitation of the AAUW and at the expense of the Institute of International Education in New York, undertook a further trip across the United States in 1920, it was with the aim of promoting a future shift within the American association toward the world at large. They hoped to convince as many members as possible, even in remote localities, of the need to build an international association of academic women, inspiring them to make this task their own—and to support it financially.[87]

It becomes clear from this sketch of the IFUW's formation that the women who managed to carve out a professional livelihood for themselves within universities, science, or higher schooling before 1914 had pursued international paths in academic training to reach their positions. Like the founders themselves, the majority of the women interested in international networking and in building up a national organization of academic women in their own countries had previously studied abroad (in Germany, the United States, Canada, France, or Britain),[88] or for other reasons felt dedicated to the liberal ethos of a civic world community regarded at that time as "Western" or even as Anglo-American.[89] Nicole Fouché has illustrated this strikingly for the case of

France: there, the establishment of a national organization of women academics was tantamount to a move toward America and Britain for women's and educational policy—something which the mainly Catholic women graduates initially refused to countenance, for denominational and cultural reasons. At least in the founding phase, the organization in France was driven solely by Protestant women, who made up a minority in the French female educational elite, but were favorably disposed to America and its women's organizations.[90]

The history of women's academic mobility shows that a transatlantic female elite had been forming in the universities of the Western world since the late 1880s, although it had not so far defined itself or explicitly coalesced as such. This elite shared many values and opinions, so that the circle of the IFUW founders quickly discovered common ground and personal sympathy. The immediate affinity that had generated such extraordinary synergetic effects during Spurgeon and Sidgwick's US tour in December 1918 characterized encounters in other countries as well. In 1919, Carey Thomas penned effusive reports from Spain and France expressing her happiness at having gained the friendship of Maeztu and Amieux: "I do not know when I have liked any strangers so much as I liked these two women."[91]

Despite all the differences in their national circumstances, the female college graduates of the first and second generation, in particular, also shared the experience of being a newcomer and an outsider. The often convoluted routes of their biographies showed similarities, as did their personal views of educational objectives, scholarly excellence, and a universal science in which it was not sex or nationality that counted, but solely the quality of research.[92] This latter aspect lent itself especially persuasively to the building of a genuine internationalism that would reach across the trenches of World War I—as I outline in chapter 4 for the case of Austria and Germany, whose associations joined the IFUW in 1922 and 1926, respectively.

3
Female Networks for Science: Programs and Politics

Internationalism, Science, and Gender

On July 12, 1920, several hundred people gathered in the auditorium of Bedford College, London, to attend the opening of the International Federation of University Women's inaugural conference. The four members of the IFUW board and 32 official delegates from 15 countries presented the new organization; also in attendance were numerous individual members of the associations in Britain, Canada, Denmark, France, Norway, Sweden, and the United States. The guest list also included prominent representatives of British liberalism and conservative members of the British government, selected protagonists of the British women's and suffragist movement, high-ranking university figures, renowned literary personalities, and official representatives of the new League of Nations—in short, the colorful spectrum of those British or transatlantic "internationalists" of the postwar period who had taken up a resolute position against Germany and the Central Powers during the war and who now placed their hopes in the achievement, through the League of Nations and similar institutions, of a peaceful global order based on mutual understanding and carried forward by an educated cosmopolitan elite.[1] Among the eminent guests were Gilbert Murray, regius professor of Greek at Oxford and England's "foremost League intellectual," who presided over the British League of Nations Union from 1922 to 1938; William Beveridge and Graham Wallas, both Fabians and professors at the London School of Economics; the writers John Galsworthy and H. G. Wells, cofounders of the International PEN club formed in 1921; the Australian medic and politician Sir John Cockburn, president of the Men's International Alliance for Woman Suffrage; Lady Rhondda, survivor of the American luxury liner *Lusitania* that had been torpedoed by German submarines and editor of the feminist magazine *Time and Tide*;[2] Lord Bryce, Regius professor of Law at Oxford, longtime British ambassador to the United States, enthusiastic proponent of the idea of the League of Nations, and author of the government's much-cited Bryce Report on German war crimes in Belgium and France; his colleague, the eminent legal historian and philosopher of law Frederick Pollock from Oxford, who had also been a member of the Bryce Committee; and Viscount Grey of Fallodon, the first president of the League of Nations, who delivered the keynote address.[3] Lord Robert Cecil, the British "soul" of the League of Nations, and Lady Aberdeen, the *grande dame* of the international women's movement and long-standing president of the International Council of Women (ICW), sent lengthy goodwill messages expressing their regret at not being able to attend.

Before this illustrious audience, Caroline Spurgeon, the first president of the IFUW, explained the aims and objectives of the transatlantic federation. She first emphasized the significance of the new organization in terms of international politics, describing the founding of the IFUW as the spearhead of an "immense process of education":

> Of education in judgment, in width of view, in knowledge, in tolerance, in a sense of proportion among individuals, and in mutual respect and sympathy and mutual help and cooperation among the peoples of the world. We believe that this is the beginning of the organized training of women to be citizens of the world, and through women, of men too,...in short, the enterprise of preparing some portion of human material for the League of Nations that is to be.[4]

As members of the new organization, Spurgeon continued, academic women bore a special capacity and responsibility for this educational task. To justify her claim, she drew on the familiar rhetorical figure of the "specific characteristics of the female sex," a trope that had been successfully deployed in all the national feminist and women's education movements of the West as a way of laying claim to and maintaining control over particular domains of social activity on behalf of middle-class women.[5] If this background was quite familiar, Spurgeon's address broke new ground in associating the special feminine aptitude for educational responsibility with a commitment to the concept of "universal science," a combination from which she derived the right and duty of the IFUW to take action for international understanding. The cause of scientific internationalism here propagated by Spurgeon had flourished before World War I and had resulted, among other initiatives, in the organizational union between European and American scientific academies.[6] The networking of the scientific world that had increased so rapidly around 1900 rested on a belief in the universal claim to truth and ethics held by the "pure" sciences, which could only, it was argued, be promoted by collaborating internationally and overcoming national fragmentation.[7] If, as Livingstone argues, scientific internationalism should be viewed as a social achievement, not the inevitable consequence of some inherent scientific essence,[8] there can hardly be an era that confirms his point as clearly as the 1914–18 conflict and its immediate aftermath. The young, almost exclusively male international scientific community that had emerged before the war had splintered under the pressure of national loyalties. If Spurgeon now recalled the belief in academic internationalism, this was fully in line with a trend that was arising after World War I among the scientists of the former Entente countries in step with the increasingly precise delineation of the idea of the League of Nations—as an emphatically Western response to victory in the war and as an aspiration to create afresh the community of the past, under transatlantic auspices and initially excluding the former wartime enemies. It was within this new North Atlantic community that Spurgeon regarded the IFUW and its academically trained members as called to fulfill a special task: "I believe," she explained, "that we, in our small cross section of the human family, are starting the great far-reaching and thrilling enterprise of bringing into being the moral forces, the knowledge, the imagination, the vision, necessary to real brotherhood."[9]

By staking this claim, the IFUW was asserting for its members an extraordinarily important role in shaping the future united community of world citizens. The board was, however, quick to reassure the assembly that the establishment of a women's organization must by no means be regarded as an attempt to pursue either "separatist" or "ultra-feminist" goals. Women and men, they argued, must work together in the international educational movement—but not enough progress had yet been made for this to be achieved without the help of an organization specifically focusing on women. The IFUW did not, then, aspire to create a separate or parallel space for itself on the international stage, but rather to open up for women the long-term opportunity to participate on an equal footing with male colleagues in the work for their shared concerns. Virginia Gildersleeve told the audience how difficult it would be to put this into practice without having an organization dedicated to women. As an example, she pointed to a phenomenon often observed in the nurturing of young scholars, "peer formation," and the recently expanding practice of academic exchanges: "We find sometimes," said Gildersleeve,

> that even the men who have the greatest sympathy with the work and aspirations of our sex occasionally forget that we are there, if the question of an exchange professorship or sending students abroad comes up. It is not because they have no interest in women, but just because they do not happen to think of us. We may not get the same opportunity to participate that we should have if we had some women in the organization or a women's committee, just as a reminder.[10]

A comparable situation prevailed, she continued, in the efforts to develop international academic networks. Male colleagues, whether in the United States or elsewhere, lacked extensive contact with women scientists or university graduates, and consequently tended to underestimate the potential of the international female educational elite.[11]

Positioning the IFUW as a cosmopolitan authority rooted in scientific universality was one of the main objectives pursued by the federation's founders. Spurgeon viewed the IFUW as an "idealist" movement with "practical" objectives; its role was first and foremost to promote international understanding, a goal that could, she argued, best be achieved through "personal intercourse" between the members and graduates of universities. Precise knowledge of national similarities and differences would emerge slowly, in small steps, especially via individual encounters based, if possible, in those locations in each country where knowledge was generated and transmitted. There was no better starting point for international understanding than that of bringing students, teachers, and researchers into contact via universities, in the hope that these positive experiences would later lead the women involved to become ambassadors in their own countries for a new internationalism—one founded on objective knowledge (objective knowledge, that is, of national differences).[12] The ambitious scope of this project becomes all the more clear in view of the fact that the officers of the IFUW were perfectly aware of the sharp contradiction between the calls for international science and the realities of science teaching, which was marked by nationalistic indoctrination. The Norwegian professor of chemistry Kristine Bonnevie, one of the three female delegates to the League of Nations, put her finger on the discrepancy: "While scientists [are] basing their work upon the interchange of knowledge between nations, the university life, with its characteristic customs and traditions, might mean for the student a development not only in a national but in a nationalistic direction."[13]

The IFUW's statutes limited themselves to the aspect of an international commitment "to promote understanding and friendship between the university women of the nations of the world, and thereby to further their interests and develop between their countries sympathy and mutual helpfulness."[14] This objective was to be achieved by a range of means: the IFUW's networking within the structures of international organizations and the exchange of women students, teachers, and researchers; the development of a network of international clubhouses; and the systematic provision of international hospitality on an individual level.

The founders' agenda spelled out further ambitions for the IFUW. The federation was to serve as an international forum for comparison and exchange of ideas, to discuss key demands regarding the politics of professionalization, and to work together to develop strategies to overcome national restrictions that impeded women's access to universities and the professions—activities like Spurgeon's 1919 protest against the gender-specific admission regulations at Oxford and Cambridge. Accordingly, at the 1920 conference in London, the presentations of the 15 national associations (like those of all the further member associations joining in later years) took the form of reports on the situation of women in their country's universities and professional environments. An optimistic conviction prevailed that women would, through comparative discussion and international networking, be able to persuade their own governments with greater inspiration and knowledge, and therefore have greater impact. Carey Thomas expressed this hope in euphoric terms: "By working together I believe we can anticipate by several centuries the progress of University women."[15]

It was the founders' concern that all the IFUW's affiliated associations should advocate four specific objectives. These were, firstly, the uncompromising struggle for coeducation in universities so as to secure and improve the quality of women's training; secondly, the call for women's "absolute freedom" to study, work, and develop their skills and potential both academically and professionally; thirdly, the demand for equal pay for equal work among male and female employees in schools, colleges, and universities; and fourthly, the right of women "to have an acknowledged right to the happiness of a family"[16] even if they were pursuing a professional or academic career.

This agenda had its origins in the politics of US and British higher education. The fourth point, in particular, not only sounded radical to Continental European ears, but was in fact disputed even among the US delegates. Reading the minutes of the first conference in London, it becomes clear that the federation's British and American initiators adopted a strategic approach to persuading their Continental colleagues of the rightness and urgency of these demands. Introduced by the resolute college president Carey Thomas, under the banner "Next Steps for University Women," Ida Smedley MacLean offered herself to the audience as a living example of how science and marriage or family need not be mutually exclusive. She spoke as a scientist and as "a woman who is married and who has gone on with her job." Being able to combine one's profession and one's family was a right, she said, and a task that needed to be handled not as a moral issue but as a logistical one.[17]

Between 1920 and 1939, the Council of the IFUW, to which all the member associations sent one delegate, met 19 times; almost every one of the European associations had the opportunity to host one Council meeting.[18] All the associations tried to arrange these intensive, three-day working meetings with the absolute maximum of lavishness, dignity, and public impact, while also providing IFUW officials with the key information about the situation in their host country. However, particular highlights in the history of the federation were the eight large conventions open to all members that, during this period, took place in Britain (London 1920, Edinburgh 1932), France (Paris 1922), Norway (Oslo 1924), the Netherlands (Amsterdam 1926), Spain (Madrid 1928), Poland (Kraków 1926), and Sweden (Stockholm 1939). Each of these conferences was attended by around 250 to 500 IFUW members from Europe, the United States, and the then British dominions. Members reported finding the meetings uniquely inspiring opportunities to meet and exchange views with other academic women. In her address at the close of the third general meeting in Oslo in 1924, Virginia Gildersleeve described the conferences as "a kind of power house of energy" for the federation's members: an energy that the women would now be able to take home and put to use "for the highest type of work on the various lines...and especially for good citizenship in the world of nations."[19] Many of the IFUW delegates and members stayed on after the meeting itself. They made the most of their visit abroad by taking part in one of the educational trips organized by the host association as a direct follow-on from the meeting, building their own contacts, or accepting other personal invitations. For example, after the Council meeting at Wellesley College, Germans Agnes von Zahn-Harnack and Anna Schönborn took up the AAUW's invitation to lecture in Milwaukee, Chicago, and Bloomington, Indiana, on the situation in Germany and the activities of the German association, the DAB. They met with German exchange students and arranged consultations with the director of the International Bureau of Education in New York, Stephen Duggan, and the head of the Carl Schurz Memorial Foundation, a body dedicated to the improvement of German-American cultural relations.[20]

The general meetings were bound by the IFUW's statutes to debate the general trajectory of the federation, set up working groups on specific topics, and hear reports on these groups' activities. Equally, the events also facilitated the board's own plans to win members over to its objectives and policy initiatives. The orientation and efforts of the IFUW committees indicate that the federation remained largely true to the essentials of the agenda it had established in 1920, and continued to work for that agenda's implementation. The committees set up in London in 1920 devoted themselves to the establishment of international guesthouses (the Hospitality Committee) and the evaluation and standardization of higher education systems and degrees across the member countries (the Standards Committee); in 1924, the Fellowship Committee and the Committee on Intellectual Cooperation took up their work. The latter was formed in response to the establishment of the League of Nations' Commission on Intellectual Cooperation,[21] and was the IFUW's closest link with the League of Nations, in that the Norwegian geneticist Kristine Bonnevie served on both bodies. She was one of only three women holding a League of Nations mandate, and had been appointed to the Commission on Intellectual Cooperation alongside Marie Curie, Albert Einstein, and 11 other respected scientists.[22] In 1920, Bonnevie helped to found the Norwegian association of academic women. She was its president for the first five years and maintained close contact with the IFUW board via her Norwegian colleague Ellen Gleditsch. It was at Bonnevie's prompting that the IFUW Committee on Intellectual Cooperation was established, under the leadership of the French scholar Marie-Louise Puech.[23] This body directly assisted the League of Nations' Commission on Intellectual Cooperation through preparatory work. Following the Commission's specifications, it carried out surveys among its own member associations on the treatment of the League of Nations in school curricula and on the social importance of children's library provision. The IFUW committee also joined in the Commission's efforts to encourage the exchange of professors, university librarians, and curators, as well as to facilitate international research by introducing a "travel card for intellectual workers," a kind of Europe-wide reader's card for libraries and archives.[24] There were also plans for a multinational service, to be run by the IFUW itself, for the translation of specialized scientific literature, along with energetic preparations for exchanges between academic women employed in libraries and archives; these plans, however, were wrecked by administrative obstacles, an uncertain political environment, and the onset of the Great Depression.[25] In the interwar years, the most time-consuming, but also most fruitful, work by the Committee, under the direction of the Paris historian Marie Monod, was an 800-page synopsis of academic terminology in Europe and the United States, designed to make it easier for women to find their way rapidly through the various higher education systems and to provide information on the academic degrees available in each country. This highly praised *International Glossary of Academic Terms* appeared in 1939, shortly before the outbreak of World War II—a dying echo of the attempt to sustain peace by means of transnational networks that bound together national scientific systems.[26]

Three further committees founded during the 1920s took up the call for the equal treatment of university-trained women in employment and marriage. The committee Careers in Industry, Finance, and Trade was set up in Brussels in 1924, Legal Status of Married Women in Amsterdam in 1926, and Investigation of the Position of University Women in Public Services in Madrid in 1928. Their practical work concentrated essentially on gathering important data, in collaboration with the individual member associations. The three committees failed in their attempt to establish robust personal networks between women university teachers and successful businesswomen, which they had hoped would enhance young women graduates' chances of finding their way in the business world; their suggestion that a transnational placement service for women intellectual workers be established also came to nothing. Instead, during the early 1930s the organization had to confront ever more forcefully the increasingly frequent attempts, in all countries, to squeeze women out of the professions. Reports on this problem were collected, and a memorandum based on them submitted to the International Labor Organization with a request for remedial action. Many of the programmatic demands that fell by the wayside in this period were victims of the devastating global economic crisis, accompanied in most European countries by political upheavals. In addition, the sheer complexity of issues like married women's citizenship meant that the international experts working for the IFUW bodies were overtaxed in terms of time and energy: the committees depended on the volunteer efforts of women who were generally already at full stretch in their teaching, research, or professional work.[27]

The numerous high school teachers in the IFUW devoted their energy to international exchanges of teachers and the comparative exchange of information about forms of secondary education. The Committee on Interchange of Teachers, appointed in Oslo in 1924 and chaired by the president of the British Association of Head Mistresses, Reta Oldham, faced problems similar to those experienced by the Committee on Intellectual Cooperation, which sought to foster exchanges between women specialists. The most troublesome administrative barriers were successfully overcome in Britain when Oldham and others managed to push the Board of Education into making two key legal changes. In 1925, the Board added a clause to employment contracts for state teaching staff that permitted them to take temporary leave of absence without forfeiting their pension claims; and in 1927, it expanded its existing schools' exchange program with the British dominions to cover all nationalities, so that teachers from Europe and the United States, too, were now entitled to obtain a work permit for British public service.[28]

International exchanges between secondary teachers were initially limited to Britain and the United States. In 1927, four American and four British women were teaching in the partner country; a year later, the first exchange took place between a school in London and another in Berlin.[29] In 1930, the number of British-American exchanges had risen to eight, while two were running between Germany and the United States.[30] The outcome of the exchange program was announced in 1932 in Edinburgh: 15 exchanges between Britain and the United States, four between Germany and the United States, and one between Britain and Germany. The 1932 figures can be taken as a grand total, since the exchange of women teachers temporarily came to a complete halt in 1933,[31] and during the period from 1936 to 1939, it was once again limited exclusively to exchanges between Britain and the United States.[32] Despite its relatively modest scale in numerical terms, this result was considered a success, and received due recognition from the League of Nations' Intellectual Cooperation Section. A 1932 report by the Section's Finnish delegate, Armi Hallsten-Kallia (who attended most of the sessions of the IFUW's Committee on Interchange of Teachers), stressed that even if teacher exchanges in Britain had remained within rather manageable dimensions, the majority of them had been initiated by the BFUW, which thus held an acknowledged pioneering role.[33] The special difficulty of organizing international teacher exchanges lay in the fact that all teaching staff had to have their salaries paid by their home country. Because most teachers in Europe were government employees, the introduction of a system of paid leave of absence in many cases involved amending civil-service legislation—changes which were called for emphatically by the various associations, but by 1933 had been implemented only in Britain and Germany.[34]

A degree of success similar to that of the teachers' exchange committee was achieved by the Committee on Secondary Education, headed by the Belgian educationist Germaine Hannevart. This committee's goal was to produce an overview of the different structures and curricula of secondary schooling for girls in Europe and the United States. Responses to a first, questionnaire-based survey were provided by all the member associations. The reports revealed such confusingly disparate conditions that immediate publication was impossible: the sole, and very sobering, common ground between the reports was the finding that in most countries, women secondary teachers did not occupy the highest positions in schools. The committee called on all members to demand remedy for this situation in their own countries.[35] Apart from this project, several years were spent in trying to find a trained researcher and the necessary funding to carry out a more systematic comparison of schooling for girls in Europe and the United States. In 1930, the Hungarian government agreed to grant high school teacher Amélie Arató, a respected scholar who had earned her doctorate in France, two years' leave of absence to carry out the project, with continued payment of her salary. The IFUW covered Arató's travel expenses and the cost of printing the study that she produced, *L'Enseignement secondaire des jeunes filles en Europe*, which was published in Belgium in 1934 and was frequently cited for more than a generation.[36] Arató's scholarly work was regarded as a great common effort of the IFUW, something that expressed the federation's spirit in an exemplary way. Funded by the government of a former Central Power, and additionally supported by an IFUW travel grant, the Hungarian teacher's research took her on a journey of more than a year across Italy, Austria, Switzerland, France, Britain, the Netherlands, Belgium, and the United States. For all of this time, she was accommodated exclusively by members of the IFUW. Federation members at her various destinations used Arató's transnational research project as ammunition in their work for the admission of women to senior positions within their country's school system.[37]

This short overview of the IFUW's committees by no means exhausts the scope of the federation's work. In the following sections, I examine in more detail the two areas that between 1919 and 1933 played the most important role in the program and policies of the IFUW, requiring of members the greatest personal and financial commitment. These were, firstly, the creation and utilization of international clubhouses and, secondly, the development of an international fellowship program. I also ask how the executive attempted to mobilize IFUW members for these initiatives in keeping with the ideal of scientific internationalism, and to what degree they proved successful.

Clubhouses and Hospitality

Personal contacts and friendships across borders seemed to offer the most important social basis for lasting academic and political internationalism. It was hoped they would solidify the foundations of the IFUW. Intersections in this web of friendships were to be the international clubhouses that the Council aimed to establish in Europe's great cities. A Hospitality Committee, appointed at the first London meeting and led by Carey Thomas, was charged with implementing this goal.[38]

As early as 1920, Caroline Spurgeon had expressed her hopes of a broad participation by IFUW members in this centerpiece of practical international networking. Every single woman, she stressed, "can do an amazing amount... towards weaving together these individual strands of friendship to form indestructible bonds which will eventually bind people together all the world over."[39] How this fabric of hospitality in the service of world peace might actually look in practice was something she could describe from her own experience: during her second trip across the United States, she had herself stayed in the IFUW's first international residence.[40] In summer 1919, the AAUW women had rented a building diagonally opposite the White House, on Lafayette Square in Washington, DC. The "beautiful old house of the square colonial type" had twelve bedrooms, several offices and clubrooms, a closed veranda that could be heated in winter, and a handsome garden. A dining room and a café open to the public were added later.[41]

This AAUW initiative was primarily a response to developments within the United States. The association's executive had previously been based in either Boston or New York, but because of the politicization of the AAUW and the general shift in the orientation of other important organizations of higher education toward the federal government, it now seemed necessary to move to the political capital instead. The Emergency Council on Education that was established in 1917, tasked with reorganizing higher education in the United States to align with national interests, continued its existence after 1919 under the new title of American Council on Education; its headquarters were in Washington, DC. In 1919, as well, a federal Bureau of Education was being developed and plans (ultimately fruitless) were under discussion to introduce, for the first time, a "secretary of education" based in the President's cabinet. "The concentration of all national movements in Washington D.C. has come to stay," wrote Virginia Gildersleeve, summarizing the circumstances which, she argued, obliged the AAUW to create a permanent, high-profile base in the US capital. She feared that "the interests of women in education will need protection as never before."[42] The AAUW had joined the American Council on Education in 1919 as a founder member. It had entrusted one of its most prominent representatives with the work of lobbying for university women's interests: Gildersleeve herself was elected general secretary of the Council for several years running, and in 1919/20 was a member of its executive committee.[43]

The AAUW building in Washington was intended to do more than to make the association physically visible within the political landscape of US education. Following the example of Suffrage House, established in Washington by Carrie Chapman Catt in 1917, the association's new national headquarters would not only accommodate the AAUW offices but also create a social center for women college graduates that additionally offered room and board for traveling members and their guests.[44] Although a university town in its own right, Washington could not compete with the academic structures and facilities—such as Women's University Clubs—that were taken for granted in Boston or New York. However, its growing federal administration meant that the capital was becoming more and more important as a place of work for women college graduates, and one of the expectations placed in the new building and its amenities was that they would tempt more of these women to join the AAUW.[45]

Even before the association had fully settled into its new home neighboring the White House, the US Department of Commerce had begun to press for the coveted premises to be returned to the Department's own use. The AAUW resolved to buy a different building instead, not far from the original location.[46] The purchase of this new headquarters, on Farragut Square, was financed through loans of around $250,000. At first, this heavy debt seemed likely to stretch the AAUW to the financial breaking point, but a national fundraising campaign soon revealed unanimous and eager support for the project among the association's membership. The campaign called on every member to contribute $14 toward the renovation work on the building in Washington. Within just a few months, $140,000 had been collected; the remainder of the mortgage was paid off within four years of the building's acquisition.[47] Besides monetary donations, members contributed furniture, carpets, tableware, and books to fit out the building to a suitably prestigious standard.

The unexpectedly wholehearted, nationwide enthusiasm for the development and maintenance of the AAUW's clubhouse and administrative headquarters in

Washington showed that having its own highly visible location in the nation's capital strengthened the association not only externally, but also within its own ranks. The building embodied personal feelings of affiliation and national presence in a tangible, easily communicable way. It served political, practical, and social purposes, but above and beyond that it was also a symbolic expression of the social and political importance of academic womanhood in the United States.[48]

The AAUW headquarters was the clubhouse of the Washington branch, as well as housing the association's growing executive and offering all members from within and outside the United States a social meeting place and overnight accommodation. Certainly, far more of the visitors from Europe applied to the AAUW Hospitality Committees in New York and Boston, and the greatest magnets within the AAUW's web of international academic contacts remained Barnard, Bryn Mawr, and Wellesley College—but the Washington headquarters functioned right from the start as a national and international advertisement for the work of the AAUW and the IFUW. It symbolized the importance that World War I had given the American association within national and international education policy.

The AAUW headquarters in Washington became the model for the international clubhouses that IFUW officials dreamed of creating—spaces where international friendship could be nurtured "through personal intercourse" in an academic context. This involved not only appropriate buildings as a physical setting, but also the psychological role of those buildings in integrating members within the IFUW. The clubhouses across the world were to represent the living, practical core of the organization, something that was accessible to all members and for which, in turn, all members would take responsibility. It is in fact remarkable how rapidly and successfully the establishment of additional international clubhouses proceeded in Europe in the early 1920s. The American women's engagement and influence, as well as their skill in acquiring funding and large-scale gifts, were decisive factors in that process.

Soon after the opening of the AAUW's Washington headquarters in 1919, an opportunity emerged that would enable a similar center to be created in Paris at a single stroke. This opening did not originate in France itself but in the United States, through Virginia Gildersleeve and her superb New York connections. An alumna of Barnard College, Helen Rogers, had taken up a post as private secretary to the New York heiress and philanthropist Elizabeth Mills Reid after graduating in 1903; in 1911, she married Reid's son Ogden, who ran the daily paper owned by the family, the *New York Tribune* (the unofficial organ of the Republican Party). After the birth of her two sons, in 1918 Helen Rogers Reid took over the advertising department of the *Tribune*, which was one of the earliest and most important sources of support for America's entry into the European war.[49] With Gildersleeve's assistance, Rogers had been appointed to the Barnard College board of trustees in 1914, and she succeeded in persuading her mother-in-law to support the establishment of an international residence for women college and university graduates in Paris.[50]

Elizabeth Mills Reid had inherited her large fortune from her father Darius Ogden Mills, who followed the Californian gold rush to Sacramento in 1849 and quickly became rich as a retailer and banker. Mills founded the Bank of California in San Francisco, amassing assets of around $40 million, which he left to his daughter in 1910.[51] She had married in 1881, and in 1889 moved to Paris to accompany her husband, Whitelaw Reid, when he was named official American envoy to France. An art-lover, Elizabeth Mills Reid soon became acquainted with the lifestyle of American bohemian society in Paris. The moral conduct of these circles generated rumors that made headlines in the *New York Times*, causing worry among the concerned parents of daughters studying art in Paris.[52] Reid came into contact with members of the US Protestant community in Montparnasse who were trying to set up a morally irreproachable club for impoverished young women artists. When a prettily situated Protestant school close to the Jardin du Luxembourg went bankrupt, she bought the building and established the "American Girls' Club" there. Reid's hope was that an institution of this kind—supervised by a woman dean, with 50 single and double rooms, a kitchen, a library, studios, exhibition space, and initially with obligatory church attendance—would make it possible for ambitious American daughters to take a part in the intellectual and artistic life of 1920s Paris without putting their good names at risk. The American Girls' Club, also known as the "Women Artists' Club," operated very successfully, and gradually became the artistic and intellectual center of the growing community of American women artists in Paris. Offering regular exhibitions and the opportunity to purchase the work of the residents, the club attained an international reputation.[53]

In 1922, Reid signed the extensive rue de Chevreuse premises over to the AAUW management to be used free of charge for six years; she also had the complex completely refitted and modernized. Only three years after the IFUW's founding, the 1922 biannual convention could thus be held in the freshly renovated, fully furnished rooms of the new center in Paris. On the occasion of the conference, the building was dedicated to its new function with great fanfare.[54] The structure served as a residence for American women students based in Paris, as an international clubhouse for the IFUW, and as the headquarters of the French association of academic women. American women in residence carried out Reid Hall's administration.[55]

The American University Women's Club in Paris quickly attracted attention through its outstanding amenities and excellent service. Jane Harrison was a British archaeologist who, upon retirement, had moved from Cambridge to Paris with her young companion Hope Mirrlees and found accommodation at the new IFUW residence for two years, through the mediation of the American Alys Russell. She praised Reid Hall in the highest terms, as having "the best bedrooms, room service breakfast, unlimited hot baths, and admirable cooking of the best French kind (touched by Americans), and more than that, a personal care and kindness that goes beyond one's heart." To appreciate fully the options that Reid Hall offered academic women in 1920s Paris, it is worth noting how protracted and fruitless Harrison's previous attempts had been to find acceptable, affordable Paris accommodation where she could work and receive guests. The scholar wrote enthusiastically of her spacious living and reception room in Reid Hall: "I can breathe and work and even think, if the delights of Paris would leave me a free moment."[56]

The stroke of luck in obtaining Reid Hall gave an important boost to the IFUW Hospitality Committee's efforts to create a transnational network of meeting places for academic women. In 1922, visions of similar halls in major cities around the world seemed, at least in the American women's eyes, to be not unrealistic dreams but perfectly realizable projects—assuming, of course, that their foreign colleagues would come to master the "tricks" of successful fund-raising that had been applied to such good effect in the United States.[57]

As it turned out, the British initiative to set up a clubhouse in London became the heart of the IFUW's complex of international hospitality. Although the British women were initially far less optimistic than their American colleagues about the chances of collecting the necessary donations through their own efforts,[58] members and sympathizers in Britain in fact showed an unexpectedly high degree of commitment, as did every one of the IFUW's affiliated associations. This was partly due to the special historical glamour and exceptional location of the building that was to house the IFUW's future London residence: Crosby Hall, a medieval town house with an impressive ballroom. The house, built in 1446 in the heart of London, had served as a residence for Richard III and the humanist Thomas More and a backdrop for Shakespeare's plays. It was one of the very few buildings in London to have survived the Great Fire of 1666 unscathed. The Hall's historical significance had preserved it from demolition in 1910; it was taken apart, stone by stone, and rebuilt by the University and City Association on an excellent site in Chelsea, with a view onto the Thames. The original plan was to use the Hall as a dining room for a new University of London residential college, but during World War I the building had been adapted to accommodate Belgian refugees. It had stood empty since 1918.[59]

It says much for Caroline Spurgeon's London connections, her powers of persuasion, and the general popularity of the idea of academic internationalism after World War I that the BFUW ventured to take on the great financial risk of buying Crosby Hall and fitting it out as an international hall of residence.[60] The acquisition and rebuilding of the premises required an investment of £50,000, of which £25,000 had to be paid up front before building work could begin.[61] To collect the necessary funds, the BFUW pursued a rigorously organized fund-raising campaign directed not only at its own members but also at the IFUW as a whole and at the British general public.[62]

After the addition of a new annex for living quarters and offices, the plan was to make Crosby Hall an international IFUW residence with a library, dining room, and clubroom. It was hoped that during the summer months, up to 50 "university women studying in London" and traveling IFUW members could be offered accommodation for a fee, the income from which could be used in winter mainly to subsidize affordable lodgings for women students from Britain and all over the world. A brochure giving details of the idea and illustrated with pictures of the building was distributed by the British women at the 1922 convention in the newly opened Reid Hall in Paris; they also presented a scale model of the London premises for members to admire.[63]

The fund-raising strategy was based on collecting the necessary resources room by room. A gift of £1,000 would entitle the donor to give one room in the new wing the name of her choice. For just £200, a donor would be commemorated for the furnishing of a room.[64] Contrary to the fears of the British organizers, the campaign to finance Crosby Hall was greeted with an enthusiastic response quite comparable to the case of the AAUW's clubhouse project in the United States. At the Paris conference, US and Canadian delegates were already pressing $100 bills on the British treasurer to help realize the planned center in London.[65] Members of the BFUW in Manchester, Birmingham and the Midlands, Glasgow, and London raised the considerable sum of £1,000 per region; further £1,000 donations came from the United States (twice, in 1924 and 1927), Australia, Canada, India, and New Zealand. The only Continental European organization to raise £1,000 for a room was that of Norway, while the organizations from Belgium, Czechoslovakia, Denmark, Finland, France, Italy, Sweden, and Yugoslavia (and later Austria and Germany) each paid £200, earning the right to be named as donors of furnishings.[66]

The collection of donations for Crosby Hall was in the hands of Caroline Spurgeon. She entrusted the organizational side to a fully paid employee who came with all the relevant experience: up to the beginning of World War I, Miss Arnold had organized highly successful events for the women's suffrage movement, and from 1919 to 1921 she had been responsible for arranging housing for British war widows. The BFUW membership as a whole also entered into the project with great energy. Alys Smith Russell, for example, an American by birth and the niece of Carey Thomas, gave the campaign invaluable support and "never stopped begging for a moment."[67] Russell had studied at Bryn Mawr before her family moved to England. She married the young philosopher Bertrand Russell and worked with him on German socialism.[68] The Russells' childless marriage broke down, and they divorced in 1921, but Alys Russell continued to dispose over her own fortune and benefited from an American lack of embarrassment in asking for donations as well as from her excellent connections to the British moneyed and aristocratic classes. She herself gave £1,000 for a room in Crosby Hall, which she dedicated as a memorial to her aunt Carey Thomas; she was also the driving force and treasurer of a group of members collecting funds for a room in honor of Thomas More's learned daughter Margaret Roper, whose Latin correspondence with Erasmus had spread her fame far beyond England.[69] Alys Russell also managed to persuade Lady Agatha Russell, the elderly daughter of the first Earl Russell, to donate £1,000 for a room to commemorate Alys's now long dead, progressive parents-in-law, John and Kate Russell, the Viscount and Viscountess Amberley, who were early proponents of religious freedom, birth control, and free love.[70]

The list of donors to the Crosby Hall project amounts to a roll call of British liberalism. It includes representatives of the women's suffrage and women's education movement; graduates of the first women's colleges in Cambridge and Oxford, Girton, Newnham, and Somerville, an important source of members for the early BFUW; and industrialists and philanthropists like Sir Otto Beit, director of the British South Africa Company and benefactor of an important fellowship for medical research. Not least, the wealthy residents of Chelsea and several large London businesses made their financial contribution to the history of the Hall. Just two years after the fund-raising campaign began, Alys Russell was able to inform the IFUW members assembled in Oslo that the British association had already collected £17,000 of the £25,000 required to start construction.[71] Work began in 1926 with a ceremonial dedication, and one year later, on July 1, 1927, Queen Mary opened the completed building amid great festivities, attended by high-ranking guests and numerous IFUW members from Britain and abroad.

The fund-raising for Crosby Hall was, then, a success story, and the same proved to be true of the use of the building, which now housed both the BFUW offices and those of the IFUW. In the first year alone, the Crosby Hall guestbook listed "several hundred" women from almost all the countries affiliated with the IFUW, although visitors from the British dominions and Continental Europe predominated. A large number of women tourists booked short stays; the majority of the longer-term visitors, who stayed for up to ten months, were researchers carrying out work at London's libraries, museums, archives, and hospitals.[72] Large receptions, workshops initiated by the guests themselves, and study groups lent Crosby Hall an open, international, intellectual, and sociable ebullience that very soon earned the "spirit of Crosby Hall" a legendary status not unlike the "spirit of Geneva." Female academic internationalism, it seemed, had become a living reality here.[73] Even if the impact of the global economic crisis drastically reduced the number of guests, especially those from overseas, and Crosby Hall constantly teetered on the brink of financial ruin, the communally funded model of an international social and academic meeting point in London became a living emblem of the goals that the IFUW had pledged to pursue on its foundation in 1919.

A vivid personal glimpse of this atmosphere can be gained from a report by Viennese physicist Berta Karlik, who spent a year in 1930/31 as a Residential Fellow at Crosby Hall. In 1932, Karlik gave a precise account of the endeavors to nurture personal contacts between the guests in and around Crosby Hall to members of the Austrian association.[74] The social high point of each day was the shared dinner in full evening dress at the long tables of the medieval hall, which began at seven every evening. New guests were seated at the "High Table" set crosswise at the front of the room. They were presented to the other diners, and specifically introduced to selected residents of the Hall. After dinner, the evening was spent together in the two great communal rooms or in the library, giving the guests time to get to know each other. Once on more familiar terms, they exchanged invitations to each other's rooms. "The atmosphere that prevailed in the whole building was a particularly congenial one," reported Karlik. "All the residents were very friendly, helpful, cheerful, almost more than I can describe. Little study groups arose as if of their own accord. We often spent time sitting with an Indian woman, for example, who answered our questions about her homeland." As well as their international origins, the women residents' interdisciplinarity exerted a very inspirational effect, Karlik said. A British radium researcher was resident at the Hall at the same time as herself, a former student at Stockholm, Paris, and New York. A French historian had made use of her stay to work at the British Museum on the influence of the papacy in the Middle Ages. An Indian archaeologist was writing up her excavation reports from the Sahara. An exchange of information and ideas on the League of Nations was offered by a Munich woman who worked in Geneva as a secretary to the League's press section. Karlik herself taught German to a small group of colleagues once a week, and offered a very successful "physics class." The unique atmosphere of Crosby Hall, in her view, arose from the fact that "many of the foreign women were spending just the one year in England, and...as a result felt this year to be one of the greatest experiences of their lives. For this reason the majority of the Crosby Hall residents lived at an enormous pitch of intensity, lifted out of their everyday habits, and this, above all, was what shaped the intellectual life of Crosby Hall." Her own experiences went beyond Crosby Hall and her own field of scientific work. Under the auspices of the Crosby Hall Hospitality Committee, Karlik traveled across all of England in the course of her fellowship, without, she stressed, ever having to spend a night in a hotel—she was always privately accommodated by members of the BFUW. A Christmastime invitation from Ida Smedley MacLean had introduced her to South Wales and to many local BFUW groups along the route. On returning to Vienna, even after several months she found it difficult to give adequate expression to the "incredible good luck" of her year-long fellowship. "Only when you feel giddy at the abundance of things," she noted at the beginning of her lecture,

> will you be able to form a correct idea of my study year in England. Then you will understand what an extraordinary experience it was for me, and...how exhausted I was on my return by the many impressions. That was only to be expected, considering that I had been offered the opportunity to become closely acquainted with more than thirty scientific institutions, meet certainly more than a thousand people (and have genuinely intense interchange with many of them), and travel across more or less the whole of England.[75]

Karlik's report emphasized again and again that the richness of her impressions was by no means restricted to scientific matters, but extended especially to cultural and social experiences. She felt "deeply indebted" to the IFUW for the year she had spent studying in Britain.

The IFUW clubhouses in Washington, Paris, and London remained the three most frequently visited, though smaller residences were also set up during the 1920s in Toronto, Montreal, and New York. American and British initiatives to create similar establishments in Rome and Athens came to nothing, despite promising beginnings.[76] In most of the countries of Continental Europe, the membership of the national associations was too small—or, in cases like Germany, lacked the financial muscle—to set up institutions of this kind. Nevertheless, the call for every country to offer opportunities for visits and encounters among academic women met with an enthusiastic and efficient response from all the IFUW's affiliated associations. In most cases, discounts were negotiated with particular, centrally located hotels and guesthouses in the capital city or important cultural centers, and national or local hospitality committees were appointed that, among their other tasks, served as points of contact to arrange private accommodation for traveling academics. In Italy in 1927, for example, women with the appropriate references could find low-cost accommodation in Bologna, Fiuggi, Florence, Siena, and Rome; in Brussels, Grenoble, and Stockholm, IFUW members could make use of University Clubs and women's student residences; in Berlin they had access to the well-appointed guestrooms of the Lyceum Club.[77] In 1932, the IFUW officially listed 36 hostels, guesthouses, and hotels, in 12 countries, that offered special rates for IFUW members. Adding to this the increasingly numerous offers of private "hospitalities," the period between 1919 and 1933 saw the weaving of a dense fabric of international interchange that was wide-ranging, complex, and diverse. The reports of the individual associations bear eloquent witness to the fact that this transnational network was not only used by the small minority of women who were actively working in university settings but also by the professional women who made up the majority of the federation and who were looking for affordable vacations or trips motivated by educational improvement, language learning, or simply pleasure. The activities of the Italian association make this expanded function particularly obvious. In 1932, the association listed guesthouses and hotels in 18 towns and cities, including such attractions as a small, family-run hotel in Cannero on Lake Maggiore "with garden and magnificent view over the lake" or a "delightfully quiet mountain resort" near Merano.[78]

With its three large international halls of residence in Washington, Paris, and London and its network of hotels, clubs, and guesthouses in those countries where the establishment of an IFUW residence was not feasible, the IFUW could, from the mid-1920s onward, build on permanent structures that enabled and nurtured exchange and encounter among its members to a very important degree. To do full justice to the scale of the new opportunities for international networking, however, other important aspects must also be considered: the educational tours, lasting several weeks, that were organized in a different country each year; the IFUW conferences; and the private hospitality that all members committed themselves to provide.

Within just a few years, the IFUW built itself the prerequisites for the promotion of personal contacts between academic women of many nations, and thus for an important contribution to international understanding. The particularity of these efforts becomes even more evident if the IFUW is set against the other institutions that arose during or after World War I with a commitment to academic interchange. Most of these, such as the American University Union in

Europe, the Institute of International Education in New York, or the German Academic Exchange Service (Deutscher Akademischer Austauschdienst, DAAD), established advisory offices to assist students abroad, or else they gathered, synthesized, and published information designed to help students find their feet while visiting foreign universities. The gender-specific network of the IFUW did provide this kind of information for women—but it also, and especially, attended to the accommodation and social integration of its members. As a result, the IFUW was a transnational organization uniquely useful for women scholars. That usefulness increased in direct proportion to the federation's success in complementing its halls of residence and hospitality program with a genuine promotion of women's research.

Promoting International Research

Promoting the scientific advancement of women in all its affiliated associations was the IFUW's most ambitious goal. During the federation's constitutive phase it was not yet clear how such an undertaking could be funded from the organization's own resources and upon what criteria it should be based. An initial idea, proposed by Carey Thomas, was that the IFUW's responsibility might be to serve as a "women's Academy of Arts and Letters and Science," in which "we shall honor and praise each other, for praise is a very important thing in success."[79]

In 1923, the IFUW Council decided to launch its own fellowship program. In this, the federation was following the example of its American mother organization, which had begun funding postgraduate studies in 1890 and whose system of postdoctoral fellowships had made it one of the very first associations worldwide to award funding of this kind.[80] From the beginning, the AAUW grants had not been restricted to the United States. On the contrary, they formed part of a "creative philanthropy" that set its sights on an important trophy inside Germany: the admission of women to the German doctoral examination, which American women hoped could have an important signal effect for the United States as well.[81] The first three American women to gain their doctorate in Germany, in 1895 and 1896 at Göttingen and Heidelberg, received financial support from the ACA.[82] Right from the start, the founders of the ACA encouraged talented women to undertake high-level scholarly and scientific work; enabling women to take an active and assertive part in scientific endeavor was also the declared aim of the pioneers who shaped the fortunes of the AAUW in the early 1920s. Carey Thomas had set up a graduate program at Bryn Mawr (the only such program at a women's college), and it was well known that Virginia Gildersleeve gave her utmost to nurture academically ambitious Barnard graduates.[83]

Including the Rose Sidgwick Memorial Fellowship, by the early 1920s the AAUW was awarding seven generously funded grants every year, two of them for study visits to Europe. The *AAUW Journal* regularly informed its readers about successful applicants, printing detailed reports by the fellowship recipients of their activities. It was hoped that the Fellows' reports would remind members of the success and vital importance of one of the AAUW's crucial goals, "the encouragement of graduate study by fellowships."[84]

Thanks to the US and British associations, which initially took on the main responsibility for awarding grants, the IFUW was able to offer its members fellowships of various kinds in the first years of its existence, even before it disposed over financial resources of its own. The principle soon emerged that a balance of consideration should be granted to representatives of both the natural sciences and the social sciences or humanities, and to both young postgraduates and established researchers. The British association offered its first travel grant in 1922, an award that was open to all IFUW members. Of the 25 grant applicants, the BFUW selected the gifted Swedish archaeologist Hanna Rydh, who planned to spend time in France carrying out research on the Stone Age.[85] The British women also raised the necessary funds to establish a fellowship at the University of Manchester for an American postdoctoral researcher and arranged for four women students from Czechoslovakia to visit London with all expenses paid. Similar projects were carried out on an exchange basis with the Italian association.[86]

As the number of affiliated countries grew, so did the number of applicants for the fellowships, with a substantial rise as early as 1923. There were 38 applicants for the AAUW's International Fellowship in 1923/24; the BFUW, which that year had been unable to raise the money for a travel grant and instead offered a "Prize Fellowship" not tied to a stay abroad, received 44 highly qualified applications.[87] Choosing the best candidates proved a "most difficult task" for both organizations.[88] The high number of applicants revealed the existence of an impressive pool of academically active women—but it also showed how paltry the available resources were in comparison with the huge need for funding. The British committee responsible for awarding fellowships lamented that "only too frequently scholars of first-rate quality have had to abandon work they were specially qualified to perform, in order to earn their livings in ways which waste their time and exhaust their energy."[89] But despite the sobering insights into the material hardship suffered by many women in the academic sphere, the fellowship committees maintained rigorously to the principles already established by the American association, which laid down that decisions be based on the highest intellectual standards and without reference to an applicant's personal situation, however difficult. This often enough meant that existing privileges were perpetuated: "to her that hath shall be given."[90]

For its International Fellowship in 1923/24, the AAUW decided to give priority to a younger natural scientist in the area of important basic research, and awarded the funding to the Viennese biologist Leonore Brecher. Brecher had earned her doctorate in 1916 in the field of experimental zoology, and had subsequently published a remarkable number of papers on developmental mechanics and color formation in butterflies.[91] She planned to use the AAUW funding to visit the Kaiser Wilhelm Institute for Biology in Berlin, in order to experiment on the biological heredity of different insect species' pigmentation in adaptation to their environment.[92]

With the principle of balance in mind, the BFUW awarded its own prize in the domain of the humanities, selecting the philosopher Cecilia Dentice di Accadia of the University of Naples in recognition of her highly respected book on Tommaso Campanella.[93] In view of the high quality of the applications, the BFUW raised some additional money and awarded five further, smaller prizes worth £25 or £20 each. Among the recipients of these were two Austrian, one French, and two Finnish women. At the top of the list was the president of the Austrian federation, Elise Richter, whose innovative studies in Romance linguistics the BFUW sought to commend. She was followed by the literary scholar Christine Touaillon, whose work the BFUW committee regarded as equally outstanding. Touaillon's groundbreaking and comprehensive history of the German women's novel in the eighteenth century was published in 1919. It was rejected that year by the University of Graz as the basis for a *Habilitation* (the university teaching qualification, requiring a kind of second doctoral dissertation), but was accepted in 1921 by the University of Vienna.[94] The fellowships and prizes mentioned so far by no means exhausted the spectrum of funding opportunities offered by the IFUW in 1924. However, the remaining grants—such as the Rose Sidgwick Memorial Fellowship, the Residential Scholarship at the University of Manchester, or the grants awarded by Reid Hall in Paris—were not open to all IFUW members but directed at individual bilateral exchanges between Britain, the United States, and France.

Accordingly, most of the women who received funding in 1923/24 were American, British, or French. However, of the seven international fellowships and prizes offered to the wider membership, three went to Austrians—indicating that five years after the end of the war, the IFUW was moving across the previous enemy lines and utilizing its funding resources to reintegrate academic women from the former Central Powers. This step was taken by the IFUW significantly earlier than by most other international academic associations.[95] The federation's deployment of a genuine internationalism, based strictly on scientific excellence, was highly effective. In Austria, the gesture was acknowledged with considerable gratitude. "I felt that the International Federation of University Women knows no difference and hostility between nations," was peace activist Christine Touaillon's praise for the organization; "every one of the women entrusted with the award of the Fellowship had no other ideal than the unity of science throughout the whole world."[96] Similar, if less spectacular, signals were sent by the awards to the two Finnish women: as Scandinavian and Czechoslovak members pointed out, academic women from small countries were dependent to a very particular degree on recognition and exchange with other countries, as a way of overcoming their dual isolation as female scholars and as representatives of what were perceived as the margins of academia.[97]

Despite their modest scale, born out of necessity, it was precisely these minor prizes that revealed how much could be achieved through the applications process and the award of fellowships and prizes. It was not necessarily important to insist on the large-scale, expensive, one-year international fellowships: even without these, the IFUW was genuinely able to take on the role of a "women's academy" that Carey Thomas had proposed in 1920. The international accolades gave recognition to the scholarly work of women, arousing substantial interest in the IFUW's agenda and goals around the world, and thus deserved to be repeated in the future: "Public recognition of the work of a woman of one nation by the women of other nations cannot fail to stimulate international good feeling."[98]

These initial experiences of the difficulties and opportunities associated with the international promotion of science strengthened the IFUW Council's resolve to accord its highest priority to the distribution of fellowships and prizes. In 1923, the Council decided that the most important task of the coming years would be the creation of an International Fellowship Fund, based on an endowment of a million US dollars.[99] The revenue from this capital would allow the award of 30 travel grants and several prizes every year. The women conceded that such a sum would not be easy to raise; each individual member of the IFUW would have to contribute $40. However, buoyed by the success of their efforts to acquire donations for the clubhouses, the Fund's initiators felt this was attainable. Only part of the necessary funds was to come from member donations: large-scale gifts were hoped for, and there was considerable optimism that a munificent public could be galvanized to support the initiative. "If the scheme grows," predicted the British biochemist Ida Smedley MacLean enthusiastically, "this fund will be one of the great endowments for research of the world, and we want to build it up out of small sums."[100]

The convention held in 1924 in Christiania (now Oslo) ratified the Council's ambitious proposal, on condition that the national affiliates would not be obliged to call in donations of a particular sum from their members. For most of the European delegates, the amount specified appeared exorbitantly high, far beyond what they would be able to afford.[101] The Fellowship Committee, chaired by Ida Smedley MacLean, made vigorous efforts to counter this mood.[102] But by undertaking the obligation to raise one million dollars by its own efforts, the federation was, in fact, committing itself to a strategy that demanded a high degree of engagement from all members. Unlike the one-off effort to establish a clubhouse, the development of a fellowship program called for long-term, systematic financial input.

In the US context, members had been successfully acquiring private donations for the promotion of scholarship since the late 1880s, a process that had functioned impeccably in drawing together the association's grassroots—made up of college graduates without higher degrees—and its leadership, mostly women who had enjoyed a full university education and held professorships. By acquiring donations, or making donations themselves, to promote women's research, members expressed their support for the ACA's (later the AAUW's) policies and their sense of belonging to the country's female educational elite. In 1932, the large number of fellowships that the AAUW was funding through its own efforts itself was widely considered a "symbol of growth," emblematic of the association's strength.[103]

For the IFUW, research funding was evidently intended to work in a very similar way, providing a material anchor for an intellectual mission, external momentum, and internal coherence. The initiators of the "Million Dollar Fellowship Fund" emphasized that the program would stand or fall on the faith of each member—regardless of her nationality—in terms both of her own financial contribution and of her personal willingness to search for other sources of money.

The Fellowship Fund was also driven by the practical politics of professionalization. The Council argued that if the IFUW was to stand for democratic, coeducational university training for men and women, equal access to academic training and degrees, and equal pay with equal opportunities to undertake university-level research, then it would have to ensure first and foremost that enough women were represented on the teaching staff of the universities. Ellen Gleditsch, a Norwegian professor of chemistry, explained to the members assembled in Amsterdam in 1926 that women in universities had hitherto made their mark primarily in low-status positions as badly paid teachers, far less so as researchers. This lamentable situation applied to all the affiliated countries.[104] Proving herself in the field of research, argued Gleditsch, was a necessary condition for a woman to assert herself vis-à-vis her male competitors in the university system. Research practice was, however, far more difficult to come by than teaching experience, and indeed could only be acquired by women who had the opportunity to dedicate their entire energies for several years, "free of cares and worries," to a scholarly project. Only then could a woman build up the necessary skills and determination to enable her to remain in research for the long haul and to deliver first-class results even under conditions that were less than ideal.[105] Top-level research, Gleditsch continued, was increasingly taking place within international networks. This was one reason why it was of utmost importance to create more opportunities for women to pursue research abroad; another was that women required such opportunities in order to compete more successfully with men, on a national level, for the scarce and sought-after full university professorships. As Ida Smedley MacLean added, this was the only way to ensure that equality of educational opportunity for men and women might actually be achieved in the universities. To reach that point, a significantly higher proportion of women must attain well-paid and appropriately resourced positions, so that they in turn could serve as teachers, researchers, and mentors for women students. "We want to increase the supply of highly qualified women available; to do this, we must encourage research."[106]

The minutes of the early IFUW conferences and Council meetings reveal that the federation's officers devoted considerable energy to persuading the assembled members of the need to offer research fellowships in an international framework. Their argument was that promotion of this kind was crucial not only for the individual researchers who stood to benefit directly from funding, but also for the federation as a whole and all its members. The women of the Council, all fully accredited as academics, regarded themselves as a vanguard within the IFUW and as role models for its members. They thus often began by describing their own experiences, as a way of underlining the benefits of a study trip abroad—benefits that could go far beyond science and help promote the wider cause of internationalism. Caroline Spurgeon portrayed her recent visit to the United States as an intellectual and emotional turning point in her life:

> It opened to me the vision of a new world, it was the most stimulating, inspiring and wonderful experience possible, as to the friends that I have made there, and the kindness I have met with, well, I can never hope even to begin tell the tale of it. In the same way my days of study in Paris are a most happy and precious memory to look back upon, and the friends I have made in France a most treasured possession.[107]

Ellen Gleditsch offered a similar line of argument. Scientific work abroad was, Gleditsch maintained, an important component of the international promotion of research; just as important, however, was "a stimulating acquaintance with their fellow workers of other lands... they will have a store of memories of those countries.... Such memories will more than repay the International Federation for all its work, for giving the money, thought and love to the foundation of a Fellowship Fund."[108] Ida Smedley MacLean, who had spent the years 1913 to 1915 in the United States with the support of a Beit Fellowship, told delegates that this honor had been "a most thrilling moment of encouragement and stimulus" for her, one that the university world offered only at rare moments, and especially rarely to women.[109]

All the members of the IFUW, said Gleditsch, could "contribute to the progress of research by women, by giving research students the chance of spending some time in laboratories and research institutions of other countries, where they will meet men and women of other nations who are doing the same kind of work and who may open their eyes to a whole new field of possibilities."[110] As for the objection that an expensive fellowship program would require the IFUW to commit itself to a measure that offered little to the mass of the members, MacLean countered with some verve: "However successful university women may be in their various professions, however great the social service they render to their community, our standing body of university women will be judged by our contribution to the furtherance of knowledge. The standard we maintain in research is not a matter that concerns only a few individuals. It affects the position of every one of us."[111]

The new program had barely been resolved before the first donations arrived, during the Christiania conference itself. Professor Gleditsch told the members about a group of elderly gentlemen who, following Norwegian tradition, had met in Oslo to celebrate the fiftieth anniversary of their graduation from university, collecting a sum of money to mark the occasion. They had planned to use this to fund a study grant for a Norwegian woman student, said Gleditsch, but had now decided to transfer it to the IFUW's International Fellowship Fund as the very first contribution.[112] In the course of the event, the 68 US participants, "so inspired by the conference," put together $1,000 for an international fellowship. They did not add this money to the general appeal, but passed it directly to the IFUW so that a grant for 1925/26 could be awarded immediately, in honor of the Scandinavian hosts.[113]

Despite this encouraging start, however, the reality of the fellowship program lagged far behind the high-flying ambitions of the early 1920s—in stark contrast to the clubhouse appeals. The hope that the Fellowship Fund would once again inspire wealthy individuals to make generous gifts remained unfulfilled. Collection efforts among the members themselves, too, yielded less than had been expected. As the 1920s drew to a close, a somber mood prevailed: it was only with great difficulty, and with the help of a large donation by the Czechoslovak president Thomas Masaryk, that the Fellowship Committee was able to scrape together the trust capital of £6,000 required to create the first IFUW International Fellowship.[114] Whereas in 1926 talk persisted of four international fellowships, by 1930 it had become clear that the actual achievements would be more modest in scale. Among the reasons for this state of affairs was the fact that, until 1927, the IFUW's two most important programs, research promotion and international hospitality, had operated in competition. While the Fellowship Fund was asking for donations, British members were putting all their efforts into the appeal for the redevelopment of Crosby Hall; similarly, the Italian association's members were concentrating their fund-raising energy on the provision of hospitality in Rome. The large-scale gifts that had been envisaged failed to materialize.

The sparse funds that the other European affiliates managed to gather were often used to award small grants within each association's national framework or to administer international fellowships on its own account. For example, in 1929/30, the Spanish association offered its own international research grant, while the French club raised the money for an "international vacation fellowship" that enabled the Austrian psychiatrist and neurologist Martha Brünner-Ornstein to spend one month researching in Professor Lapique's laboratory in Paris. Examining the rationale behind the selection of candidates for grants, it seems that national concerns were often accorded more importance than international ones. Thus, a report by the French association noted that Brünner was a fortunate choice because her aim in visiting the Sorbonne was to learn to apply a new neurological measurement procedure: a French invention. The association was anxious to see this invention recognized and disseminated among scientists in the German-speaking world.[115] In 1928, the Spanish association awarded an international fellowship, funded by the Spanish Ministry of Foreign Affairs, to the Austrian historian Margarethe Mecenseffy, who was studying the links between the Austrian and the Spanish Habsburgs;[116] in 1929 the same fellowship went to the US historian Dr. Lillian Estelle Fisher for her work illuminating the "true mission of Spain in America."[117] In 1929, the German association DAB presented 1,000 reichsmarks to a fellowship fund maintained by the Reich Ministry of the Interior.[118] When, in 1930, the DAB received a gift of 2,000 reichsmarks, it preferred to set up its own "German Fellowship" for a woman scholar from abroad instead of contributing the money to the IFUW's Million Dollar Fellowship Fund.[119] Such priorities slowed the accumulation of the required capital for the fellowships; they also fragmented the already meager funding available to women, making it hard for applicants to stay abreast of the different options. In some cases, the result was a failure to award funds due to a lack of suitable applicants.

Outside the United States, the largest contributions to the IFUW's International Fellowship Endowment came from Australia, Britain, Denmark, New Zealand, Norway, Sweden, and Switzerland. The commitment of these associations, on the one hand, quite simply reflects the fact that their members were the ones most able to afford the international promotion of science.[120] On the other, however, it seems that these were countries where the practice of donating money was more deeply rooted in academic culture, and also that the membership of the associations there included a significantly higher quota of women who were themselves professionally active in science and academia. But the disproportionate efficiency of the US fund-raising remains striking: while the European donations came primarily from charity bazaars, fund-raising events, and the sale of IFUW postcards and notepaper, thus differing little from the general culture of charitable giving, the American women succeeded in appealing to their members' sense of belonging to a responsible educational elite—for example, by asking them to donate at least one day's wage per month to the endowment fund.[121]

Also important is the fact that, in contrast to Crosby Hall or other clubhouses, the funding of which had proceeded with such élan, an "International Fellowship Endowment" was a rather abstract notion, poorly suited to attracting heartfelt and enthusiastic donations. The AAUW, which had the most experience of carrying out vigorous appeals for donations and had long acquired private funding for fellowships, quickly identified this problem and began instead to raise money for four individual International Fellowships. This seemed more concrete, easier to explain, and more capable of being achieved than the huge and abstract target of "one million dollars." Even more effective proved to be the idea of attaching specific names to the individual fellowships. In 1930, the BFUW's executive committee in Britain decided to award an international "Caroline Spurgeon Fellowship," worth £100, for the following year, as a signal of gratitude to the former president for her great efforts in helping to open Crosby Hall as an international residence. The members greeted this initiative with such a "generous response that a scholarship of 100 pounds a year for two years will be awarded for the years of 1931–1933."[122]

In 1927, the international project of the Million Dollar Fellowship Fund passed into American hands. As a result, the international orientation of the Fund's original purpose now shifted in favor of promotion efforts within the United States. Of the 25 one-year fellowships that were to be funded from the endowment in 1932, only ten were still advertised as "international."[123] Optimism regarding the feasibility of collecting such an ambitious sum through membership dues was not dented in the United States until the end of the prosperous 1920s; quite to the contrary. The AAUW also benefited considerably from its international activities and the increase in female college graduates, as reflected in a constant rise in membership figures. Between 1918 and 1924, the number of AAUW members grew from twelve thousand to nearly twenty-one thousand. The association's international orientation, reported Ellen F. Pendleton of Wellesley College in 1924, "fired the imagination and enthusiasm of the younger college woman, whose interest it was particularly desirable to secure as soon as they graduated."[124] In 1924, the AAUW succeeded in going beyond membership dues and donations alone to acquire large appropriations from charitable foundations, a move which fundamentally altered the association's financial circumstances. Grants from the Laura Spelman Rockefeller Memorial, the Carnegie Foundation, and the Carnegie Institute of International Education totaling more than $100,000 prompted a surge of professionalization in the association's labors.[125] Skilled and specialized secretariats were put in place, responsible for dealing with administration and member relations. They also began to provide systematic information on the importance of national and international association activities.[126] Additionally, voluntary work for the association was offered by individual members not in paid employment. Dorothy Bridgman Atkinson launched the Million Dollar Fellowship Fund campaign from St. Paul, Minnesota, with the help of her husband's secretary.[127] The campaign initiated by Atkinson was the AAUW's first nationwide appeal for donations. Each of the 25 sections created by the association for this purpose was charged with raising $40,000 for the endowment fund, a sum that corresponded to the capital required for each one-year grant. In 1932, five years after the campaign began, the assets collected for the Fellowship Fund amounted to just over $200,000, one-fifth of the million envisaged by the organizers.[128]

Up to 1933, the IFUW's actual achievements in the international promotion of science remained far behind the ambitious goal set by its Council in the early 1920s. The reasons for this may be found partly in the great scarcity of private and public financial resources in Europe,[129] partly in cultural differences within the federation.[130] At least for Continental European states, it proved impossible to establish a proactive approach to procuring funds. The degree to which these countries' reserved attitude should be interpreted as a form of skepticism toward the IFUW's program among individual national associations will be discussed in chapter 4, with specific reference to the German case.

Despite these difficulties, progress was made up to 1933 in the transnational exchange of ideas between women scientists and scholars through the IFUW network. In the first 13 years of its existence, the organization and its national affiliates offered 45 international grants, prizes, and fellowships.[131] One-year travel awards, allowing women to pursue scientific research abroad, accounted for around half of this funding. From 1928, support was offered annually from the endowment of the Million Dollar Fund. The fact that this amount was modest in comparison to the demand for international fellowships among women researchers, and that further efforts were therefore urgently required, was something the leadership never tired of impressing upon the federation's membership. It was partly with this in mind that, from 1928 onward, the Fellowship Committee supplemented the IFUW's own funding opportunities with a detailed booklet listing all the international grants for which women could apply.[132] The Council called on members to make up for the lack of paid support by offering "complete hospitalities" for women researchers, and in the early 1930s several associations took up this challenge. In 1936, the South African association offered six months' hospitality at the Cape of Good Hope, while Austria and Sweden took in two British researchers each for several months. Paris, Brussels, and Copenhagen, too, reported success in accommodating researchers from abroad.[133] Individual colleges in the United States created their own international visiting scholar programs and extended invitations to members of the network. First and foremost among these colleges was Barnard, which invited Marguerite Mespoulet, Caroline Spurgeon, and other prominent IFUW members to New York as visiting teachers.

Studying the reports and curricula vitae of the women who received IFUW funding between 1922 and 1933 reveals that the federation's support measures provided new momentum to their recipients' careers, exerting a decisive and lasting positive influence. For these women, the IFUW's promotion of research opened up new perspectives. The first recipient of the International Junior Fellowship, awarded out of the IFUW's own endowment fund for the first time in 1928, found the idea of a funded research trip abroad so far beyond the limits of her imagination that she had to be pressed into applying by her sister members.[134] The Swiss biologist Anne-Marie Du Bois studied in Neuchâtel and Geneva, earned her doctorate in embryology in 1927, and then overcame great financial difficulties to continue her studies at the Sorbonne. The IFUW funding enabled her to research for one year at the Kaiser Wilhelm Institute for Biology in Dahlem, Berlin, with zoologist Richard Goldschmidt, a world leader in the field of genetics who was also well known for his unprejudiced support of female scientists. For Du Bois, fresh from her doctorate, the one-year visit to the KWI in Berlin turned out to be a breakthrough in her scientific career. Under Goldschmidt's guidance, she was able to experiment on the heredity of sexual characteristics in a butterfly species and acquire the foundations of her later specialization; in addition, she found in Goldschmidt a mentor who helped her network within the new field of genetics. When her IFUW funding ended, Du Bois remained employed at the KWI laboratory for an additional six months. Goldschmidt then helped her find a two-year grant from the Carnegie Institute of Washington that allowed her to pursue her research at respected US laboratories. The biologist's nearly four years of scientific work abroad smoothed her path into a career in Swiss science. Upon her return from the United States in 1933, she received the entitlement to teach biology at the University of Neuchâtel and, in 1937, was appointed senior laboratory scientist at the dermatological clinic of Geneva University. In 1940, she stepped in to replace a colleague who had been called up to military service and took on the teaching of his histological laboratory courses. Du Bois continued to teach at Geneva University until her retirement in 1974.[135]

Similarly, the Viennese physicist Berta Karlik's one-year fellowship at Crosby Hall in 1930/31 laid the foundations for an extremely successful scientific career. In 1928, Karlik had completed her studies in physics and mathematics in Vienna summa cum laude.[136] She wrote her doctoral thesis while serving as an intern at the Institute for Radium Research in Vienna, an internationally renowned institution whose director, Stefan Meyer, was particularly supportive of women students. Meyer encouraged and helped Karlik throughout her studies, giving her—like many female colleagues at the Institute—the opportunity to pursue her own independent scientific work.[137] Upon completion of her doctorate, Karlik first sat the teaching examination and subsequently completed her probationary year of teaching at a Viennese technical high school. During this time, she enjoyed access to the laboratories of the Institute for Radium Research. Karlik wanted, if at all possible, to continue with her scientific work; however, she was all too well aware how rare it was to find a permanent appointment in the world of science, and decided that training as a teacher was an absolute practical necessity, a conclusion reached by almost all women with scientific ambitions during this period.[138]

Karlik published her first important works on the scintillation process in 1929 in the papers of the Institute for Radium Research, and these studies paved the way for her selection as the recipient of a young researcher's Residential Fellowship at Crosby Hall for the academic year 1930/31.[139] Armed with Meyer's letters of recommendation, Karlik spent her fellowship year acquainting herself with the important laboratories and scientists of western Europe active in the field of radium research. In Britain, she worked at the Royal Institution in London with William H. Bragg, the 1915 winner of the Nobel Prize in physics and, like Meyer, an energetic promoter of women.[140] While in London, Karlik worked closely and productively with one of Bragg's many female colleagues, crystallographer Isabel Ellie Knaggs. It was the radiographic analysis of the structure of cubic metals, developed in Bragg's laboratory alongside Ellie Knaggs, that would establish the foundation of Karlik's international reputation.[141]

At the Cavendish Laboratory in Cambridge, Karlik carried out research with Ernest Rutherford. Like Bragg, Rutherford was among Britain's leading nuclear physicists and was in close contact with the Institute for Radium Research in Vienna.[142] Karlik also visited many London hospitals, acquiring familiarity with the practical applications of radium research in the treatment of cancer. In Paris, she observed the work of Marie Curie, visiting the Pasteur Institute and the laboratory of Louis de Broglie.[143] When she returned to Vienna in 1931, it was not yet clear whether Karlik would succeed in pursuing a scientific career, but her collaborative work with numerous eminent scientists in Britain and France upon completion of her doctorate certainly played a decisive part when, after a year of unpaid work at the Viennese Institute for Radium Research, she was offered one of the few salaried research assistant posts there in 1932. Thanks to her scientific experience abroad and her command of foreign languages, the 28-year-old Berta Karlik joined what was still a rather small international community of physicists. As Otto Hahn remarked with some respect, she had "been able to expand [her] scientific horizons far further than is possible in the course of normal scientific life."[144] Karlik's scientific reputation was reinforced by further research visits, this time funded by the Austrian state, to the oceanographic commission in Bornö, Sweden, the results of which appeared in a widely read paper on measuring the uranium content of seawater.[145] To be sure, the status and salary of Karlik's appointment by no means matched her scientific reputation and, like so many other female scientists, she faced considerable delays and impediments to her scientific career. Even so, her case can be described as one of the first "linear" female scientific careers in twentieth-century German-speaking Europe, since it was not significantly disrupted by Germany's *Anschluss* of Austria in 1938, by the war, or by the upheavals of the postwar period. In 1937, Karlik, then 33, received her university teaching credentials; named an associate professor in 1940, two years later she published her much-noted discovery of element 85. In 1947, she was appointed director of the Institute for Radium Research in Vienna and professor extraordinarius, and in 1956 she became a full professor—the first woman to hold this position in Austria. When, after World War II, the Institute gradually managed to rebuild its prewar renown, this was due in large part to Karlik's international connections.

During the 1920s, the IFUW's insistence that the international promotion of women in science was one of its central tasks helped establish a reputation for nonpartisan internationalism, which brought the organization a high level of credibility. Even if the actual promotion activities were modest in scale, these grants sent important signals: they lent visibility to women scientists' internationally relevant research, and offered quite a few of those women a completely new range of opportunities to establish themselves in research. In this respect, the women's network of the IFUW stands out as having promoted science with a powerful impact in the few fields where influential male professors were prepared to support women and enable them to carry out independent research work in their laboratories: the new disciplines of nuclear physics and genetics.[146]

Above and beyond these personal advances, the funding efforts gave expression to the IFUW's aspiration to make science and scholarship key elements in the formation of a transnational female elite. In selecting women to receive its grants, the organization set high academic standards, demonstrating that it had a sure eye for the potential of the next generation of women academics. The IFUW's international promotion of knowledge and science helped it to define a very specific profile, easily distinguishable from that of all the other international women's and professional organizations. This explains why the IFUW called on its members to commit themselves to the promotion of research as the most important contribution to a global community of graduate women. When, in the early 1930s, hopes died that the IFUW could achieve its aim of contributing to the maintenance of peace through collaboration with the League of Nations, the organization's web of hospitality and its promotion of science came to the fore as a cohesive grounding for the diverse choir of representatives of national elites that were assembled in its ranks. Through the trials of the 1920s, the IFUW had laid the foundation for extraordinary efforts to rescue academic women from Germany and the whole of German-occupied Europe when the international community collapsed in ruins.[147]

4
Reactions in Central Europe: The German Case

Science, Internationalism, and the Women's Movement

In the immediate aftermath of the world's first global conflict, there was no question of admitting Germany to the IFUW. The fact that the IFUW steered clear of the Central Powers, and especially Germany, arose from the initiative's orientation on work within the Entente, which lasted well beyond the armistice. The IFUW did not differ in this respect from the academies of science and the other international scientific and professional associations. Germany was also excluded when the new International Research Council (IRC) was formed, under American leadership, between October 1918 and spring 1919, with the task of intensifying the inter-Allied scientific cooperation that had begun during the war. Early on in this founding phase, the IRC passed a resolution ruling out both official and personal contacts with the Central Powers. Before these countries could be readmitted into the international academic community, said the IRC, "the Central Powers must renounce the political methods which have led to the atrocities that have shocked the civilized world."[1]

The response of German academics and their organizations was to dig in their heels and present a counter-claim of their own: they would only be prepared to join the international community if they received an unconditional invitation. The result was a "war of the academies" that the contemporary German historian Margarete Rothbarth considered to be still raging in the early 1930s. Rothbarth had been working at the League of Nations' International Institute of Intellectual Cooperation in Paris since 1926 and observed the slow progress of political and academic rapprochement from there. In her view, the war had caused more devastation for Germany in "cooperation between scholars" than in almost any other sphere.[2]

Recent studies confirm Rothbarth's evaluation,[3] with the caveat that it was largely the Germans themselves who bore responsibility for the persistence of the devastation. A stance of self-imposed isolation, particularly marked among German professors, meant that the process of reconciliation unfolded significantly later in the academic world than in the political sphere, and had not fully taken root by 1933.[4] Overall, as Gabriele Metzler has argued, Germans "squandered" their chance of academic internationalism during the Weimar period.[5]

The history of how the German federation of academic women, the Deutscher Akademikerinnenbund or DAB, was founded and admitted into the IFUW reveals a very different pattern. Understanding was reached between the German women and the new female international academic community by the mid-1920s. The DAB was formed in 1926 and it immediately joined the IFUW. By 1932, the German body's position within the IFUW had become strong enough for the Edinburgh convention to accept German as the third conference language alongside English and French—a concession that the IRC never offered German academic organizations, to the latters' chagrin. The strength of German academic women's ties with the IFUW also becomes apparent in the fact that, in Edinburgh, the international umbrella organization accepted the DAB's invitation to Berlin: the sixth international assembly of the IFUW was to be held there in 1936. This process of rapprochement was certainly not free of tension in the period, and in 1933 it began to take a very different course from the one that had been envisioned.[6] Nevertheless, in light of the relations between the IFUW and the DAB, it is fair to say that Weimar's university women—unlike their male colleagues—made full use of the opportunity offered by membership of the IFUW. This chapter begins by exploring the reasons for their embrace of scientific internationalism and the manner in which it took place.

The first step toward closer rapport came from the IFUW itself. Compared with the majority of male-dominated academic associations, the IFUW moved away earlier from the policy of categorically excluding the former enemy nations. Gradually, it began to steer a course of integration, a change that found expression in the IFUW's research promotion policy. The admission of the first successor state of the former Central Powers, Austria, in 1922 played an important role in this process. Although Austria's admission to the IFUW took place concurrently with the new Republic's admission to the League of Nations, and to this extent harmonized with official Western diplomacy, Austrian membership sent a message to the world of science. After all, in 1919, the Austrian Academy of Sciences had joined the cartel of the German academies, committing itself fully to its political line and thus isolating itself internationally in just the same way.[7] This made the founding of an Austrian association of academic women and its accession to the IFUW in 1922 all the more striking. The importance of this advance in the politics of reconciliation was underlined by the association's success in attracting Austria's most renowned female scholar as its president.[8] This was the 57-year-old Elise Richter, linguist and specialist in Romance literatures, who had been appointed to the University of Vienna in 1907 as the Habsburg monarchy's first female university lecturer and in 1921 as the new Austrian Republic's first female professor extraordinarius. Her personal decision to become president of a national organization of women academics was probably accompanied by considerations of global politics. In Habsburg times, Richter had spent long months traveling in western Europe—especially France and Italy, but also England. Elise Richter spoke several languages, and had a low opinion of the provincial nationalism that was then cultivated in German-speaking universities. She was part of the liberal, cosmopolitan Viennese intellectual elite of the Habsburg era, which was prepared to take up a committed political position when the monarchy collapsed and helped to build the Republic of Austria within the new community of nations.[9] Richter herself said she believed "the international idea of cooperation and mutual assistance among those with a common cause" to be a self-evident and logical correlate of university work.[10]

A politically active representative of the new Austria and a scholar committed to internationalism, Elise Richter proved the ideal president for an IFUW member association. These advantages evidently outweighed the fact that she had, on her own admission, previously avoided contact with other women at the university, let alone with the women's movement, in the hope of attracting as little extra attention as possible to her gender. Richter had made her way at the University of Vienna as an individualist and an exception, with the help of certain male patrons. In her memoirs, she portrayed the IFUW's invitation as the moment when it dawned on her "how completely unfamiliar I was with absolutely all those women who had trodden the same path as I, and perhaps had suffered the same experiences. That piqued me to take on the task."[11]

It would be interesting to investigate whether Richter's male colleagues had anything to say about her new office and the admission of Austrian women graduates and researchers to the IFUW. We do know from Richter's memoirs, written in 1941, how difficult she found it to organize the female representatives of Austria's politically polarized academic sphere into a nonpartisan, internationally oriented association. Establishing the Austrian university women's association, the Vereinigung der Akademikerinnen Österreichs (VAÖ), was, as she put it, like stirring up "a hornet's nest." It meant "walking on eggshells" and an attempt to rise above politics to the lofty attitudes that very few of her newly assembled colleagues were "able, or even willing, to reach":

> The Catholics were implacable enemies of the Protestants, the nationalists of the Social Democrats, and everybody of the Jews—even at this stage following the idea of race. They had been strongly attached to these positions from their student days, wanting on no account to fall behind their male colleagues in the robustness of their views, often surpassing them in the narrowness of their prejudices and in their timidity toward the international idea.... It was considerably easier to tie the knot of friendship with sisters in Japan or Australia than with the Czech women or one's colleagues from the very next street.[12]

Links with the IFUW and the opportunity to make international acquaintances and receive fellowships were, in Richter's opinion, the crucial factors in bringing together and keeping afloat an association so deeply riven by political disagreement. The IFUW's research grants and prizes, awarded to Richter herself and two other Austrian women in 1923, were immensely important for the survival of the young VAÖ, but their luster reached as far as Germany, too, probably not least because they honored the research of German-speaking women.

With regard to integrating Germany itself, the IFUW—again following the political line of the former Entente—was less forthcoming. Having said that, there were members, especially among the American delegates, who advocated admitting the German women. Physiologist Ida Hyde, who had earned her PhD in Heidelberg in 1896 with the help of an ACA grant, made no secret of her stance in this respect. She contacted the Germans on her own initiative, a move that would ultimately do more harm than good. During the run-up to the Paris convention of 1922, Hyde was staying in Heidelberg, near her friend from student days, economist Elisabeth Altmann-Gottheiner—now professor extraordinarius at Mannheim's commercial college and one of the few German women university lecturers to play an active part in the national and international women's movement.[13] Acting without a mandate and without the knowledge of the IFUW Council, Hyde erroneously told Altmann-Gottheiner that the Paris conference might resolve to admit the Germans to the IFUW. She propounded this view so convincingly that, in spring 1922, Altmann-Gottheiner called two meetings of German university women to discuss the possibility of joining the IFUW. A petition was drafted noting "that no doubt the University women of Germany would be glad to join the Federation, and probably also be willing to send delegates, if the invitation were extended to them in the same way as it has been to the women of other nations."[14] A German federation numbering thousands of university women could, said the signatories, be established at very short notice; in the meantime, Elisabeth Altmann-Gottheiner and Gertrud Bäumer were nominated as unofficial delegates to the Paris event.

Just weeks before the conference began, Hyde passed the petition to the IFUW board, causing consternation. The question of admitting a German federation of university women had been discussed at the margins of the first members' convention, in London in 1920, but no conclusions had been reached. A more thorough discussion was planned for the Paris meeting in 1922. Put under pressure by the German petition, Caroline Spurgeon asked the French host association whether, under these circumstances, it might be justified to invite the German women to attend as unofficial visitors. This would, she said, depend on their presence being "acceptable for our French hostesses. None of us would wish to have them unless their presence was likely to contribute to international good feelings rather than the reverse."[15]

The French women's answer was unequivocal. In a series of letters, they recounted their firsthand "expériences deplorables" of the German occupation, and refused to contemplate an ad hoc visit by the German academics. They insisted that the question first be discussed on principle, as planned—and without the German guests. The letters were accompanied by two brochures of a "Ligue pour perpétuer les crimes allemands," recalling the atrocities committed by the German occupiers. Unfortunately, Spurgeon's diplomatic dispatch to Altmann-Gottheiner has not survived, but Altmann-Gottheiner's answer has. If, she wrote, she had known that the question of admitting a German federation was to be discussed in Paris, she would never have put forward her request, "as we should never apply for admission if we did not know that we should be absolutely welcome. As matters stand, we should of course not be able to attend a conference which is to discuss our admission."[16]

It is quite conceivable that these deliberations would have turned out more favorably for the Germans if it had not been for the irritations preceding the convention. Nevertheless, after exhaustive debate in Paris, the IFUW's Council settled on the wording: "The German university women, when they are properly organized in a national federation and apply in the usual way, shall ultimately be admitted to the International Federation, probably at the time when Germany has been admitted to the League of Nations, if not sooner."[17]

This closing phrase could be interpreted as indicating a more accommodating stance toward Germany. The German women themselves, however, interpreted it as an insulting affront and as perpetuating the policy of exclusion.[18] Hence the bitterness with which the leaders of the German women's movement noted in their monthly review, *Die Frau*, that the IFUW now seemed to have aligned itself completely with the general "loathing of Germany as a country unworthy to join the community of nations," or at least was pandering to those feelings. Gertrud Bäumer and Helene Lange, the most high-profile representatives of the bourgeois women's movement in Germany, were outraged that such a position could even be contemplated "in an association of women who epitomize the highest educational standard of their countries, an association that, moreover, claims to represent the ideals of international peace."[19] Bäumer and Lange felt that the internationalism propagated by the IFUW was discredited by this resolution. For German university women, they concluded, the question of founding a national organization and possibly joining the IFUW was off the agenda for the time being.[20]

The IFUW's Paris resolution reinforced the view, widespread among German academic women, that Germany could do without international links—indeed, that the national interest demanded they be forgone.[21] In several pieces for *Die Frau*, Bäumer devoted extensive discussion to the rationale behind this stance, which she initially described as the only defensible response to the Treaty of Versailles.[22] The new community of nations that had arisen through the treaty in 1919 was, she argued, "founded upon the trampling of German honor," and it was only through Germany's absence from the international stage that the world could be suitably alerted to this fact.[23] In Bäumer's view, internationalism had been invalidated by Versailles, and must therefore be rejected not only by male office-bearers and dignitaries, but also by their female counterparts—all the more so in that women now had the right to vote and, with it, adult political responsibility. For Bäumer, gaining the vote in 1919 cast sweeping doubt on the previous practice of nurturing international relationships between women. She argued that since women had begun to bear responsibility for political processes to the same degree as men, it was no longer possible to sustain the previous division of the world into a political sphere, defined as male, and the feminine idealism that had been "peace, international understanding, human community, and...the source of cordial sympathies." The "irresponsibility" with which, before they attained political adulthood, women had been able to create networks "in areas where politics had interests to preserve" was now, in view of the hated peace treaty that had been forced on Germany, nothing less than "wicked superficiality." If women wanted to be taken seriously as actors in the international arena, they would have to abandon the "play of sweet feelings" and pursue "the *politics* of peace."[24] Bäumer was not alone in believing that rebutting international relationships was the only political act appropriate to the situation: for large segments of the German population, the hostility to foreign triumphalism that had marked the immediate postwar years hardened into an isolationist ideology as the Weimar Republic slowly gained its footing.[25]

However, it was typical of the Weimar period's dynamism that the defensive attitude toward international work soon softened somewhat, at least for some liberal members of the women's movement. This was due to the markedly more conciliatory attitude toward Germany displayed by the International Council of Women (ICW) and the International Woman Suffrage Alliance (IAW)[26] than by most other international organizations, and especially than the academic ones. The umbrella organization of German women's associations, Bund Deutscher Frauenvereine or BDF, was invited to the ICW's annual congress in Norway as early as 1919, and in 1920 the IAW invited the venerable German association Allgemeiner Deutscher Frauenverein or ADF (later known as the Deutscher Staatsbürgerinnenverband, the Association of German Women Citizens) to Geneva. At that stage, German women were not yet willing to accept the invitations, but in 1920 a German delegation, for the first time, attended a meeting of the Women's International League for Peace and Freedom (WILPF) in Rome, and another traveled to The Hague in 1921 for the international congress of the ICW.[27] This latter group included Gertrud Bäumer and Elisabeth Altmann-Gottheiner.

In 1921, overwhelmed by the warm reception the German women were given in The Hague, Bäumer noted with obvious emotion that before World War I no one had fully realized "how much moral force really lies in these women's organizations of the world, in their collaboration and the bonds between them." It turned out, she wrote, that the will of all women to join together in overcoming the war's catastrophic consequences was stronger than wartime hostility itself. As a result, it had been possible in The Hague "to find important lines of common aspiration" even while "preserving our own standpoint intact." Through the encounter with women of other nations, a generally shared "sense of the female destiny" had emerged that, in turn, vividly reminded Bäumer of "the active task of women," also beyond the political sphere—something that "had a very profoundly uniting effect."[28]

From this point on, relations between the German women's organizations and their international counterparts were rapidly normalized. At the IAW's Rome congress in 1923 and the ICW's Copenhagen meeting in 1924, no particular controversy surrounded the German women's attendance despite the renewed political tensions generated by the Ruhr crisis,[29] and German was reestablished as the third conference language in the international women's associations.[30] In 1926, Bäumer's reawakened belief in a specifically female contribution to international understanding found poignant expression at a ICW rally in Paris: after her speech, she embraced her French co-speaker to the tumultuous applause of the audience.[31]

It may have been this optimistic turn in the process of reintegrating German women into the international women's organizations during the early 1920s that prompted Altmann-Gottheiner and Bäumer to react so eagerly to Ida Hyde's advances. Against such a positive backdrop, their outrage was all the greater when they read the text of the IFUW's 1922 resolution. Certainly, the IFUW was a new body; admittedly, it was looking to position itself on the international map of academic organizations and molded its attitude to Germany to match theirs. But in Germany, the IFUW was regarded primarily as an organization of women and, accordingly, was measured against other women's organizations in terms of its will to achieve political reconciliation.

Given this rift, how did the international organizations nevertheless manage to grow closer, a process culminating in the foundation of a German federation of university women and its admission in 1926 into the international umbrella organization? Personal networks between the IFUW and the international women's movement proved critical. The international women's congresses offered neutral ground for first encounters between academic women from Germany and members of the IFUW, and as such were a decisive factor in the founding and integration of the DAB. As Gertrud Bäumer noted in 1921 apropos of the ICW congress in The Hague, participants at the international women's conferences included significantly more academically educated women than had been the case before the war,[32] and many were also members of the IFUW. The meetings of the international women's associations enabled German and non-German university women to become close on a personal level, and some IFUW members also deliberately used these gatherings to promote the IFUW and as a diplomatic platform. When the IAW met in Rome in 1923, the Italian university women's association organized an informal get-together for all women graduates attending the congress. The association concluded from the success of this occasion that IFUW member federations "may soon be formed in Germany, Hungary, Switzerland, and the Ukraine."[33]

There are strong indications that Gertrud Bäumer—an economist, Liberal Democrat member of the Reichstag, newly appointed ministerial official, and doyenne of the bourgeois women's movement in Germany—was tentatively approached in 1921, at the ICW meeting in The Hague, about the possibility of establishing a German association of university women.[34] In spring 1923, Bäumer became the first woman in Germany to speak out publicly in favor of founding an association of this kind.[35] The women who endorsed that idea and forwarded its progress in the subsequent years had also attended the international women's congresses of the postwar period, along with Bäumer or at her behest. Germany differed from most of the other western or northern

European countries, and indeed from Austria: in Germany, it was ultimately not women actively engaged in research work who ensured that an association was founded and admitted to the IFUW, thus reestablishing academic internationalism. Instead, that task fell to a small number of politically active university and college graduates with links to the international women's movement. These women were the driving force in founding a German federation of university women.

Specifically, the initiative to open a public debate and change the mood among German academic women originated with a small number of women students. At least two of them belonged directly to the circle around Bäumer: Irmgard Rathgen, a student of economics in Hamburg, and Gabriele Humbert, a student of German literature in Berlin. In 1925, Rathgen coedited a fiftieth-birthday homage to Gertrud Bäumer from the "third generation" of the women's movement.[36] A member of the German Democratic Party (Deutsche Demokratische Partei or DDP) and, from 1925, executive secretary of that party's cultural committee, Rathgen attended the ICW congress in Copenhagen in 1924.[37] Shortly afterwards, she traveled to the third convention of the IFUW, in Christiania, Norway, in the company of Emma Alp, a budding art historian and committed pacifist from the southern German university city of Freiburg.[38] In Christiania, the two women introduced themselves as the unofficial representatives of German university women. It is probable, though not verifiable, that this was done in consultation with Bäumer; what is certain is that Alp and Rathgen had gone to Norway to gauge the chances of a German application to join the IFUW being approved by all members. Rathgen herself claimed already to have received personal assurances from "leading women" at the ICW congress in Copenhagen that "all representatives in Christiania, including the French delegation, would accept a German application for membership."[39] However, these assurances seem to have been insufficient to convince Rathgen's colleagues that such an application would be approved without embarrassing dissent.

According to the IFUW's conference reports, Emma Alp took the floor in Christiania to express her regret that it had not yet been possible to found a German association of university women. Younger women, especially, were keen to join the international community, she said. Alp did not leave it at that; she also asked the assembled members for an assurance that the IFUW would unequivocally welcome the membership of the German women.[40] The transcript reveals a youthfully naïve and vehement Alp reiterating an appeal that had been made before by many German associations to international organizations, but mostly without success (especially in the case of academic bodies), causing continued and profound German resentment. In Christiania, too, the initial response to Alp's call was that the statutes did not provide for an invitation of this kind. All new associations were required to submit an application for admission, based on which the Council would come to a decision. The Germans must abide by this rule like everyone else. However, after lengthy discussions, the assembly found a very diplomatic compromise, showing how much the process of the Germans' rapprochement with the IFUW benefited from the interest in harmony and the political experience of women who had long been active in the international women's movement. It was none other than the much-admired Margery Corbett Ashby—a 1901 graduate in Classics from Newnham College, Cambridge, and a member of the BFUW, but best known as a high-profile British suffragist and as the newly elected president of the IAW—who managed to persuade the 1924 conference to set a signal that would satisfy the German women without being regarded by the other member organizations as inappropriate special treatment.[41] Corbett Ashby suggested that the IFUW send a message of greeting to all the national associations currently in the process of forming, a proposal finally approved by the members present. Such notes were to be sent out not only to Germany but also to Bulgaria, China, Estonia, Greece, Luxembourg, Poland, Mexico, and Romania.[42]

In Germany, the events of Christiania were accompanied by carefully staged media attention. In the run-up to the IFUW conference, Gabriele Humbert—as the editor of a new nationwide monthly magazine for female students, *Die Studentin*—had already launched a debate on the arguments for and against a German federation of academic women. Though Humbert herself made no secret of her skeptical view of the IFUW, she allowed Emma Alp to put her points in the journal's very first issue. Alp provided a brief outline of the goals of the IFUW, and followed it with an impassioned appeal to her sister students not to fritter away their energy on the international boycott. Women students must themselves realize how much their interests were "intertwined" with those of their European and American colleagues, and how much they needed "these other comrades." Alp called on students not to make the narrow limits of their own nation the measure of all things. The effort of expanding their horizons would not be "too great"—certainly not, she stressed, for "the scientific thinker, the truly cultivated human being."[43]

On her return from the IFUW conference in Christiania, Alp wrote a further article in *Die Studentin*, again propagating academic internationalism. She now explicitly appealed, quite in the style of the IFUW, to her readers' sense of duty as female representatives of the German educational elite. Because of their sex, she argued, women had a special vocation to combat the "degeneration" of the nation's sciences, "the highest goals of which should and must be truly international, supranational."[44] This was exactly what she had experienced in practice at the IFUW conference, she added. In terms reminiscent of Bäumer's upon her return from The Hague three years earlier, Alp too underlined the "warm and heartfelt goodwill" that the delegates of the IFUW had shown Germany and its two representatives. "Again and again," she noted, delegates had expressed a desire "to see German academic women join the federation's ranks in future."[45] She herself had been asked to tell her colleagues at home of the delegates' "wish and hope to be able to work together with the German women in years to come." Emma Alp's article in *Die Studentin* was directly followed by a full-page item by the IFUW's secretary, Theodora Bosanquet, propounding the organization's program and aims. Although Bosanquet's piece did not close with the promised invitation to the German women, it emphasized the successes of the IFUW and predicted that the federation would "without doubt... in the very near future encompass university women of all those countries in which the higher education of women has become an established fact."[46]

Irmgard Rathgen supported Alp's comments in *Die Studentin* with further arguments in favor of German membership of the IFUW.[47] While studying economics, she had, she explained, come to understand that the international exchange of experiences was an absolute necessity. There were specialist questions that could not be resolved by remaining within a purely national framework. Over and above that, female academics and researchers were especially likely to draw personal and collective benefit from meeting their international colleagues. Rathgen was particularly impressed by the academic self-confidence of colleagues from the British and American women's colleges. "With regard to the creation of a tradition of female intellectual work," she wrote, they were "well ahead" of the German women. In Germany, the very different structure of the higher education system meant that "different paths" must be found to build a collective consciousness of this kind, but nevertheless there was much to be learned from American and British women.

Alp's and Rathgen's energetic advocacy of the international female academic community did not remain uncontested in the journal. Some opined that the time was "not yet ripe" for such a move,[48] or continued to insist on the "pride" of German women researchers, calling for them to maintain, like their male colleagues, their voluntary isolation and refrain from joining any international organizations.[49] But these voices fell silent when, in late 1924, the IFUW's promised message of support arrived in Germany and was printed in both *Die Frau* and *Die Studentin*. The IFUW's official note, which explicitly welcomed the founding of the German association and expressed the hope that "we may expect it to join the international federation in the near future," met with a gratified response. In the women's organizations and among women students, the IFUW greeting was regarded as the hoped-for unequivocal invitation for Germany to join, and as a basis upon which the German women could now officially seek admission to the IFUW without losing face.[50] From this moment on, there was no more debate about whether a federation of German university women should be established, only about the form that federation should take in preparation for joining the IFUW. This new debate, oriented primarily on the context within Germany, is analyzed in more detail below. First, however, I retrace the nature of the relationship of the German women with the IFUW from their admission into the international federation in 1926 until the end of the Weimar Republic.

The "Language Question," 1926–32

It seems unlikely that the German delegates could have imagined a more dignified form for the official admission of the DAB at the fourth IFUW convention in Amsterdam, 1926. The German group joined the IFUW along with Poland, Hungary, and Estonia, as the twenty-sixth national association.

Virginia Gildersleeve, elected president of the IFUW to succeed Caroline Spurgeon, cordially greeted all the new member associations. The opening words by Estelle Simons, a lawyer from Utrecht and president of the Dutch association, were both far-reaching and personal; they were directed only at the German women and, moreover, read out not in the federation's official languages English and French, but in German. It was only with the admission of the German women, said Simons, that the IFUW had truly fulfilled its claim to be international:

> I have always felt very strongly that [the IFUW] was not a genuinely international forum for as long as the academic women from a country so important in the sphere of knowledge were missing. It is therefore a great pleasure for me to welcome you now.[51]

At the end of the opening ceremony, Agnes von Zahn-Harnack, the first president of the DAB and the leader of the German delegation, asked to address the assembled women, and thanked them for their friendly welcome. For her and her sister delegates from Germany, she said, it was difficult to express all they felt "on this auspicious occasion." She drew on the authority of Goethe to invoke a happy and successful cooperation between the associations gathered in the IFUW: "Goethe...said that we all belonged to a tribe that sought for light. Light meant understanding, confidence, admiration, and friendship. The International Federation will help us to find that light."[52]

With its almost 4,000 members, the DAB was the second-largest national group in the IFUW, after the American association.[53] As the Germans observed, it not only had almost as many members as the British and French associations put together, but also boasted by far the highest proportion of women with doctorates.[54] The German delegation in Amsterdam represented the Weimar Republic's female educational elite in impressive breadth. Alongside the DAB president, historian and leading German feminist Agnes von Zahn-Harnack, it included the DDP parliamentary deputy and future DAB president Marie-Elisabeth Lüders; Rosa Kempf, an economist and the director of a women's professional social-work school in Frankfurt; Maria Schlüter-Hermkes, the association's general secretary and a Catholic high school teacher; Anna Schönborn, principal of the girls' high school Uhland-Oberlyzeum in Berlin; and the plant physiologist Margarete von Wrangell, who in 1923 had been appointed Germany's first female full professor at the agricultural college in Hohenheim.[55] Among the nine other German women who attended the Amsterdam conference were Maria von Linden, a parasitologist at the University of Bonn who had become Germany's first titular professor in 1910; and Elisabeth Altmann-Gottheiner, who had earned her university teaching qualifications at the commercial college in Mannheim as early as 1908 and, in 1924, had been appointed there as an associate university professor.[56]

Most of the women in the German delegation in Amsterdam were later, at the IFUW Council meeting in Vienna in 1927, appointed to one or other of the federation's standing committees, and from then on took an active part in the IFUW's practical work: Elisabeth Altmann-Gottheiner in the International Fellowship Award Committee; Marie-Elisabeth Lüders in the Committee on Careers for Women and, when that was dissolved in 1929, the Committee on the Legal and Economic Status of Women; Anna Schönborn in the Committees on Exchange of Information Concerning Secondary Education and Interchange of Secondary School Teachers. In 1927, the physician Ilse Szagunn offered her services to the IFUW Committee on Standards, and Agnes von Zahn-Harnack took up a place on the Conference Committee in 1929. After the early death of Elisabeth Altmann-Gottheiner in 1930, physicist Lise Meitner replaced her on the International Fellowship Award Committee.[57]

However, a shadow soon fell on the spirit of cooperation, understanding, and friendship that Agnes von Zahn-Harnack had evoked in Amsterdam. The Germans felt that neither the status of their association nor the importance of their country was being appropriately acknowledged within the international federation, despite the cordial welcome in Amsterdam. If the Dutch president had briefly greeted her German colleagues there in their native language, only English and French were in use as conference languages at the 1926 convention. The German delegation promptly submitted a request for German to be used as the third conference language of the IFUW, alongside English and French; it was a demand that would preoccupy the Council and conferences until 1932 and contribute to considerable tension within the IFUW. For the Germans, the confrontations over the "language question" in the IFUW became the pivotal issue for their full recognition and a focal point that crystallized their notions of how German academic women should be representing the political interests of their country on an international level, in light of the Versailles Treaty and the attainment of women's suffrage. For the IFUW, in turn, the increasingly vehement German demands became a difficult diplomatic challenge to mediate between the national interests of its member countries and its own objectives.

When the IFUW Council met in Vienna in 1927, the German academic women—supported by their Austrian colleagues—made sure that the language issue took a prominent place on the agenda.[58] Together with the Austrians, they succeeded, after lengthy discussions, in persuading the Council to approve a resolution that

> English, French, and German may be spoken at the Conference. Translations, if asked for, will be given in French and English, since these are the languages spoken and understood by more than two-thirds of the present members. The secretarial work of the Conference will be conducted as heretofore.

This formulation shows that complete equality between German and the other two conference languages was impossible from the outset. Moreover, the Council added a caveat that the resolution was by no means final, and that its practicability must first be tested during the next meeting, in Geneva, in 1929. The majority of the Council members wanted a third conference language only if this did not cause any added obstacles to day-to-day business.[59]

In Geneva, Agnes von Zahn-Harnack took advantage of the Vienna agreement to use her native language for her 20-minute plenary lecture on the IFUW's relationship with the women's movement. A short English summary was circulated among those present, based on the printed version.[60] However, this was far from drawing a line under discussion of the appropriate status of German vis-à-vis English and French; on the contrary, the debate took up substantial space in Geneva as well. The German women were hoping for a specification or revision of the Viennese declaration, especially with respect to the question of whether or not German contributions must be officially translated into the other two conference languages and, if so, whether the IFUW would bear the cost of such translations. For the Germans, the logic of equal status between all three languages seemed to indicate that translations were not necessary at all. This intervention caused considerable annoyance in Geneva, and revived the emotionally charged debate over the principle of which languages to admit as conference languages. The rule applied by the League of Nations, allowing representatives of each nation to use their own official language as long as they offered English or French translations, was rejected as too costly and time-consuming. The Bulgarian delegate Raina Ganeva suggested approaching the problem not as a question of prestige but as a pragmatic issue, and that genuine interest in international understanding would be best shown by making an effort to express oneself in the languages that were understood by the largest number of women present. Like the appeal by the British professor Edith Morley "for a little more lightness and less solemnity in the discussion of a question which was really only concerned with getting through business quickly and avoiding the tiresome process of double translation," Ganeva's attempt to find agreement failed to convince the assembly.[61] What was at stake for the Germans here, and for that very reason encountered resistance on so many levels, was not in fact a practical problem calling for pragmatic solutions, but a fundamental question of cultural politics. It became all the more explosive when, after an emergency session discussing various possible revisions of the Vienna resolution, a clear majority of the delegates in Geneva voted for a version that—while expressly confirming the status of German as a third conference language—made it secondary to English and French more decisively than the Vienna declaration had done.[62] German delegates took particular umbrage at the sentence: "In view of the fact that English and French are the languages understood by the greater number of members present, it is understood that wherever possible the business meetings will be carried on in these two languages." They called this an "extraordinary surprise" and a "substantial and unacceptable change for the worse" in comparison with the resolution approved in Vienna.[63]

The conflicts around the language question, both within the DAB and between the DAB and the IFUW, escalated in the wake of the Geneva resolution. In Germany, the Association of German Women Philologists (Deutscher Philologinnenverband) was most forceful in its claim for equal status for the

German language within the IFUW. With its 2,150 members, this was by far the DAB's largest member association and could exert considerable influence on its stance.[64] However, the call for German to receive equal status was not simply a hobbyhorse of the philologists; it was supported by the broad majority of DAB members. In June 1930, Frieda Kundt, a high school principal in Berlin, proposed a motion to the DAB's board in the name of the women philologists' association asking for the IFUW to be required to "place the German language on an equal basis with English and French in every respect at all events organized by the federation." This was passed by 33 votes to 7.[65] For the philologists, this result logically implied a further demand: if the next conference of the IFUW, to be held in 1932 in Edinburgh, did not prove willing to accept the German women's request, the DAB must cease its work within the IFUW until a satisfactory solution had been found.[66]

In 1930 and 1931, Agnes von Zahn-Harnack used the IFUW's Council meetings in Prague and at Wellesley College near Boston to stress the urgency of the German concern over languages "in the course of various personal conversations."[67] Finally, the British professor of medicine Winifred Cullis, who had been elected president of the IFUW in Geneva in 1929, agreed to visit Berlin in fall 1931, in order to engage the German women personally and to seek a joint solution to the language question. Talks between the DAB's board and the IFUW president took place in Berlin on October 31, 1931. They were held in German, as the board had agreed at a preparatory meeting just before its guest arrived.[68] Only DAB president Marie-Elisabeth Lüders ignored the agreement on her own initiative, using English in her welcoming address to Cullis.

Essentially, Lüders's speech put forward two arguments. Firstly, she said, the IFUW's guiding motivation and the dominant content of its work consisted in "cultural policy objectives." These shared cultural goals could only benefit "from the intellectual and scientific specificity of such a large and many-faceted cultural sphere as the German one," whose language predominated "throughout all of central and eastern Europe." Secondly, the DAB's desire to see the German language accepted in the discussions of the IFUW was supported by the fact that in the largest international women's organizations, such as the ICW, the IAW, and the Medical Women's International Association, "all three languages coexist on an equal basis, and this arrangement has so far not resulted in any difficulties for their daily business."

In her response, Winifred Cullis set out the Anglo-American view: the only inequitable aspect of the treatment of German compared with the two other languages was that "it is not possible to require a translation into German." She justified this in financial terms related to improved efficiency, but did concede to her hosts that it meant asking a sacrifice of the German women. The German language, she said, was certainly acknowledged as a language of culture and of science. However, because the IFUW was not a purely scientific association, this fact could not be taken into account at the cost of carrying out business as efficiently as possible. Among the other national associations, she added, there was little sympathy for the German women's appeal, and if

German were recognized as a third and equal language, it could be anticipated that the Spanish and the Italian associations would immediately put forward identical demands. In fact, she said, at the Viennese meeting in 1927 the Italian women had threatened to make a claim of this kind if the Germans' petition succeeded, and if the federation rejected an Italian request there was a danger that the Italian leader Benito Mussolini himself might pressurize the association to leave the IFUW.[69] Although personally sympathetic to the German women's concerns, Cullis thought the "mood in the other national associations" to be, unfortunately, unfavorable.[70]

In the course of the lengthy and heated discussion that followed, the offended German women rejected the comparison with their Italian and Spanish counterparts, citing the size of their association and the global reach of the German language. They emphasized that worldwide intellectual developments made promoting the German language a "cultural task" in terms of both science and internationalism. The women present drew up an alternative to the Geneva decision on the spot. It did not include the hoped-for general principle of equality between German and the other two conference languages; Cullis had been able to convince the meeting that such a demand had no chance of success. Instead, the Berlin version of the resolution, drafted together with Cullis, ran as follows:

> English, French, and German may be spoken at the conference meetings. Translations from these three languages will be made into French and English. At the formal meeting English, French, German, and the language of the hostess country may be used. There will be no interpretations, but the speakers will be required to provide written résumés either in English or in French. To make our work more effective, the IFUW asks its members to simplify the proceedings of the conference as much as possible by reducing to the minimum the necessity of translating.[71]

To be sure, this new draft differed from the Geneva resolution at most in shades of meaning. The only real distinction was that the appeal to reduce the necessity of translation was now directed less specifically at the German language, and instead, at least in theory, now applied to all members, irrespective of which of the three languages they preferred to use. For the DAB members gathered in Berlin, this sufficed for the new text to be pronounced a success—providing, of course, that the Edinburgh conference of the IFUW was willing to pass the resolution unchanged. Cullis could not promise that, but she did undertake to do her best.

In a slightly altered form, the Berlin version of the resolution was presented for approval to the 1932 conference in Edinburgh.[72] Fearing that the resolution could still fail at the last minute, the German delegation had orders to walk out of the congress immediately if the resolution was rejected.[73] But it did not come to that; their proposal was advocated so persuasively by Winifred Cullis, the Dutch botanist Johanna Westerdijk, and BFUW president Ida Smedley MacLean that, as the German delegates noted with satisfaction, it was "supported by all sorts of different nations and then approved unanimously, with Italy abstaining."[74]

At the IFUW conference, high school principal Anna Schönborn, standing in for the indisposed DAB president Marie-Elisabeth Lüders on the Council, gave a lecture in German on the question of whether girls attending higher schools were overworked—an issue much debated in educational policy at the time—and attracted an audience of more than 200 women. This pleased her greatly, as did the fact that she had "never been addressed so often in German" at an IFUW conference. Her impression of Edinburgh was that

> compared with Geneva, and I should also say compared with America, the atmosphere had changed completely, something that the wider circumstances would hardly have permitted us to expect. In fact, an international spirit prevailed, free of all other, political, influences. I experienced a great amount of cordiality and at the last plenary session, when the president made her farewells, I was asked to express thanks in the name of the whole assembly. I must say that I have never gone to a congress as unwillingly as I went to this one, but I have never come back so satisfied.[75]

Schönborn's personal impressions may be taken as a reliable barometer of the DAB's growing rapprochement with the international community of women academics, given that she had been present at all the international conferences and Council meetings of the IFUW. Six years after the DAB joined the IFUW, the ice had finally been broken between the Germans and their international sisters, at least as regards the leadership elites of the associations. The process was consummated by an invitation to Berlin, where the next convention, open to all members, was to be held in 1936.[76]

The German delight at the outcome of the Edinburgh conference requires some explanation, given that the oft-repeated goal of equal status between German, English, and French as conference languages had still not been achieved. Their demands regarding the "language question" had cast the Germans as awkward and obstreperous newcomers, yet in the end they bowed to the Anglo-American dominance in the IFUW and the new hegemony of two world languages behind which German must now, unlike in the prewar period, accept an important but subordinate position.

The language debate reveals that German academic women had adopted a new approach to politics on an international level, as Gertrud Bäumer had demanded in the early 1920s. The hairsplitting negotiations over nuances of wording in the various versions of the language arrangements may be read as an object lesson in the diplomatic pursuit of German interests, something that became the core stipulation of the DAB's international policy—in analogy to and close association with Germany's wider cultural policy abroad.[77] Reinstating the international standing and influence that Germany had lost through the war was a task that encompassed claims to linguistic hegemony, as reflected in the 1925 establishment of the German Academy (Deutsche

Akademie) and its numerous, publicly funded German language courses abroad.[78]

An important role was also played by a specific dynamic inside Germany. The founding of the DAB and its admission to the IFUW had initially been advocated and driven only by a few liberal feminists. However, in the mid-1920s, interest in international links began to grow among university women outside the liberal leadership of the women's movement. This facilitated the formation of the DAB, but it also created new problems, since the international political activity of many of the new association's officers was by no means always motivated by a belief in liberal internationalism. In fact, the liberal representatives of the Weimar Republic's women's movement also played their part in this nationalist turn within international engagement: they advised those members who continued to oppose internationalism to act internationally but "starting from a strictly national standpoint," so as to "carry a German outlook into the sphere of international work."[79] Nevertheless, some liberal figures were worried about their conservative colleagues' desire to participate in life outside German borders, feeling that a carefully considered balance between preserving German interests and reaching international understanding was more important than simply "representing Germanness abroad."[80]

One thing, though, did unite the liberal and conservative-nationalist academic women in the DAB: their repudiation of Germany's sole war guilt and a radical rejection of the Versailles Treaty. As a result, the desire to restore the sovereignty of what they called the "high-ranking German cultural nation"—expressed among other things in their espousal of the "language question"—became a central feature of the international policy pursued by the German university women's organization. Liberal members such as the economist and German Democratic Party deputy Rosa Kempf[81] supported this policy just as strongly as did sports physician Edith Lölhöffel, an early follower of National Socialism, and her protégée Ilse Szagunn, who played an active part in the IFUW as the deputy president of the DAB.[82] For the majority of the DAB board's members, the aim of "safeguarding German interests" was the highest priority. The conflicts over the language question do, however, indicate that until 1932 the DAB's representatives in the IFUW were pursuing their national interests with a strong sense of what was practically feasible.

For its part, the IFUW emerged from these debates as remarkably unforthcoming in its reaction to the Germans' requests when compared with the other international women's organizations. Its founding members were, almost without exception, academics from the Entente countries. Like most of their male colleagues, these women had carried out war work with particular enthusiasm during World War I, and had conceived of the IFUW as an Anglo-American competitor to the Central Powers.[83] If the "war of the academies" survived World War I less stubbornly in the IFUW than in the other academic organizations, it smoldered longer there than within the other international women's organizations. To be sure, from 1922 onward, the IFUW gradually abandoned the boycott against Germany. But when the Germans brought up the "language question" immediately after being admitted to the federation, the IFUW's American- and British-dominated Council faced the challenge of disengaging from the organization's original conception, which also had a linguistic component. In its earliest days, the future IFUW had been intended as an international exchange program for female graduates in the "English-speaking" world, as a way of strengthening the networks between those women.[84] The internationalization of the IFUW initially took shape in English; French was added as a concession to the Continental allies. Early experiments to make more use of Esperanto as an international language were quietly dropped in 1926.[85]

The appeal to all members to take a pragmatic attitude to the language question, put forward especially by the British women, must be interpreted at least in part as a rhetorical move that concealed a reluctance to undermine the linguistic hegemony of English (and French) in the IFUW.[86] If the vote in favor of German ultimately passed off so smoothly in Edinburgh, this should probably be attributed to a changed attitude to the language question in the inner circle of the IFUW's Council, a shift most likely prompted by Winifred Cullis's visit to Germany in fall 1931. The two sides moved closer together in a process of political negotiation during which particular national interests were weighed up against the principle of internationalism on the Anglo-American model, and were calibrated anew.[87]

A Tradition Takes Shape

The debate within Germany about the value and objectives of an association encompassing all university women sheds interesting light on crucial social issues that affected not only the Weimar Republic's higher education and graduate professions, but also the culture of the women's movement and academic culture more generally. Research in women's and gender history, social history, and the history of science has analyzed these issues, but has not yet addressed them as an interrelated complex—whether in the shape of the oft-lamented lack of a new generation in the women's movement or the difficult situation of women students, graduate women professionals, and women based in German universities and research institutions. Like other political problems in the early Weimar period, they were exacerbated by the era's relentless economic pressures and the harsh collision of social interests. In this context, the founding of the DAB was an important attempt to pursue new paths by establishing a sense of female solidarity between German women graduates within and outside the universities, across disciplinary, professional, and generational boundaries. It was a solidarity that had existed in North America and Britain for several decades and had molded the IFUW's ideal of female academic internationalism. It would be hard to overstate the impact of this international organization on the thinking of the women who gathered in the DAB. The IFUW's influence on these Weimar women is analogous to that identified by Karen Offen and Carol Miller for the international and national women's movements of the 1920s. Offen observes a "new and unprecedented surge of activism" among academically educated women at the international level, focused on the goal of instigating and supporting national initiatives to strengthen the position of women.[88] The national organization of university women in Germany, as I will show, suggests that this upsurge in activism applied in equal measure to the sphere of academic networking between women, with a similarly intense interplay between the national and the international level. This finding contradicts the long undisputed claim by Richard Evans and Brian Harrison that international feminist politics and networking developed a momentum of its own, largely independent of specific national contexts.[89] It also casts doubt on the established image of a "stagnating" women's movement during the Weimar era.[90]

The close interaction between national and international trajectories, and the intimate connection between science and the women's movement, is indicated by the article that launched the German discussion on how to build an association uniting all the country's university women under one roof and representing their interests in the IFUW. In 1923, Gertrud Bäumer published "The Plight of Women Intellectual Workers," which did not in fact bewail the difficult situation of professionally active university graduates and argue for better representation of their professional interests through an umbrella organization.[91] Rather, it offered an incisive and movingly sympathetic portrait of the highly ambivalent social and political situation in which, especially, the young generation of women academics found itself during the Weimar Republic's early years. In Bäumer's view, just because women had won the vote, and some individual academics had been granted their "habilitation" qualifications for university teaching, the post–World War I period was anything but a brave new start for female students. On the contrary, it was an "immeasurable misfortune for the fate of German women in intellectual professions that the second generation after university studies were opened up for women" had been born into an era that faced them with such extraordinary "economic obstacles." Their situation was made doubly difficult by "the widespread shock to Germany's sense of its culture, continuity with its past, and faith in its future." Neither had women in Germany yet securely or completely "conquered their part of the intellectual universe." In Bäumer's view, the women students of the postwar period were

> to a certain extent still pioneers. The pioneers of the elite have done their work, but these students are the pioneers of the average, and in some ways that is the far more testing task. It is based on their evidence, and not the evidence of the first generation, that the case for university studies for women will stand or fall.[92]

Bäumer had earned her doctorate in 1904 at the University of Berlin, and without doubt counted herself among the "pioneers of the elite." She reiterated that since the end of the war, female students had encountered enormous and growing difficulty in delivering the evidence of their capacities. Their often "pitiful" financial situation meant that while these students, like their predecessors, still had no female teachers as role models, they now also lacked economic and social backing. Under these conditions, Bäumer found it unsurprising that many young women, discouraged, gave up their aspirations. The older women's expectations of their younger colleagues—the hope that the youngsters would now find their feet in the intellectual professions and infuse those professions "organically with the feminine spirit"—were bound to seem unreasonable and to prompt a hostile rejection.

Bäumer called on the women's movement as a whole "not to desert the younger generation" but, instead, to try to address women students' social, intellectual, and economic plight. She found graduate women to be under a particular obligation; they must, she said, "join together in taking an interest in their intellectual successors." In Bäumer's view, the first generation of women university graduates must take a share of the blame for the prevailing circumstances. They should always have considered it their duty to help close the social and intellectual "chasm" between women students and the university, and to build connections between the generations. These tasks, in particular, could not be achieved by means of the academic professional organizations, but only through

> an association covering all German university women, something that could be constituted as a cartel of existing organizations and one of whose most important tasks would be the care of women students: providing accommodation and hot meals, setting up clubrooms, and so on. If, for once, all university women were to act together, it would become obvious that they are actually quite a large and effective social stratum.[93]

Today's perspective confirms Bäumer's diagnosis that the early Weimar period found the female educational elite in a state of profound crisis. The occasional award of a habilitation qualification, and the remarkable careers of a few individual women in university research and teaching, tend to distract from the negative evidence that Bäumer analyzed so keenly. It is too tempting to evaluate the hard-fought battles of women as a generally positive trend, whereas their impact on the universities and research of the Weimar Republic was actually minor.[94] At least in the first half of the 1920s, women students were forced to fight their way in universities under very difficult economic circumstances. The threatening economic climate led seasoned pioneers to lose sight of the solidarity they needed to offer their younger female colleagues. The pioneers' complaints of an alleged lack of seriousness and determination among the growing band of women students, and those students' "ingratitude" toward the vanguard generation, began with the typology of female students set out in a 1917 commentary by Marianne Weber;[95] it continued undiminished in the early Weimar period. An example is a 1921 article by Elisabeth Altmann-Gottheiner on the professional outlook for German women graduates, claiming that only the women "of the early days of women's university studies represent a true intellectual elite."[96] Even greater weight accrued to this dismissive statement in that Altmann-Gottheiner, like Weber and Bäumer, was among the exponents of the bourgeois women's movement, the highest goal of which had been to open up universities and research to women. Altmann-Gottheiner's verdict echoed the sentiments of many of her older comrades, who felt that "in recent years, certain elements had found their way into the ranks of women students that seriously lowered the general standard."[97]

Altmann-Gottheiner's position is replicated in many statements by the first generation of graduate professional women. However, its political charge became particularly intense when the professional associations of women university graduates took the same judgment on board. Four of the five largest professional groups of graduate women had formed associations during the final decade of Wilhelm's empire. These were the college-educated teachers' section of the German women teachers' association, formed in 1908;[98] the Deutscher Juristinnenverein, an association founded in 1914 by 28 women jurists in Berlin;[99] the women economists' society Vereinigung der Nationalökonominnen (VdN), formed in 1916 at the wartime conference of the BDF to safeguard the interests of the "growing number of academically educated women economists" in response to their "constantly improving professional prospects";[100] and the organization of women chemists (Verein deutscher Chemikerinnen), founded in 1918.[101]

Given the dismal economic context after World War I, the new specialist associations of women graduates were far from encouraging young women to take up university studies. They welcomed the falling numbers of female students and if anything considered it their duty to warn young women against pursuing a degree.[102] In her 1921 essay, Altmann-Gottheiner cited these professional associations when arguing that the economic setting made it advisable to carry out "the most rigorous selection among the candidates for graduate professions."[103] She saw no need for an exception even in what was traditionally the most feminine of all the occupations requiring a college degree: in view of the "immense overcrowding" of the high school teaching profession, combined with the low public demand for such teachers, Altmann-Gottheiner recommended that only outstandingly gifted women "with a genuine enthusiasm for the educational vocation...should still dare to embark on the thorny path of the senior mistress's career."[104] Similarly off-putting advice had previously been given by the president of the German women chemists' association, Toni Masling. After the association's second general meeting, held in Würzburg in September 1919, Masling issued an "urgent caution" against studying chemistry.[105] The president of the jurists' association, Margarete Berent, put forward a similar argument. She worried that the recently granted permission for female candidates to sit the articled clerk or assessor examinations would tempt increasing numbers of women to consider studying law, and warned, according to Altmann-Gottheiner, that "as long as the careers of lawyer and judge remain closed to women, one can only advise young women not to study jurisprudence."[106] Women doctors made the same point, and Altmann-Gottheiner added an exhortation with respect to her own subject, economics, although her professional association, led by Marie-Elisabeth Lüders, initially hesitated to follow suit, finding that "the economist's profession is still in need of a new generation of particularly talented younger women, especially if those women's financial situation permits them to wait a few years after graduation for a suitable position to arise."[107] To offer support during this difficult transition period, the association set up its own agency helping young university graduates to find their first position.[108]

Gloomy as circumstances were in the early years of the Weimar Republic, Altmann-Gottheiner regarded this difficult phase as promising a useful catharsis for the new generation of the female educational elite. In the long term, she hoped, women's opportunities on the academic labor market would improve. The new democratic state needed a female elite and its highly trained brainpower—all the more so since women now had the vote. Without "intellectual leadership," and that included leadership by women, the task of rebuilding the country would be impossible. "University study for women," Altmann-Gottheiner concluded, "thus does have a future. It will shine all the more brightly the more carefully we select among the women flocking to take it up."[109]

This kind of discourse puts Bäumer's 1923 appeal for intergenerational solidarity between all university women into very sharp focus. Her purpose was to build wide-ranging networks and a sense of common responsibility among academic women of every generation, discipline, and profession in Germany, as a way of securing the survival of the female educational elite beyond the lifetimes of the individual pioneers. Bäumer believed this goal necessitated special efforts to create a female academic tradition that went beyond personal, professional, or disciplinary ties and that must draw its identity-building power from the community of female university graduates as a whole. Her article made no specific reference to the source of inspiration for her deliberations—in all likelihood a deliberate omission, since the heated debates about whether German women should join the IFUW were then still in full flow. However, it is evident that the IFUW and its founding associations served Bäumer as a model for her vision, as a speech she held at the Lyceum Club in Berlin, celebrating the foundation of the DAB in 1926, underscores. There, she argued that the new organization must

> make all the women involved in university education more strongly aware of the unified creative force of feminine intellectuality [*Geistigkeit*] in scholarship, the professions, and personal style, beyond the boundaries of professional affiliations...and give expression to the solidarity of the female cultural stratum which that force makes possible.[110]

By taking this line, the new organization unequivocally adopted a strategy of building a female academic tradition. That strategy had been launched successfully in the United States in 1881 and continued in the framework of the IFUW. The resulting networks of college and university graduates, initially supraregional and after 1919 supranational, declared a university degree to be the common feature that defined a female "class entirely new." In Germany, too, the connecting and sustaining component of the female educational elite was to be not women's professions, paid employment, or marital status, but solely their gender and academic training. Networks between such women were to form the foundations of a female academic tradition into which later generations of women students would be able to grow.[111]

To grasp the genuine novelty (and the Anglo-American provenance) of this principle in the Germany of the early 1920s, it is useful to draw a comparison with the beginning of academic women's networking and its social and political context in the United States. The Association of Collegiate Alumnae (ACA, later AAUW), which provided the model for the DAB's constitution in 1926, was born in 1881 out of a broad-based bourgeois educational movement championed by men as well as women. After the American Civil War and the subsequent reconstruction phase in the southern states, this movement was reacting in part to a growing demand for female teaching staff; indeed, Barbara Solomon describes the call for women's education in the United States between 1850 and 1900 as the "demand of the age." This was the context for the ACA's success in establishing a nationwide network of women college graduates. The association's efforts to promote social acceptance of a model of academic womanhood was further facilitated by the fact that, during the second half of the nineteenth century, the demand for women's education became both a catalyst and a common denominator of the US liberal reform movement as a whole. A demand upon which both men and women, proponents and opponents of abolition or women's suffrage could agree, it offered shared ground that in some ways compensated for activists' other divisions.[112]

In Germany, calls for women's education were first voiced during the revolution of 1848 and gathered strength from the mid-1860s onward.[113] Unlike the US (and British) situation, the goal of providing women access to higher education did not become a broad-based, unifying reform objective of the liberal middle class either in the period following the 1848 revolution or in the course of Germany's militarization before, during, and after the foundation of the German Empire in 1871. On the contrary, it developed as a politically polarizing demand put forward by social democracy and the women's movement.[114] For prominent representatives of the bourgeois women's movement in Germany, comparisons with other countries were a constant and painful reminder of the impact of this political polarization on girls' and women's actual experiences of education. After visiting Girton College, Cambridge, in the mid-1880s, Helene Lange reported that the "friendly helpfulness of men, free of that patronizing superiority," had left her unable to suppress a "sense of bitterness" when she thought of the battles around girls' education in Germany. The "keen interest that women in the leading circles of society felt for the whole educational movement and backed up with offers of rich resources" impressed her as a "stark contrast to circumstances at home." In Germany, with only "very few exceptions," backs were firmly turned on this aspect of the women's movement, and "the slightest contact with the idea of 'emancipation' in general" was shunned.[115] As social democrat Lily Braun complained, at the turn of the twentieth century, when almost all the neighboring countries had accepted women physicians as public employees, in Germany even the admission of women to universities was still "feared like an act of revolution."[116]

The politicization of the women's educational movement in Germany, along with its lack of a broad base of support from society and the state, was one important reason for the explicit exclusion of women from university education formulated in 1879 and the tardy granting of regular admission to university degrees, which was achieved only between 1900 and 1909 in the German Empire (varying from state to state).[117] At the same time, the political and social configuration inimical to women's education also impacted negatively on the relationship between the women's movement, women students, and the academic world. Networks based on gender appeared dangerous in an academic context exclusively controlled by men. Drawing on the autobiographies of the first generation of German women students, Patrizia Mazon has reconstructed this field of tension; she observes that the women pursuing academic vocations were the first to distance themselves from the women's movement. This was even true of women who, like the later parasitologist Maria von Linden, received periodic funding from the Allgemeiner Deutscher Frauenverein. In her student years, von Linden adopted an insistently androgynous style and conformed, as far as possible, to the male-connoted academic conventions at the university, making a special point of rebuffing the hopes placed in her by the women's movement. She saw herself as an individual blazing her own path, and refused to be co-opted as proof of women's capacity to work in science without losing their femininity.[118]

Of course, this first generation also included students who were already active in the women's movement when they entered university, women who took a firm stand on female education and access to the graduate professions. According to Mazon, these women developed a fierce ambition to fulfill the academic standards of their university programs, but shied away from the arduous attempt to gain access to male student culture and to become, as far as this would have been possible, part of the academic milieu.[119] In Mazon's view, it was precisely their background in the women's movement that "swept" students like Alice Salomon and Marie-Elisabeth Lüders out of the university immediately after graduation despite their talent for research.[120]

Only rarely did a woman move to and fro between the two worlds. These were scholars committed to the interests of academic women and unafraid to associate with the women's movement, such as the physicist Elsa Neumann, who died tragically, aged just 30, in a 1902 laboratory accident. In 1899, Neumann had become the first woman to earn her doctorate at the University of Berlin, ten days after being appointed by her powerful patron Max Planck as the first, and for many years sole, female member of the prestigious Physical Society.[121] She continued to pursue her scientific interests outside the university after completing her doctorate,[122] but also took an active political role in support of the women's movement: in April 1900, she founded an association to provide interest-free loans to women students, which by 1914/15 had accumulated enough donations to fund several grants to women students.[123] Looking at the people who joined Neumann in her efforts, it becomes clear that in early twentieth-century Berlin a group of women researchers, along with academically ambitious graduates and professors, had taken up the cause of the new generation of university women. The board of Neumann's association included the bacteriologist Lydia Rabinowitsch-Kemper, who had earned her doctorate in Berne, then taught in the United States, and was now working with her husband Walter Kemper at the Prussian Institute for Infectious Diseases (the later Robert Koch Institute); the historian Hildegard Wegscheider, who had studied in Zurich, taken her doctorate in Halle, and spent one year teaching on Helene Lange's college-preparatory courses in Berlin, but was dismissed upon marrying the physicist Max Wegscheider and in 1900 founded the first private girls' high school in the Berlin district of Charlottenburg, with support from Minna Cauer's Frauenwohl organization;[124] Franziska Tiburtius, who studied medicine in Zurich before becoming the first woman doctor to establish a practice in Berlin; and the scholar of German language and literature Helene Herrmann, Berlin's first married doctoral candidate (her doctorate was awarded in 1904). Herrmann's husband Max, a specialist in theater studies, was also a member of the association, as were Elsa Neumann's most important male mentors.[125] Initiatives like Neumann's association or the academic section of the Lyceum Club in Berlin[126] were high-profile but isolated and locally restricted activities where university women from various disciplines met and worked together for the benefit of the next generation.[127] But Germany certainly did not have a nationwide or, more importantly still, an intergenerational organization of graduate women comparable to those initiated in the United States in 1881 and Britain in 1907, which were so successfully deployed to improve educational opportunities for girls, create employment opportunities for college and university graduates, and campaign for the broad social recognition of academic women.

The early success in networking among US and British women, unlike their German sisters, was greatly aided by the existence of women's colleges. Different though these institutions were in the American and British contexts, when compared with circumstances in central Europe they indicate the long-term and significant advantage that was gained by beginning with women-only higher education (even if it was often dismissed as "second-rate") as opposed to the coeducation that was the only option offered by German universities. The segregated system favored networking and professional activity for women, both of which arose much earlier in the women's colleges than in the established universities. Likewise, the women's colleges fostered the invention and societal anchoring of a model of academic femininity that could be conveniently reconciled with traditional notions of "womanhood."

In this respect, the key issue was less the relevant institutions' curricula than their representation of what Sophie Forgan, in her study of the history of educational architecture in Britain, has described as a kind of academic domesticity.[128] The architecture of women's colleges differed substantially from that of the older educational establishments for men, in both ground plans and exterior aspect. On the one hand, the women's colleges were more protected from the outside world through enclosed courtyards; on the other, in the nineteenth century their facades were marked not by an intimidating neo-Gothic but by variations of the "Queen Anne" style that cited the domestic architecture of the early seventeenth century. In Margaret Vickery's view, this cozily agreeable architectural style helped to soften criticism of women's university education in Britain by effectively countering the impression that educational institutions would rob women of their femininity or turn them into unnatural "bluestockings."[129]

The academic domesticity staged architecturally since the nineteenth century—in both Britain and the United States—was echoed in the interiors of those universities where young men and women studied together. In the late nineteenth century, British and, especially, American coeducational universities had set up "women's rooms" or "women's halls." While male students amused themselves in smoking rooms fitted out with spittoons, billiard tables, and durable furnishings, the decor of their female colleagues' common rooms exuded the same domestic charm as that of the women's college facades. Fresh flowers, dainty wicker furniture, good carpets, expensive tea sets and large mirrors, pianos and other musical instruments all made it clear that the academic training being enjoyed by young women in the institutions of higher education did nothing to compromise their femininity.[130]

Women's exclusion from the established educational institutions of men and the efforts to make women's academic education socially acceptable (at first through separation) led to British and US students creating spaces within the higher education system where women could find solidarity and communication among themselves. In other words, their sequestered academic and domestic life in the women's colleges, or in the women's halls of coeducational institutions, made it easier for women students to establish a gender-specific academic identity within the universities.[131] That was not possible for women students in Germany in even the most rudimentary form.

Considering the external face of academic domesticity that prevailed in late nineteenth-century British and American women's colleges and universities, it is perhaps surprising that so little significance was initially attached to the discussion of the "special character of women" in relation to women's capacity for academic study and work.[132] Reporting on her 1880s trip to the women's colleges in Oxford and Cambridge, the German pioneer of girls' secondary education Helene Lange remarked that, when it came to academic work, nobody in England

> talked about the "special character of women." . . . The general opinion was this: The appropriate path to science was initially the one that men had trodden. It was dubious whether two separate routes toward science existed; but if, in the course of time, different methods and paths were to arise out of the viewpoints of women, these would be all the more surely discovered and all the more effectively distinguished from their masculine counterparts if the viewpoints of men first served as a foundation. This was particularly advisable in that the general, formal principles of science were indubitably human, not masculine.[133]

Many representatives of the German bourgeois women's movement supported this approach wholeheartedly,[134] but the conclusion they drew from it was that women must, at all costs, study alongside male students at the existing universities and with the same professors.[135] The international comparison shows that this rigorous German insistence on coeducation as a way of demonstrating women's capacity for academic work arose from the particular structure of the German educational system. In the late nineteenth century, the mission of higher education in Germany (almost always publicly funded and government regulated) had been resolutely and exclusively directed at the future male educational and state elite. This may explain the strength of women's commitment to coeducational academic training. German university women firmly opposed the segregated advanced education for women that was common practice in many medical schools, and in part also at Oxford, Cambridge, and Harvard, well into the twentieth century.[136] Its consequence, the German women argued, was a second-rate education—a fear that seemed confirmed by their travels in Britain and the United States. A "second-class education" of this kind could only play into the hands of the opponents of university studies for women, who denied that women were capable of "real science" or of practicing graduate professions.[137]

However, by taking this stance in their struggle for admission to the universities, the German women's movement implicitly committed itself to the notion that the proof of women's academic aptitude depended on how well they made their way as individuals within a masculine university culture, without the gender-specific models, structures, and resources that would have enabled them to assert their position as new, female citizens of the "academic state."[138] Women students on the foreign and often hostile terrain of the university faced immense pressure from the dual expectation of having to stand their ground intellectually while also remaining "feminine." In Germany, the lack of an infrastructure of women's colleges and "women's halls" like those in Britain or the United States, and more generally the lack of social acceptance, in many cases led to a sense of alienation between the women's movement and women students—that is, the female academics of the future.

The first local and national academic networks among German women emerged shortly before the turn of the twentieth century. Germany differed from the United States, and later Britain, in that these networks were initiated not by the alumnae of specially renowned women's colleges, but by the women students who had only just attained permission to attend coeducational universities as auditors. In a range of different ways, those women tried to compensate for their marginal position within the university by closing ranks and creating rudimentary infrastructures to help them cope with their studies. The first of these networks was a Berlin-based women students' club named the "Verein studierender Frauen in Berlin." Founded in 1896, it was clearly modeled on older organizations from Switzerland.[139] The club maintained an information office offering guidance on conditions in various programs at the University of Berlin and passing on information about accommodation for women students in the city. It also set up contacts with the societies of women students in Swiss universities, where many German women had gained their first university experience and where it was still easier for a woman to complete a doctoral degree.[140] Women at the universities of Bonn, Freiburg, Heidelberg, and Marburg followed the example of the students in Berlin, and, in 1906, the new groups were federated nationally as the Verband studierender Frauen Deutschlands. This liberal league initially focused on agitating for women's right to matriculate at German universities; from 1909 on, when universities were legally obligated to accept women as regular students, its priorities shifted to social and subject-related activities designed to promote the "lively understanding of university studies and womanhood." The Berlin club arranged readings and evening debates in the university's "ladies' room" and organized weekend excursions. In the wartime conditions of 1915, the league, now named Verband der Studentinnenvereine Deutschlands, joined the country's umbrella organization of women's associations, the BDF.[141]

In 1909, a rival organization was founded in Berlin: the Deutsch-Akademischer Frauenbund an der Universität Berlin, firmly oriented on the values of "Greater Germany," anti-Semitism, and the traditional student corps. The resulting polarization among organized women students in Germany was a political one, but was also reflected in the degree to which the organizations tried to comply with the dominant masculine and fraternity-based model of student life. The majority of the liberal league's members were "free societies" that functioned like associations. In contrast, the nationalist Deutsch-Akademischer Frauenbund an der Universität Berlin insisted on its status as a "community of moral education," tinged with nationalism and closely modeled on the male student fraternities. It had a sorority house, gave itself a coat of arms and corporate colors, and established strict hierarchies and rituals; its goal was female students' "organic integration into the academic state."[142] In 1914, it joined with similar new groups in Münster, Göttingen, and Greifswald to form the Deutscher Verband Akademischer Frauenvereine or "German league of university women's associations" (DVAF).[143] The spectrum of female student organizations was completed by two initially smaller groups, the Catholic and the nationalist Protestant associations of women students. In Germany, the structure of women's student organizing, based on individual associations, offered scant room for an overarching and connective female academic tradition to develop as it had, in fits and starts, in Britain and the United States. Moreover, according to Claudia Huerkamp, these associations were not particularly attractive to women students. In 1918, only around 700 women students in the German Empire—just 10–15 percent—belonged to one of the organizations mentioned.[144]

In 1923, when Gertrud Bäumer called for a common organization to unite all Germany's university women, what was on offer in terms of academic female networks and organizations was thus firstly a number of rather small organizations of women students splintered along political, ideological, and religious lines, and secondly women graduates' professional organizations, divided by discipline. Although attempts had been made during the 1914–18 war to draw the two umbrella organizations into one federation,[145] the necessary momentum had dissipated in the political and economic turbulence of the 1918 revolution.[146]

It says much for Gertrud Bäumer's influence and the persuasiveness of her idea of a new umbrella organization, as well as for the negotiating skill of the jurist and Reichstag deputy Marie-Elisabeth Lüders, that the situation changed so thoroughly between 1923 and 1925. During those years, after some hesitation, all the women's associations of students and graduate professions agreed to form the DAB. In July 1923, Bäumer had invited the representatives of all these organizations to her office in the Interior Ministry, in search of agreement on the structures and goals of a future federation of university women. Supporting the new cohort of women students and building "closer relationships between the older and younger generation" were settled upon as the two key goals of the new organization; furthermore, the DAB was to "enable German university women to represent their interests within the IFUW by offering them a unified representative body."[147]

The paucity of documentary evidence precludes a full account of why the DAB was not officially founded until two years after that meeting. Up to the end of 1924, when the IFUW's eagerly awaited invitation arrived in Germany, the process of reaching agreement had been paralyzed by disputes over whether Germany should join the IFUW at all; other compelling reasons also played a part. In the early 1920s, financial worries alone militated against establishing an additional organization for university women. In the turbulent years of defeat, revolution, and hyperinflation, all the associations were facing enormous difficulty in sustaining their existing activities: many members were barely able to pay the dues for their own individual club. Contributing to an additional umbrella organization would have entailed a further financial burden, one that many associations considered too much to ask of their members.[148]

There were also concerns about the substance of the proposed new organization, issues touching on the very core of the IFUW's objectives. Many university women in Germany considered it anything but wise to define themselves as part of an organization not through their membership of a profession, but solely by virtue of their completion of higher education and their gender, as advocates not purely of professional interests but of university education and womanhood in general. Women physicians articulated these concerns with particular clarity. When their own organization, the Bund Deutscher Ärztinnen (BDÄ), was constituted in 1924 on the prompting of the Medical Women's International Association, many members expressed unease. They feared their position within the organized medical profession would be weakened if they spoke up specifically as a group defined by gender. In the graduate world as a whole, it seemed injudicious to draw more attention to one's gender than absolutely necessary; furthermore, the difficult labor market of the early 1920s left members anxious to avoid the impression that organizing as women was part of some militant formation of the sex. Even after the federation of women doctors was formed, its members frequently reiterated that they had not the slightest intention of trying to "compete with our male colleagues."[149]

If individual groupings were hesitant to organize as women within their disciplines, their worries were redoubled when it came to a cross-disciplinary and cross-party organization of all Germany's university women, a body that would be defined purely in terms of its members' university degree and gender. On the other hand, the bleak state of the graduate labor market in the 1920s was increasing societal pressure not only on specific groups of graduate women professionals, but also on the female academic elite as a whole. Renewed doubt was being cast on the "returns" to be expected from women's participation in intellectual life and research; women's right to continue claiming a share of society's scarce resources for higher education, the professions, and academic research was disputed. The most prominent critic was the liberal minister of education and culture in the state of Baden, Willy Hellpach. Running for the office of Reich president in 1925, he pronounced that academic education for women had brought "absolutely no enrichment of the nation's culture or science" and should thus be considered a failure.[150]

The movement to unite all the organized university women took shape as an assertive response to this crisis, with the declared objective of enhancing the visibility of university women's "cultural influence." It was a response that indicates the profound impact of the IFUW model and its lasting contribution to changing the way that graduate women in Germany saw themselves. Most strikingly, women working in German universities and research institutes now decided to create an organization of female university teachers, the Verband der Hochschullehrerinnen Deutschlands, in the run-up to the DAB's formation—and convinced 20 of the existing 25 German female university lecturers to join it in 1925.[151] In 1929, the organization's membership already numbered 36, the great majority of the 42 women in Germany who had completed their habilitation.[152] The members included the highly respected, 60-year-old parasitologist Maria von Linden, who had so firmly refused to be associated with the women's movement during her student years some decades before. Further prominent members were the chemist Margarete von Wrangell, who had earned her habilitation in 1920 at the agricultural university in Hohenheim and became Germany's first full professor there in 1923; the economist Elisabeth Altmann-Gottheiner, who had worked unpaid as an associate professor at Mannheim's commercial college since 1924; the physicist Lise Meitner, a researcher at the Kaiser Wilhelm Institute of Chemistry in Berlin since 1913 and a qualified university lecturer since 1922; the economist Charlotte Leubuscher, who, in 1921, completed her habilitation at the University of Berlin as the first of her discipline in Germany; the Indologist Betty Heimann, who had earned her doctorate in 1919 in Kiel and her habilitation in 1923 in Halle, where she had taught since then; the archaeologist Margarete Bieber from Giessen; and the geneticist Elisabeth Schiemann from Berlin.

The membership register of the women university teachers' association is now lost. We can only identify the members indirectly—for example, through their active participation in the DAB's executive committee, as in the case of Margarete von Wrangell, or in one of the local groups, as in that of the biologist Paula Hertwig (the first woman to earn her habilitation in zoology at the University of Berlin, 1919) and the bacteriologist Gertrud Meissner (a lecturer in Breslau since 1927), both of whom were on the board of the DAB's Berlin chapter. The first president of the lecturers' association was biologist Rhoda Erdmann, who had taught at Yale from 1913 to 1918, obtained her habilitation at the University of Berlin in 1920, and built up an independent institute for experimental cell research at the Charité teaching hospital in Berlin.[153]

The founding of the association of women university lecturers and its admission to the DAB illustrates the innovatory power of the university women's networking that Gertrud Bäumer, encouraged by the IFUW, had called for in Germany in 1923. Although by far the smallest member of the German federation in numerical terms, the association without doubt attracted those women who were to play the greatest part in shaping external perceptions of the DAB's status. It was they who emphasized that the new federation did indeed represent university women as a whole and demonstrated the full range of women's "cultural contribution." For the university lecturers, this meant abandoning their previous strategy of playing down their gender and keeping their distance from the women's movement. By establishing an association of women university lecturers, they were now expressly defining themselves as women working in the academic world—a fact that can only be explained in the context of the IFUW, for the crucial trigger was provided by that organization and the academic internationalism it propagated. The academics active in the IFUW leadership were a stimulating role model for their German colleagues. At the same time, IFUW networking opportunities and research funding programs offered these women particularly practical incentives to make use of the network's resources.[154]

The importance that the DAB's officials attached to building bridges between women academics and the women's movement becomes evident in their decision to provide an important forum for the university lecturers' association so soon after it was founded. At the festive and well-attended inaugural meeting of the DAB in May 1926, Charlotte Leubuscher was asked to speak to the audience on "the professional situation of women university lecturers."[155] Leubuscher's presentation was the first attempt to draw general conclusions on the circumstances of female university lecturers in Germany. Although the overwhelming majority of these women had many years of experience in research behind them, it was only between 1919 and 1923 that they had been permitted to complete their habilitation. Leubuscher noted the particular difficulties that inflation and recession had caused for women lecturers, who, in most cases, worked without pay. However, she stressed that the association did not plan to react to this fact politically by posting its own demands. The situation of female lecturers, she said, in many ways did not differ from that of their male colleagues, and should therefore be represented "in cooperation with our male peers." Remaining prudently reticent on the matter of the ferocious competition with male colleagues, Leubuscher was more forthright in criticizing the women's movement, which she accused of having failed so far to take any interest in the concerns and achievements of women researchers or to address the situation of women university lecturers. "No historian in the future," said Leubuscher, would be able to deduce from the minutes of meetings of the women's organizations "that the years after 1918 saw the first women attain their right to lecture at German universities."[156]

Certainly, the publications of the Weimar women's movement only very rarely featured articles by women academics, at least until the mid-1920s, but Leubuscher's accusations seem to have been somewhat exaggerated.[157] More important, and more surprising, is her declaration at the DAB's inauguration that lecturers wished to see the women's movement give more attention not only to their scientific achievements but also to their economic circumstances. She was happy to note that the organized university women had now—by asking to hear about women university lecturers at the DAB's very first meeting—set a clear signal that they were not "completely indifferent to the situation of women lecturers." These women, continued Leubuscher, were not expecting "an interest in their personal destinies." They were, however, convinced

> that for the entirety of academically educated women, it is of paramount importance to ask whether the profession of the university lecturer will remain open to women in future, not only in theory but also in practice, and how the position of women within that profession will develop in the long run.

It was for this reason, she added, that an association had been formed and that it asked for support from the women's movement. Regarding the promotion of the new generation of academic women, something that the DAB had made its most important concern, Leubuscher stressed the special contribution of university lecturers. The association of women university teachers aimed to encourage long-term support for young academics within the framework of the IFUW.[158]

Leubuscher's speech to the gathering of university women focused on the alarming financial situation of women university lecturers and their difficulties in earning a living from their teaching work. Surprisingly, she made no mention at all of the public debates instigated by Hellpach's comments querying the scientific productivity of women. Yet it was precisely this public discussion that had spurred women university teachers to seek a hearing and assistance on a national and international level. This point was made by the association's president, Rhoda Erdmann. At the 1926 meeting, the 56-year-old biologist chose to make a very critical public review of her own academic career. Her talk addressed head on Hellpach's aspersions regarding the social benefit of women's education: in view of the tiny number of women who had been appointed so far, she argued, it was absurd to ask at this stage whether women researchers were capable of contributing substantially to the progress of the exact sciences.[159] Without analyzing in detail Erdmann's impressive autobiographical description of the pitfalls of a scientific career and her situation as a researcher, it is interesting to ask at this point what circumstances enabled her to give such an outspoken account of the genuine obstacles to the academic productivity of women in Germany.

Erdmann earned her doctorate in Munich in 1908, and went on to work in experimental cell research at the Robert Koch Institute, Berlin. In 1913, she was invited by a respected biologist at Yale University to work on his material for one year. Caught unawares by the outbreak of World War I, Erdmann was initially unable to return to Germany, and received an offer to continue researching at Yale. During the year from 1914 to mid-1915, Erdmann had

> complete freedom to work on whatever I wanted, the most excellent working conditions that one can imagine, sufficient staff, etc. Added to this, I was made an Associate of the Rockefeller Institute.... I received a large salary and my research work was facilitated in every way, in the large laboratory of that just, kind, and highly distinguished researcher Prof. Harrison. Students applied, doctoral researchers asked for advice, and so it appeared that, if I remained in America, my scientific career would develop to my satisfaction.[160]

However, Erdmann decided to return to Germany. Despite her international renown, once in Berlin she could not find a full, publicly funded professorship and was never again able to work under anything like the conditions she had enjoyed in Yale—even though in Germany, too, she was an acknowledged authority in experimental cell research. Erdmann emphasized that her short autobiography was designed to "show only the fact that there is so much productive power among women that is not exploited, that is suppressed and cannot come to full fruition because the very few excellent positions that men have created—and probably *for* men—are only with great reluctance awarded to women."[161] For herself, she was left with

> the bitter feeling that if I had been offered all the opportunities at an earlier stage, and if I now possessed an excellent laboratory, I would easily have been able to compete with the famous laboratories of the world and contribute to the promotion of German science.[162]

Under the aegis of the IFUW and as part of the preparations for founding the DAB, in the mid-1920s a gender-specific organization of women university lecturers thus formed for the first time in Germany. Its objective was not only personal networking, but also the creation of a platform for a public, political discussion of women's marginal and disadvantaged position in German academia—although the newly organized women lecturers still took great care not to provoke even more resistance to their presence within universities and research institutions. The IFUW offered opportunities for withstanding or countering such pressures on women. International networking also opened up access to additional resources abroad; and in Germany, it was certainly a source of prestige that enhanced the willingness of university women, in particular women researchers, to define themselves as members of a female elite and make their voices heard in public. For the small group of academic women who held the habilitation qualification, this meant ceasing to regard themselves as lone fighters within the *civitas academia*, keeping a more or less conspicuous distance from the women's movement. They could now take a stand as members of a transnational female network that had its roots in the women's movement and that went beyond the boundaries of the universities. To this extent, the DAB's foundation heralded a dynamic new departure. Given the difficult economic and political circumstances of the Weimar Republic, the effect of its clarion call should not be underestimated.

In many respects, the DAB represented an ideal of how an organization of all university women could come about under German conditions. Its foundation seemed to inaugurate a new self-image among university women, apparently overcoming the oft-lamented splits between the women's movement, professional women with degrees, and women academics.[163] The DAB had the potential to become the quintessential Weimar women's organization. Unlike all the IFUW's other member bodies, the DAB was itself an academic umbrella organization. It had very few individual members, and was based almost entirely on the corporate membership of the women's graduate professional associations. Well suited though this organizational form might be to representing the greatest possible spectrum of academic womanhood in Germany, it did also have its drawbacks. Especially problematic was the structure and membership of the DAB board, to which, according to the bylaws, each member association must send a delegate. As a consequence, the political antagonisms between the member associations—a paralyzing factor for other women's organizations in the Weimar era as well—were soon reflected in the DAB executive.[164] Alongside Marie-Elisabeth Lüders, Agnes von Zahn-Harnack, and Margarete von Wrangell, who represented the executive's liberal democratic majority, Ilse Szagunn (president of the Deutsch-Akademischer Frauenbund and the DAB's vice president) from the outset spoke for the federation's nationalist and conservative, revanchist, and anti-Semitic groupings.[165]

At first, the emergence of political factions within the DAB board made itself felt primarily in the conflicts around the "language question" in the IFUW, as discussed earlier in this chapter; after 1933, it would contribute crucially to the federation's alignment with the Nazi regime. But in the first years after its foundation, the DAB experienced a short period of florescence. Twenty-seven local groups formed in quick succession, some of them entering into the new task with great enthusiasm.[166] Judging by the wide spectrum of professions represented by these groups, on a local level the DAB had rapidly became a genuinely interdisciplinary and cross-professional network. In Freiburg, the jurist Erica Sinauer founded a group that attracted numerous women doctors and high school teachers; Leipzig women benefited from the commitment of meteorologist Luise Lammert, who had spent 1928 researching in Australia on an IFUW International Fellowship; in Emden, Elisabeth Weber, a senior teacher, drew together a female academic milieu; and in Berlin several members of the DAB's executive took up work with a number of university lecturers.

The general aspiration was to follow the Anglo-American pattern in defining the federation not by professional sector but by academic status, as can be seen in the fact that married women, not active in the professions, were also persuaded to participate actively at local level. It was not unusual for assistance in creating local chapters to come from graduate women who had given up their professional or scientific ambitions upon marriage; in Munich, for instance, the local group was led by a professor's wife.[167] Some surviving correspondence from the late 1920s documents networking between the groups above local level. For example, women newly arrived in an area might ask for the names and addresses of established members as a way of overcoming their loneliness.[168] Lecture evenings, small receptions, and socials with women students were regular features in most of the local chapters.

At a national level, the DAB's first campaign was an initiative to improve the income of women university lecturers. An information sheet setting out standardized fees for academic lectures by women was published immediately after the DAB was founded.[169] Politically, the federation objected in strong terms to the increasingly aggressive discourse, gathering pace in the late 1920s, that called for the removal of graduate women from the professions and married women from senior civil-service posts. The DAB collected comments in the press, regional parliaments, and ministries, and reacted with statements of their own.[170]

An ambitious tradition-building project was also initiated, led by Agnes von Zahn-Harnack in cooperation with the German State Library in Berlin. With additional funding from the Emergency Association of German Science (Notgemeinschaft der deutschen Wissenschaft), work began on the "memory of the women's movement," an annotated bibliographic record of all publications (books and articles) relating to the woman question; its aim was to provide a black-on-white documentation of the achievements and struggles of women's cultural contribution in the preceding 100 years, as well as the cultural capital it had accumulated. Two women librarians worked on this complex project at the State Library for more than six years. Because each individual book was checked by the compilers, the listing of titles and the short annotations was very labor-intensive, and aimed to fulfill the highest academic standards.[171] Despite severe funding problems and political uncertainties, especially in the later years, the project was successfully completed and published in 1934.[172] The bibliography on the woman question was long regarded as a model of scholarly bibliography and annotation.[173]

Finally, an important component of the organization's work was fostering contact with colleagues from abroad and taking care of them during visits to Germany—a task growing to such dimensions, particularly in late-1920s Berlin, that the DAB's executive struggled to keep up. For example, numerous delegates from Norway, Spain, Yugoslavia, and Greece arrived in Berlin after the IFUW's conference in Prague, and were welcomed there by many of their German colleagues. When a group of Bulgarian university women came to visit, the DAB leadership organized a small conference on the woman question in Germany for them; welcomes were also extended to other members of the IFUW from Argentina, England, the Netherlands, Romania, Switzerland, and the United States, some of them women who held great influence in their home countries. According to the summary of the DAB's activities for 1930/31, "all members of the executive committee assisted with great dedication in this work, which makes very heavy, often personal, claims on their energy."[174] Berlin members may occasionally have found it burdensome to offer hospitality as required by the IFUW. However, the women hoped their efforts would earn them "the entitlement...for German members to enjoy similar hospitality in other countries in return."[175] Furthermore, the women were well aware that their commitment was helping to lend political weight to the DAB. The Foreign Office had, noted the end-of-year report, acknowledged and supported the DAB's work as "especially important in terms of cultural policy."[176]

The DAB acted primarily to protect the interests of working graduate women, who made up the majority of its membership. But it also aimed to draw the new generation of academic women into the task of nurturing a female academic tradition in Germany. This process can be retraced through the commitment of female researchers, working graduates, and politicians in the field of women students' welfare.

Nurturing the Tradition

Gertrud Bäumer's discussion of the "plight of women intellectual workers," the 1923 publication that inaugurated the process of founding the DAB, may be interpreted as part of a dialogue with a group of women students close to her. The first of these representatives of the new generation of women academics to respond to Bäumer's diagnosis were the same individuals who had instigated the reconciliation between German women and the IFUW. Irmgard Rathgen was one of them. Deeply impressed by the British and American university women at the ICW's 1924 world congress in Copenhagen, Rathgen took a very similar view of the German generation conflict to that presented by her mentor. In the essay "Generations," published in 1923 in Bäumer's honor, she noted that while the "older generation" might complain of the dearth of suitable successors, the students themselves felt "abandoned in these boundlessly troubled times." Rathgen considered this a dilemma that endangered not only the continued existence of the female intellectual elite but also the women's movement as a whole: in her opinion, it was the ranks of women students that, "by internal necessity," would yield "a significant proportion of the leaders of the women's movement." The lack of a "tradition for university women" was thus by no means simply "a gap in the masonry like any other gap, but rather the absence of foundation stones and mortar."[177]

In her address to the DAB members, assembled for the first time in 1926, on the "wishes of women students" and ways to develop intergenerational support—that is, the transfer of values, experiences, and personal connections—between university women, Rathgen's first point was the need to create "clubs in university cities." These could, she said, become "a genuine gathering point, a center for university women."[178]

In fact, university women in various towns and cities had already begun to experiment with clubs of this kind. In Munich and Frankfurt, for example, the economist Rosa Kempf had initiated a local gathering of university women across faculties, age groups, and marital status as early as the end of 1923. The aim of these groups, which served as the models for the formation of new local DAB chapters soon afterwards, was to tap the "schooled intellect that women had gained through their university studies" in the exploration of general issues "that specially affect women within the national community." Its initiators were confident that such intellectual exchange between "working women and married women who are not in employment, younger and older women" would enrich the intellectual and spiritual life of both the individual and society.[179]

However, Irmgard Rathgen had a rather different goal in mind. In contrast to the DAB local groups that were soon to be set up, most of which met just once or twice a month, she called for fixed venues that were fully tailored to the needs of women students in big cities and would constitute a daily point of contact for them. Especially in Berlin, where almost a thousand women were studying in the mid-1920s,[180] Rathgen identified an urgent need for facilities close to the university offering women students a homelike retreat and relaxation, practical assistance, and intellectual stimulation. A location of this kind could relieve these students' private lodgings of the functions of

> sociability not permitted by the landlady. Given our straitened means—and in the coming period of economic depression, they will be even more straitened—a facility like this makes a very simple room tolerable by allowing us to take refuge in a cozier space. It will create human contacts, build bridges across the disciplines, provide advice and assistance in matters of our studies and personal life. It may be equipped with books and journals.[181]

Creating a permanent venue near the university where women students could go between lectures to rest, read, and chat, perhaps also enjoy cheap meals, and to benefit from the advice, experience, and contacts of the older generation of academic women became the practical vision embodying the ideal of an intergenerational tradition among female academics. It was a mission taken up the DAB, and especially its Berlin chapter, shortly after they were founded. The new federation's executive was not pursuing a genuinely new idea—the Ottilie von Hansemann House in Berlin's Charlottenburg district had been established in 1915 as Germany's first student residence for women, funded entirely by private donations. It offered space for 95–100 women in single and double rooms, but its monthly rates of 135–200 reichsmarks for a room with breakfast and lunch were beyond the financial reach of the great majority of students in the immediate postwar years.[182] Even before World War I, there had been attempts to set up a publicly funded student facility in Berlin that would offer dormitory accommodation for both men and women, as well as common rooms and recreation space specifically for women students. Plans of this kind were originally thwarted by the authorities' objection to student self-administration, and after the war it proved impossible to find reasonably priced premises close enough to the university.[183]

A renewed attempt, under the aegis of the DAB and in cooperation with Berlin's student services organization (founded in 1923), was more successful in establishing a clubhouse or "day home" for women students. The process bears the unmistakable stamp of the international links with the IFUW and the influence of Crosby Hall in London, which evidently served as a model for the Berlin project and lent it political momentum.[184] The international clubhouse in London had opened on July 1, 1927, in the presence of the Queen, numerous British dignitaries, diplomatic representatives of more than 20 countries, 21 IFUW delegates, and numerous guests, "many of whom had traveled thousands of miles to attend the ceremony," as the DAB's vice president Ilse Szagunn reported in *Die Frau*.[185] Together with four other representatives of the DAB, including Marie-Elisabeth Lüders, Szagunn had attended the opening ceremony and the subsequent reception, which appears to have made a deep impression on all the German women. "If the ceremony symbolized the close connections between the women's movement and female academia," wrote Szagunn, "then the illustrious assembly that met to celebrate Crosby Hall's opening showed the prominence of the public role played by British university women, and with them the British and the international federations of university women."[186] At the same time, the opening of Crosby Hall also gave the DAB a chance to garner political capital at home. To honor the occasion, the Foreign Office in Berlin donated £200 to furnish one room in the London clubhouse's new residential wing—the German government was officially underwriting the significance of the DAB and its international networks.[187]

Officialdom again showed itself willing to support the DAB and its German projects when the first women students' clubhouse was established in Berlin. In 1926, Reichstag deputy Marie-Elisabeth Lüders, physician Ilse Szagunn, and school principal Anna Schönborn, all three of whom served on IFUW committees and had excellent connections in Berlin's local offices and authorities, succeeded in finding high-ranking support and substantial donations for the project. Shortly before the Crosby Hall opening ceremony, the Ministry of Education and Culture had agreed to make available, at no cost, a spacious apartment in the "apothecaries' wing" of Berlin's old City Palace, with eight large rooms, a kitchen, and several additional utility rooms. The palace was just minutes away from both the university on Unter den Linden and Berlin's commercial college, opposite the stock exchange in Burgstrasse—the two institutions where the majority of Berlin's women students were enrolled. A further advantage of this central location was its immediate proximity to the student cafeteria, which was housed on the first floor of the apothecaries' wing.

To fund the required rebuilding and furnishing of the new "day home," the organizers issued a public appeal very similar to the earlier calls for donations to Crosby Hall.[188] The success of the campaign also echoed, if on a smaller scale, the enthusiastic support that had been mobilized to establish the London clubhouse. Against all expectations, the necessary twenty thousand reichsmarks were collected within three months, thanks to the "highly sympathetic" responses of ministers and ministries, professors' wives, the presidents of professional bodies, senior civil servants, and members of the women's movement, as well as several of Berlin's large companies and department stores.[189]

On May 4, 1928, the Berlin day home for women students was opened in the presence of the donors and various dignitaries, and dedicated to its goal of "becoming a hearth and home for women students in the Reich capital in a form that has not yet been provided in any other German university."[190] One of those attending the ceremony was Helene Lange, the 80-year-old teacher and *grande dame* of the German bourgeois women's movement in whose honor the new facility was named. A bust of the home's eminent namesake was donated by the Berlin section of the association of German women philologists and unveiled on November 27, 1928.[191] The carefully staged opening festivities indicate that the Helene Lange Home was designed to meet more than simply practical requirements. Not unlike Crosby Hall, its objective was to build connections between the women's movement, women academics, and female students across the generations, and to help establish a female academic tradition within the universities. The management of the home was entrusted to Gertrud Hamer-von Sanden, Gertrud Bäumer's later life companion. Hamer-von Sanden had spent many years in England and was a member of the Women's International League for Peace and Freedom.[192]

The Helene Lange Home in the City Palace had three large common rooms with 14 tables, 48 armchairs and 10 benches, numerous magazines, and a small but steadily growing library. It had a bathing area, a kitchen, and a scullery, but the greatest public attention was drawn by the three relaxation rooms, with their 31 daybeds allowing students to take a midday nap. The home was open on weekdays between ten in the morning and ten at night, and offered space for around 70 students. Access to the home cost two reichsmarks per semester, with a small extra contribution to cover beverages and baths.

The women students' home in Berlin was a success from the very start. Of the 2,495 women students enrolled at universities in Berlin in the winter semester of 1928/29, 350 purchased a semester ticket for the home, and the manager also issued several hundred day tickets. At the busiest times, from midday to 4 p.m., more than 100 women were using the home's rooms every day, with increased demand in winter.[193] The home quickly became a meeting point for a range of different study groups organized by the students themselves. Cooperation with the "old-established" academic women also moved ahead: Agnes von Zahn-Harnack, who, in 1926, had been appointed to represent the DAB on the executive committee of the German National Academic Foundation (the committee's only woman), arranged several gatherings for female recipients of the scholarships, taking the opportunity to publicize the work of the IFUW. The home's Advent and Christmas season festivities met with an enthusiastic response, attracting more than 80 women.[194] Over the first three years of the Helene Lange Home's existence, the number of evening events increased steeply. The manager's 1931 report noted that "especially in the past year, the organization of lectures, discussion evenings, and so on, has resulted in an intellectual focus emerging, and with it a strong mutual influence between the students and the established university women who are already making their way in the practical world."[195] The home also ran teas and receptions for colleagues from abroad. In 1930 alone, it welcomed "guests from Australia, India, Argentina, England, Poland, Sweden, Austria, and Switzerland" and held a reception to honor the IFUW's vice president, Professor Johanna Westerdijk from the Netherlands.[196]

However gratifying the developments at the day home, economic and political turmoil cast increasing doubt on the continued existence of this women's space in the early 1930s. The original plan had been to expand the day home into a full residence for women students, but this was dropped. The economic crisis meant fewer student visitors, many young women now being unable to afford even the low price of a semester ticket. Although the facility's schedules remained almost as crowded up to 1933 as during the first six months of its existence, the number of women using the home began to decline overall. By early 1933, the little store of capital available to cover expenses arising beyond the home's income had shrunk to a residue of just 312 reichsmarks. Further doubt was cast on the home's financial basis by the fact that the supporting association's numerous Jewish members were no longer able to pay their dues.[197] It is not clear whether or in what form the Helene Lange Home in Berlin survived beyond summer 1933. By that time, other similar initiatives, such as the smaller "Bettina House," a women's student residence in Marburg, had been taken over by the National Socialist women students' organization.

Given the very short history of the Berlin day home for women students and its liberal supporting association, it is difficult to evaluate the success of this attempt to establish a female academic tradition in Germany by means of a venue near the university where women students could relax undisturbed by their male counterparts, cook, talk, work, and meet up with older academic women. Some of the key figures on the DAB executive committee may well have taken special pains to organize events at the Helene Lange Home and to make their presence felt there. That said, there was certainly also some truth in the frequent complaint that Berlin's female researchers did not put into practice their professed aim of contributing actively to the promotion of a new generation of women scholars—that they did not fulfill their high-minded promises. Lise Meitner and Paula Hertwig, for example, belonged to the home's supporting association, and their financial donations testify to their approval of the initiative,[198] but they did not make an active commitment to the role of mentoring.

The Helene Lange Home enabled the DAB's officials in Berlin to create an oasis of intergenerational dialogue, an institution that seems to have resulted in successful communication between the generations until 1933. But it ran clearly counter to the general trend. The sharply declining numbers of organized women students, and in particular liberal women students, in the early

1930s shows that the DAB was not in a position to halt women students' widespread drift away from the women's movement and toward right-wing ideals. In the early years of that decade, the liberal associations of women students gathered under the DAB umbrella forfeited large parts of their membership.[199]

Despite all the efforts to set up and sustain clubhouses for women students, and despite the popularity that the Helene Lange Home, in particular, enjoyed among students in Berlin, in the late Weimar Republic it proved impossible to bridge the gap between the women students of the day and the older university women represented in the DAB. When the DAB organized a "women's *dies academicus*" at the University of Dresden in 1930, several of the presentations highlighted a profound shift in the profile of women students, a trend identified by Marianne Weber as early as 1917. "It seems," observed Gertrud Jung, "as though nothing has been passed down from the woman student of the first two decades to the woman student of today....Today's student does not feel distinct from the ranks of her non-student contemporaries."[200] Women now expected university to give them "the best possible preparation for the profession they have chosen to pursue," and, ran the frequent complaints, "this profession is nowadays far from always being sensed as an inner vocation."[201] The idea of belonging to a female elite, or, as the student Hildegard Gallmeister put it, of fostering "a far-reaching, shared academic women's spirit and women's will," was something that no longer convinced most women students as the 1930s began.[202]

At the time, many representatives of the DAB reflected on this development, which they could do little to counteract. It would, however, be wrong to interpret their deliberations solely as an expression of crisis, bearing in mind the large amount of money and time dedicated to welfare for women students by university women not only in Berlin, but also in the other local groups around the country. The extensive commentary on the "crisis of university studies for women," filling ever more pages in feminist publications as the Weimar Republic neared its end, does demonstrate beyond doubt that the pressure of worsening circumstances was altering the attitudes and values of the young generation of women students—so rapidly that it had become difficult to win these students over to the older hopes and ideals of the women's movement. At the same time, however, the very frequency of such laments articulates an IFUW-honed awareness that it was absolutely necessary to establish a female academic tradition in Germany, at the heart of which was support for the new generation of university women.

5
World Community under Threat

The year 1933 brought a radical break for both the Deutscher Akademikerinnenbund (DAB) and the international network of academic women. When the Nazis came to power and immediately began to impose the "Führer principle" onto German society, voluntary and interest groups found themselves under massive pressure either to close down completely or to bow to the Nazi Party's demands, which included the expulsion of all Jewish members. Neither academic nor women's organizations were spared. The federation of German women's associations (BDF), led by Agnes von Zahn-Harnack, resolved its own dissolution on May 15, 1933; several of its member bodies quickly followed its example or had anyway already ceased to exist.

The history of the DAB, however, assumed a different course. The organization rejected the idea of voluntary dissolution, and instead voted on May 18, 1933, to line up with the Nazi state in an act of *Gleichschaltung* or Nazification. The DAB was not dissolved until three years later. This chapter begins by examining how the DAB's transnational connections figured in its Nazification, then dissolution, in the "Third Reich," and how the international umbrella organization and its German members responded to the political and moral challenge of Nazism.

The chapter's second section asks what became of the networks between university women in Germany once their organization had collapsed, and of their concern to represent the interests of a national female elite. The third section explores German women's responses to the ideology of National Socialism. It looks at those members of the women's network whose aim was to assure female academics' participation in the building of the Nazi *Volksgemeinschaft*, or racial community, by formulating feminine forms of National Socialist science and law.

The Impact of Nazification

The analysis of the DAB's Nazification offered here casts new light on a historical context that has provoked heated controversy. Research on the Nazification of the German women's organizations has hitherto concentrated on the BDF, harshly criticizing its self-dissolution in May 1933.[1] Early studies claimed that the BDF, as the umbrella organization of German women's associations, resolved its own dissolution only in order to expedite its members' integration into the new structures of the Nazi state. This view has endured, supported by the fact that several individual associations were incorporated into the large-scale Nazi professional organizations directly after the BDF's dissolution. However, relevant evidence has remained sparse.

Examining the Nazification of the organization that gathered all the Weimar Republic's associations of women graduates promises to contribute new insights to this discussion. As we will see, the Nazi transformation of Germany was embedded in international contexts, and international sources are therefore indispensable for a full reconstruction of the DAB's internal politics between 1933 and 1936. Such an analysis can open up our view onto conflicts within the federation and the personal motivations of individual executive members, and can also illuminate the Nazification of German women's organizations more generally.

The process of Nazifying clubs and professional associations began a few months after the National Socialist German Workers' Party (NSDAP) came to power in January 1933. By early May, the pressure on women's organizations was intensifying. The BDF council was the first to be targeted by the Nazi organizations, but the president of the DAB, Marie-Elisabeth Lüders, also found it necessary on May 10, 1933, to invite all members to an extraordinary general meeting with new elections to the executive. The meeting's goal was to protect the DAB "from external encroachments."[2] It is unlikely that Lüders intended to use this event to dissolve the DAB—rather, she was hoping to sustain the federation's structures for as long as possible. All the statements from DAB circles before, during, and after the Nazification process are dominated by the argument that university women must refuse to be removed entirely from the positions they had achieved, out of degree courses, politics, and professional life. Lüders had battled discrimination against professional women and the rise of National Socialism not only as the DAB's president since 1930, but also as a Reichstag deputy, party politician, president of the association of German women economists VdN, BDF council member, and editor of the BDF bulletin. For her, cooperation with the Nazi state was out of the question. Like other democratic opponents of the dictatorship, however, she deemed it advisable to hold onto existing positions as long as possible, hoping that the Nazi government would not survive for long. In an April 1933 letter, Lüders told Dorothee von Velsen it was only through quite "extraordinary patience" that "one can remain on the tracks. Setting a single point wrongly out of haste will put the whole train in danger."[3] Thinking about the practical form that women's academic professional organizations could adopt in order to subvert the Nazification process, Lüders considered the experience of the large professional organizations that had already been smashed; she saw a solution in "bringing the whole thing back to the line of what used to be 'women's education–women's work,'" in other words to the only organizational format for professional lobbying that had been legal before the ban on women's political associations was lifted in 1908.[4] Lüders pursued this stalling strategy in her capacity as president of the VdN, but she found no majority for a parallel decision within the DAB leadership.[5]

The political polarization of the federation of women's professional bodies had already been coming to a head in the final years of the Weimar Republic. The DAB executive's liberal members were viewed with considerable suspicion by sections of the women physicians' association BDÄ, several members of the association of women philologists, and especially the nationalist and *völkisch*-minded league of academic women's associations Deutscher Verband akademischer Frauenvereine (DVAF), which brought together the ultraconservative clubs of women students. This antagonism ultimately put paid to Lüders's May 10 appeal for a general meeting. Her deputy, the physician Ilse Szagunn, made no secret of agitating against the plan, and initiated a coup by the board against its president. Szagunn headed the DVAF and held important roles in the women physicians' association. She had been elected to the DAB executive in 1926, and as deputy president was responsible for coordinating the work of local chapters.[6] She made good use of this office to incite opposition to Lüders among those local DAB leaders likely to welcome the federation's Nazification. At the same time as Lüders was inviting members to the general meeting, Szagunn sent out a call to the local chapters to mobilize the forces within the federation "that joyfully welcome the new Germany," gain their "active and leading cooperation," and remold the DAB in line with the "national revolution."[7] Szagunn evidently succeeded in hijacking the planned general meeting and turning it into a session of the full board, attended not by all DAB members but only by the presidents of the member associations and local chapters. This meeting took place on May 18, 1933. Three days earlier, the umbrella organization of all German women's associations, the BDF, had approved its own dissolution.

As a member of the BDF board, Lüders had voted for its immediate dissolution, but when she chaired the DAB meeting on May 18 she appears to have been hoping that the demands of the Nazi Women's Front (Deutsche Frauenfront) would be satisfied by the election of a National Socialist onto the existing executive. However, a majority of the women present advocated the reelection of the entire executive. Lüders resigned in response; Agnes von Zahn-Harnack and Anna Schönborn also decided not to stand for reelection.[8] There was a degree of compromise in that the new president, high school teacher Johanna Willich, did not actually belong to the NSDAP but to the nationalist-conservative DNVP, but three other members of the new executive were Nazis: high school teacher Friederike Matthias, physician Lea Thimm (who shortly afterwards also took on the presidency of the Nazified BDÄ), and the new secretary Luise Gelius, holder of a business diploma.[9]

Even so, apart from Thimm and Matthias, "old fighters" of the Nazi movement and newcomers to the DAB, all other members of the new executive were recruited from the DAB's Weimar membership.[10] The new president, Johanna Willich, had belonged to the Berlin local chapter for many years and was well known within the international network of academic women, especially in

Britain and the United States. Willich had spent a year in St. Louis, Missouri, as one of five German high school teachers on an IFUW international teacher exchange program in 1930–31. She had attended the festive opening of the IFUW guest house Crosby Hall in London, and offered her services as a translator during the disputes on the language question.[11] This international experience, along with her high profile in the IFUW, probably played a large part in her election as the new DAB president.

Other Berlin chapter members elected to the executive were Charlotte Möller and Ilse Balg. Möller, a university lecturer and senior assistant at the Institute of Oceanography based in the University of Berlin, had joined the NSDAP in early April 1933;[12] Balg was a student, like Szagunn an active member of the right-wing DVAF, and had been the student representative on the previous DAB board. The presidents of the Wuppertal and Halle chapters were also elected: Editha von Moers, a teacher of languages with literary ambitions in the field of regionalist poetry, and Dr. Lore Liebenam, a nationalist-*völkisch* specialist in English literature who had devoted herself to "foreign cultural work" with an allegedly pro-German England and who played an active role in the continued process of Nazifying the DAB.[13]

On May 23, 1933, Johanna Willich and her deputy, Friederike Matthias, were summoned by Lydia Gottschewski of the National Socialist Women's League (NS-Frauenschaft) "for consultations on integration into the 'Women's Front.'" At issue in this meeting was the DAB's assent to "subordination to Adolf Hitler, acknowledgment of the tasks with which the National Socialist state charges women, the removal of non-Aryan members from executive roles, and the election of National Socialist women to the prominent positions."[14] The two DAB representatives made no objections to these demands and, as the women doctors' journal *Die Ärztin* reported, granted "the required signature without hesitation."[15] The DAB was thus permitted to remain in existence, keeping its old name, until early 1934.

The new board immediately set to work to fulfill the Nazi Women's League stipulations. On June 12, 1933, the board called on its member organizations and the executives of the 30 local groups to align themselves with the regime and to ensure "that all leading positions" were held by women of "racially German origins" who "consciously affirm the goals of the national government." On October 6, 1933, the executive ordered the expulsion of "all Jewish members," on the grounds that "the DAB has joined ranks with the Women's Front."[16]

In fall 1933, conflicts arose over powers and responsibilities in the institutionalization of future Nazi women's policy, creating additional pressure to conform. Whereas Lydia Gottschewski, responsible for the German Women's Front, was content with a "discreet removal" of Jewish members—that is, women defined as "Jewish" by the new regime—from the DAB, a different line was taken by a newly created division of the Interior Ministry, the Deutsches Frauenwerk or German Women's Agency: it made the DAB's continued existence in Nazi Germany conditional on more far-reaching commitments to anti-Semitism. Just one week after the October 6 instruction to exclude "Jewish" members, the DAB chapters received another circular from their executive informing them of a "voluntary additional resolution." Henceforth, "Aryan" women married to "non-Aryans," too, could only remain in the DAB if their husbands had not yet been dismissed from the civil service under the anti-Semitic Law for the Restoration of the Professional Civil Service of April 7, 1933.[17] In the missive, dated October 14, 1933, Willich also advised the DAB local groups that the time had now come to change the federation's statutes in order to transform "our community, which once represented mainly professional and cultural interests, into a community of convictions." The supreme goal of that community could "no longer be the advocacy of special interests related to profession and status," but must become "the nurture of cultural interests within the German *Volksgemeinschaft*."[18] What that meant in practice was spelled out by committee member Lore Liebenam from Halle in a memorandum headed "The work of the federation of academic women in the New Germany," issued to local chapters at the same time. Liebenam demanded not only the exclusion of all Jewish members, but also a strict ideological line for the DAB. All the organized university women should "belong to the NSDAP in one form or another," undertake instruction in National Socialism, and focus their collective attention on topics in eugenics, racial biology, and "the Jewish question." The representation of professional interests was "part of the rather disappointing tradition" of the bourgeois women's movement, and must be relegated in favor of a "happy solution of the woman question, which is of the greatest importance for our aspirations to regain the health and vigor of our whole German *Volk*."[19]

While the newly elected DAB executive set out to restructure its organization along Nazi lines, Marie-Elisabeth Lüders—although she had resigned as DAB president on May 18, 1933, and therefore had no official authority—still put her hopes in the DAB's continued presence in the international fora of the IFUW, regarding this as a way to influence decisions on the DAB's future. Lüders correctly judged that maintaining international links would also be in the interests of the Nazis themselves, and she put all her own connections to work to achieve it. Hoping that her personal networks would bear political fruit, she wrote "open and unvarnished reports" for some senior Foreign Office functionaries of her acquaintance, detailing her activities and the general mood at the IFUW's international conferences. She passed on her impressions "with no regard for whether other observers see these matters differently."[20] Her objective was to show as forcefully as possible that German women would only be able to remain visible on the international stage if the German associations survived in their previous form, or, at least, were not replaced by Nazi organizations. For, as Lüders wrote to trusted allies in the Foreign Office in September 1933, even if "a new German women's organization" were constructed that "fulfilled the conditions for a national association to be accepted into the various international organizations," this would not alter the fact that "the German colleagues who were 'internationally' elected in the past" would retain their office until 1936.[21] Lüders had an IFUW mandate of this kind herself, and was determined not to give it up prematurely.[22] She regarded the continued presence of the democratically elected German women as offering some chance of survival for the DAB in two respects: she speculated that the DAB's international connections would protect it from complete destruction by the Nazis; and she argued that retaining such offices would prove vital to securing international support in the future. In a complacent note to the Foreign Office in early May 1933, she reported that women in the IFUW were expressing the wish "to stand by the current German members." This concern was, she added, "expressly underlined by the hope... that the circumstances of the German organizations would soon ease," until which point the existing "bonds" must be kept intact.[23]

As a German politician, Lüders regarded membership of the international women's organizations as "perhaps the only remaining way of preventing further harm to the German cause."[24] In light of this notion of damage limitation, she felt obligated to assert her position as representing the "good" Germany and to erase her international colleagues' "doubts, prejudices, and bad feeling" toward the German events of 1933. She tried "to provide information and reassurance, and in many cases also to contribute... to self-criticism." An important point she wanted to convey was that the National Socialist assumption of power must be blamed not least on a general failure of economic and disarmament policy, by all European nations and the United States, in the wake of the Versailles Treaty. To ensure this point of view was represented, she wrote, as many German women as possible must be present internationally. It was important to "miss not a single opportunity" to face up to "critical questions," even if these occasionally involved undisguised hostility. She knew of no better way to protect her own nation's reputation, only now rehabilitated after a decade of painstaking international efforts: "If we do not participate, decisions always go against us, and I do not wish that for Germany."[25]

However, Lüders's endeavors to shield her country from "damage" increasingly drew her into the dubious position of defending the policies of the Nazi state, and opened her to suspicions of sharing the new government's views. Her situation was made no easier by the fact that she had to push herself on the Foreign Office in order to obtain permission to travel. Before the World Economic Conference in summer 1933, for example, she argued that if she were allowed to travel to London, she could use the opportunity to work in Germany's favor "upon the English women"—women with whom she had been "closely acquainted, in some cases friends" for many years. Lüders also invoked the advantages that the IFUW's international hospitality could offer for her mission. She would be able to "stay in the international hostel of the international academic women's federation and thus to be in the midst of all the women, whether British ones or those traveling to Britain, who participate in public life. It would be impossible to exclude me from any enterprise or entertainment in these circles, because I am their colleague as a member of all three international organizations." Links with the other general secretariats of the international women's organizations based in London could, continued Lüders, also easily be created in this way.[26]

Lüders did her very best to stave off any official criticism of Nazi policy. Her efforts were driven by the fear that international censure of the Nazi regime and its persecutory policies might reinforce isolationist tendencies within Germany, and that "the possible repercussions could rebound upon the very people they are trying to help."[27] Lüders's forays, as noted in her reports, nevertheless sound somewhat defensive. Thus, in late May 1933, she confided to her Foreign Office contact that she had managed to ensure that the IFUW Disarmament Committee in Geneva "will not write to the German Reich Chancellor about the Jewish question."[28] She took the opportunity of a dinner with Lord Robert Cecil, a key figure in the League of Nations disarmament negotiations in Geneva, to argue that the debate on "a long French resolution" against the treatment of German Jews should be held in closed session.[29] On several similar occasions, she explained to her interlocutors that the "Jewish question" was "an internal German matter," and anyway not so different from many other events around the world "on which all those who are outraged today were silent." This type of criticism increasingly reminded her, she said, of the anti-German propaganda in the World War I; behind the international indignation, she suspected an attitude that she described as part of the "tragic guilt of the whole world" for the political debacle in the German Reich. The outside world had accorded the Weimar Republic none of the concessions that might have stabilized its political and economic situation at the crucial moment: the practice of "conceding after grueling battles, always too late, and then with ill grace" was now being continued as the world refused "to give the 'current system' a sufficient chance." She seems to have hoped that a conciliatory and restrained stance toward Nazism on the part of the international bodies would allow them to exert a moderating influence on Germany's new regime and its electorate, and would counter tendencies to seal Germany off completely from the international community and its political principles. It was probably for these reasons that, in September 1933, Lüders assented to the request by the Interior Ministry women's division to counsel two freshly commissioned female members of the new German League of Nations delegation in Geneva and introduce them into international circles. She interpreted this task as meaning to answer "the countless questions addressed to me regarding who the women are, where they come from, what they do, how they stand on the women's movement."[30]

Given the circumstances, it is little wonder that Lüders's strategy of using personal interventions and her reputation for democratic integrity to deflect harm from Germany began to falter even before 1933 had ended. The British and American representatives of the international women's organizations, along with delegates from Poland and Czechoslovakia, were increasingly reluctant to accept business as usual in view of the spiraling terror in Germany. The "treatment of the Jewish problem" and "developments in the Lutheran church," in particular, were items that the IFUW and the International Council of Women refused to consider a purely internal German matter.[31] At the same time, mistrust of the German delegates themselves was growing. In May 1933, Lüders had observed that "the atmosphere among our existing colleagues" was "markedly friendly and cordial in 'personal' terms,"[32] but this had changed by the fall. In September, Lüders was already describing the "difficulty of providing explanation and information" as "very severe."[33] In particular, she suspected that those adherents of the international peace and women's suffrage movement who were now actively assisting German refugees were playing a double game. The "great amiability on the outside, and the great interest displayed by many others in hearing details of circumstances in Germany," could not, Lüders reported, be taken at face value as expressing a genuine wish or even as assent. It often turned out to be an illusion, "quickly rectified" upon "closer knowledge of the person and when mutual acquaintances enable a verification of that person's actual attitude and statements."

The Weimar Republic's last DAB president had to give up her self-imposed mission to turn her international connections to the benefit of the "good" Germany and women's organizations when, in early 1934, she ceased to receive visas for foreign trips. Her links with the Foreign Office broke down at the same time.[34] The other Weimar-era DAB leaders who had held office in the IFUW or other international women's organizations were sidelined in the same way. Physicist Lise Meitner had already resigned from the IFUW Fellowships Committee in spring 1933, unwilling to represent Nazi Germany. In fall 1934, Agnes von Zahn-Harnack gave up her presidency of the committee preparing the planned 1936 convention in Berlin. Zahn-Harnack was unable to accept the IFUW's offer to remain in the federation as an individual member: although the Foreign Office granted her permission to join on this basis, the German Women's Agency refused, as it wished to send a Nazi delegate in her place.[35] At the Budapest meeting of the IFUW in late summer 1934, therefore, Anna Schönborn—elected president of the IFUW Committee for Interchange of Teachers in 1929 and part of the DAB executive up to May 18, 1933—was the last remaining German representative with a mandate.[36]

The reason for clinging to IFUW office may initially have been to preserve the DAB's pre-1933 structures in Germany for as long as possible with the help of its international links. But while that strategy is understandable, by remaining visible in the IFUW and defending the German case the former DAB officers in fact helped to create a false impression of continuity. In Germany itself, they had taken an unambiguous stand against the Nazi regime by resigning from the DAB board. National Socialist women turned this dilemma to their own advantage in gaining access to the international networks.

Whereas Lüders and others eschewed all national work after the DAB's Nazification, there were more receptive responses among many other DAB members, mostly the younger ones. An important cause of this seems to have been the aggressive Nazi rhetoric calling for highly qualified women to be removed from professional life—a familiar discourse that not only became considerably harsher with the Nazi accession to power, but was also put into political practice. The relevant policy measures included the Nazi "Law against the overcrowding of German schools and universities" passed in April 1933. Though primarily targeting Jewish school and college students, this measure also provided for a discriminatory quota to be imposed on female students, whose numbers declined dramatically until 1939.[37] Many 1933 sources reflect the anxiety that the Nazis' exclusionary policies triggered among female students, exacerbating the nationalist and anti-Semitic attitudes that were already as rife among women students and young female academics as everywhere else. At a DAB meeting in June 1933, Ilse Balg, the student representative on the new, Nazi-dominated board, complained about the rumors circulating at Berlin University that in a year's time there would be no more female students, and that women would range last in the hierarchy of the National Socialist German Students' League, behind even "the foreigners and Jews."[38]

These and related fears of a permanent loss of ground quickly led the DAB's "apolitical" and "temporizing" members to close ranks with its National Socialists. The Nazi women expected the DAB—as well as the BDÄ, the leading positions of which they had also begun to occupy—to improve their situation within the Nazi state, and its well-established international networks seemed particularly attractive and prestigious means to that end. The "apolitical" and "temporizing" women drew hope from the regime's recognition of their organization, which seemed to promise "that women will participate in the future reordering of the state."[39] This was the implication of the June 12 circular in which the new DAB executive instructed its membership and local chapters to line up with the Nazi regime. Affiliation with the Women's Front was seen as a way of ensuring "the positive participation of academically trained women in the organization and incorporation of women into the new *Volk* state."[40] According to a report in *Die Ärztin*, a DAB conference in summer 1933 indicated the "general depression" giving way "to new life and new hope in our association as well," thanks to a renewed confidence that professional women would be able to avoid marginalization during the reorganization of society.[41]

Exactly how many DAB members were prepared to remain in the organization under the conditions dictated by the Nazi regime and its followers is hard to say. In 1934, German women informed the IFUW that after the NSDAP's accession to power the DAB's membership had dropped from around 3,100 to an estimated 400.[42] These losses may be partially explained by the enforced exclusion of Jewish members and the demonstrative resignation of some non-Jewish women, but their chief cause was that the DAB was gradually ceasing to serve as an umbrella organization for female graduate professional bodies. The associations of women philologists and women jurists and the women students' clubs were wound up, the association of women economists dissolved itself, and the women physicians' association BDÄ, now also Nazified, withdrew from the DAB in spring 1934, leaving the umbrella without any associations to shelter.

Reports from the local chapters suggest that the executive's instructions were generally positively received and implemented. On July 25, 1933, for example, the Freiburg women reported that they had reelected their executive committee and fulfilled "the conditions for joining the German Women's Front."[43] All the committee's members were now of "Aryan descent," but the president remained the same: lawyer Maria Plum had led the group since 1931. The Karlsruhe chapter, too, fell in line with the Nazification process, though again its president since Weimar times, high school teacher Gertrud Carl, retained her office. With a letter of November 29, 1933, Carl threw out those of the chapter's Jewish members who had not already resigned—the Karlsruhe membership was halved.[44] Groups in Mannheim, Hamburg, and Munich also continued to exist. The Berlin chapter announced that in winter 1933–34 it would hold a "study group on politics" in close cooperation with the BDÄ and led by BDÄ president Lea Thimm. This would meet every two weeks for the purposes of "intensive, small-group self-instruction with the help of National Socialist writings."[45] The local chapters in Wuppertal and Halle carried on meeting under the direction of Editha von Moers and Lore Liebenam, and the longtime president of the Chemnitz group, physician Frieda Lange-Malkwitz, also remained active.[46]

However, no matter how carefully many members on both national and local level tried to comply with the Nazi requirements, their efforts did not suffice to maintain the DAB in its previous shape. When the German Women's Agency was established by Reich Minister Rudolf Hess, with Gertrud Scholtz-Klink appointed as its head in spring 1934, the disempowerment of the recently elected DAB leadership began in earnest. On April 30, 1934, Scholtz-Klink—the "Reich Women's Leader" (*Reichsfrauenführerin*)—discharged Johanna Willich from office and appointed her former deputy as the acting leader of the DAB: Friederike Matthias, a Nazi Party member since 1929 and the National Socialist Teachers' League expert on higher schooling for girls from 1934 onward.[47]

The subsequent fortunes of the German university women's federation were closely bound up with its international umbrella organization. The DAB's ideological recalibration and its policy of anti-Semitic exclusion led to open conflict with the IFUW, culminating in 1936 with the German women leaving the international network. From the very start, the numerous dismissals of female civil servants and the isolation of Jewish women graduates had been viewed with consternation from abroad. British, French, and American members of the IFUW, in particular, were outraged.[48] The national periodicals of the British and US associations, like the IFUW newsletter, ran frequent and detailed reports on the general situation of university women in Nazi Germany, with special attention to the fate of those who had lost their means of existence due to their "Jewish origins" and had been forced to emigrate.[49] In the early days, disapproval was directed at the Nazi government, not at the German sister organization itself. Even the news of the DAB's Nazification, conveyed by Lüders to some of her IFUW colleagues during the Geneva disarmament conference in late May 1933, initially elicited solidarity and dismay at the radicalism of Nazi attacks on independent women's organizations. As Lüders reported from Geneva to the Foreign Office, IFUW women found "such measures" against the DAB "especially incomprehensible in that 'science' is not considered a political or 'racial' matter."[50] The IFUW's response was not limited to personal expressions of sympathy. In London on May 31, 1933, British university women took part in a rally of women's organizations and female legislators protesting the dismissal of German women from the senior civil service.[51] Although the IFUW itself and all its national organizations were usually reluctant to make public statements on current political events, the BFUW annual conference in August 1933 passed a resolution addressed to the German ambassador in London. This condemned all kinds of discrimination based on religion, race, sex, or political convictions that deprived teachers and students of their livelihood and barred them from suitable education or training. The British women regarded such discrimination as irreconcilable with the "interests of humanity," and strongly censured the German treatment of highly qualified people who had made a special contribution to their country's reputation and honor.[52]

From late 1933 onward, however, the sense of international solidarity with the German women began to give way to a suspicion that the DAB itself was actively supporting the politics of exclusion in Germany. When the IFUW Council gathered in Budapest in September 1934, its agenda included a discussion of the German situation, and specifically the DAB's position on Jewish women's membership and academic freedom. Two German representatives attended: the National Socialist Friederike Matthias, who had been made acting DAB president that May, and Anna Schönborn, well known to the IFUW Council. She had been part of the DAB executive until May 18, 1933, had attended every IFUW Council meeting and convention since 1926, and still held a seat on the IFUW's Committee on Interchange of Teachers. In Budapest, Matthias admitted to a worried Council that the past year's turbulence had lost the German federation many members, but also asserted that she was willing and able to rebuild the DAB in future without racial, religious, or political discrimination. Of course, she said, unfortunately "non-Aryan" women could not be part of a German league of university women at a national level, since each of the individual professional associations was "automatically bound" by an "Aryan paragraph." But efforts would be made to enable members affected by these regulations to remain in the DAB via individual membership of the local chapters.[53] A draft proposal to amend the statutes accordingly was circulated by Matthias to the Council members in advance.[54]

In view of the systematic expulsion of Jewish members from the DAB's local chapters that had been completed in fall 1933, this supposed amendment was nothing more than a delaying tactic, intended to keep the IFUW Council from taking direct measures against the German body. Matthias's predecessor Willich had already realized that the official adoption of an "Aryan paragraph" in the DAB statutes might result in German expulsion from the international federation.[55] Although she had been willing to set such concerns aside in late 1933 and prioritize the need to "satisfy the demands of our German life in the National Socialist state,"[56] in fact the DAB statutes had not been touched by fall 1934 despite the drastic de facto changes to membership arrangements—in order to avoid jeopardizing the federation's IFUW membership.

For the IFUW Council's 29 members, the Budapest encounter with a representative of the "new Germany" was a dramatic political event they would not soon forget, as several later reported.[57] Especially the British and French presidents, and the president of the newly founded association in Palestine, were already well aware of the consequences of the Third Reich's anti-Semitic and misogynist policies, having received first-hand reports from émigré women.[58] Matthias further fueled suspicion by apparently attempting bribery: she underlined the DAB's desire to remain in the federation in individual conversations with promises of substantial contributions to the international fellowship fund from German state coffers.[59]

Officially, the Council maintained a polite front toward the German women. After some searching questions on how Germany dealt with its intellectuals, freedom of thought, and the preservation of world peace, it expressed its best wishes for a successful reorganization of the German federation along "more international lines."[60] Behind the scenes, fierce debates raged over how to proceed, especially between the American and British women. The US delegates made the case for maintaining links with academic women in the fascist regime for as long as possible, while the British women argued for the Germans to be excluded.[61] They succeeded in convincing the IFUW Council to put an amendment of the statutes to the vote at the next members' convention. This amendment was intended to clarify the precise circumstances under which the IFUW would be prepared to accept the continued membership of the German women: in future, no national organization that refused membership to university women for racial, religious, or political reasons would be allowed to join or remain within the IFUW.[62] The next large-scale assembly of delegates, which would decide on the adoption of the amendment, had long been planned for 1936 and, ironically, the location was to be Berlin. But because Agnes von Zahn-Harnack had given up her seat on the 1936 preparation committee and most members of the Council did not trust the other German representatives, considerable diplomatic finesse was deployed to persuade Schönborn and Matthias to give up the idea of organizing such a large event in Berlin at a time when the DAB was still in the throes of reorganization.[63] The Council in Budapest accepted an invitation by the Polish association to meet in Kraków.[64]

On the German side, the outcome of the Budapest negotiations was presented as a German Women's Agency diplomatic victory that had safeguarded the DAB's "right to exist."[65] As a Women's Agency circular explained on October 10, 1934, Matthias had conducted negotiations so well "that the damaged relationships with other countries have been rebuilt most gratifyingly, and the DAB's continued membership, previously in doubt, has now been unanimously resolved." The IFUW Council had, said the circular, accepted Matthias's proposals "relating to National Socialist principles."

The reorganization of the DAB that Matthias announced in Budapest began to take shape immediately after the Council session, though in a strikingly different form than the one she had presented to her international colleagues. In early October 1934, Gertrud Scholtz-Klink ordered that all the local-level elected DAB executives be dismissed and the entire federation be permanently restructured in line with the "Führer principle."[66] Along with this demolition of its remaining democratic structures, the organization received a new name: it was now the Reichsbund Deutscher Akademikerinnen, or Reich Federation of German University Women (RDA). In February 1935, the RDA was incorporated into the German Women's Agency as a new division.[67] Friederike Matthias was appointed to head the division, and thus retained her leadership role. The surviving local chapters, too, were absorbed into the German Women's Agency without resistance. Scholtz-Klink defined the division as a "working alliance of all the graduate sector representations and groups affiliated with the German Women's Agency," which was to continue the interdisciplinary gathering of academic women in a similar way to the old federation. "For international reasons," in other words with an eye to safeguarding IFUW membership, what was now actually merely a section of the German Women's Agency retained its federation name RDA.[68] A statute-like agreement between the German Women's Agency and the RDA on February 28, 1935, stipulated that the RDA's chief task was

> to nurture cultural and friendly relations with graduate women in other countries. It is a member of the International Federation of University Women. Through its cooperation with that Federation it serves cultural understanding and enhanced political awareness among academic women at home and abroad.[69]

This redefinition of purpose was designed to secure the federation's continued presence on the international stage, but it was far from resolving the conflict with the IFUW that had emerged in Budapest. The arrangement between the Women's Agency and the RDA drew a careful veil over the membership restrictions, and the German women were not officially excluded from the international umbrella organization because IFUW delegates failed to reach agreement on amending the statutes at the 1936 Kraków meeting. The assembly did, however, decide that the principles of the Budapest resolution should be applied immediately, even if they could not be formally incorporated into the IFUW's constitution until the next members' convention in 1939.[70] After the Kraków conference, therefore, an official exclusion of the German women would have been perfectly possible.

In fact, by this point there was no longer any need for such a step, because in January 1936 the National Socialist women themselves had announced that the RDA would leave the international federation.[71] The Reich women's leadership explained the decision in two detailed circulars of spring 1936.[72] As the primary cause it named "difficulties in the handling of the Jewish question." After the passage of the Law for the Restoration of the Professional Civil Service, one circular noted, some of the IFUW's member associations had "disloyally criticized measures taken by the German State," and the IFUW had "not considered it necessary" to provide the "rectification" requested by the German women. Clearly, the phase in which the Nazi functionaries still tried to hide anti-Semitic policy from the international public had ended with the promulgation of the Nuremberg race laws in late 1935. These laws, wrote the women's leadership, made a further tightening of membership conditions inevitable for the RDA, and it would thus no longer make sense for the German women to work for international relations within an IFUW framework. Furthermore, for the new Germany's "Aryan" university women, the situation in the IFUW had changed such that "not only can we not expect any sympathy for the racially and philosophically determined reorientation of German work for women," but "the RDA can do nothing to prevent today's IFUW from regarding support for Jewish academic women from the old Germany as an important task." The RDA believed that when the international umbrella organization distributed fellowships, it gave preferential treatment to women graduates "who have been dismissed from their posts due to the German law on the professional civil service." This was unacceptable, and showed that the IFUW, "as an organization of the old women's movement," failed to grasp the importance of "race and political outlook" and instead viewed academic excellence as the only valid basis for the demand for women's rights. The RDA decried that policy, adding that the IFUW was by no means the only channel for promoting international relationships; "other doors" were open to other countries.[73]

Nevertheless, the RDA women avoided closing the gates to the IFUW altogether. They carefully selected the arguments presented in the official announcement of their withdrawal, making no mention of the reasons they had given the German audience. The circulars pointed out that it would be diplomatically unwise to argue too openly, particularly with regard to the "Jewish question." Instead, they explained, the IFUW had been informed that the structure of German women's politics did not permit "small, specialized associations." The dissolution of IFUW membership had been carried out "loyally," enabling "the continued reciprocal provision of hospitality."[74] The Nazi women academics evidently still wanted to benefit from the IFUW's informal networks.

The German Women's Agency's laments of "difficulties in the Jewish question" referred to the committed efforts of, especially, the BFUW to find academic work for university lecturers forced to emigrate from Germany.[75] But this international policy of support may also have served the Women's Agency as a welcome pretext to withdraw voluntarily from the IFUW before its exclusion could be enforced. The frequent claims of "disloyal" attacks from IFUW ranks are strong indications that this was a calculated attempt to save face, used within Germany as part of anti-Semitic attacks on the international community, for there is actually no evidence of any specific conflict with the IFUW of the kind implied by the Women's Agency. This interpretation is further supported by the Women's Agency's dissolution of the RDA immediately after its withdrawal from the IFUW, on the grounds that its existence had been "justified solely by membership of the international federation of university women."[76]

Networks and Careers under Nazism

The dissolution of the RDA in May 1936 did not mark the end of female academic networking in the Nazi state—the destruction of the association structure did not equate to academically trained women disappearing as a distinct group of professionals in the Third Reich. Quite the contrary: although worries were voiced in the Women's Agency that the RDA's incorporation, and the double coverage of its members in the professional organizations and the Women's Agency, had increased the risk of giving rein to "egoistical special interests," less publicly the advocacy of academic women's interests began to coalesce under Nazi direction and was gradually anchored in new institutional formats starting from 1936. In this sense, the RDA's dissolution ushered in a new phase of female academic networking, this time orchestrated by the Nazis. University women within the German Women's Agency began to use or create structures within the Nazi hierarchy to enhance their visibility and consolidate their access to professional and social leadership positions. As early as July 1, 1936, a new section devoted to "academic issues" within the Women's Agency took up its duties. The Reich Women's Leader called on the former RDA members to apply for individual membership of the Women's Agency and continue their labors in this new framework, whether in the culture, education, and training sections of the National Socialist Women's League or as Reich- and *Gau*-level advisers on academic questions.[77] Several presidents of local chapters complied, again contributing to a sense of continuity, and retained their presence within the Nazi hierarchy.[78]

These developments in the RDA as a whole can also be identified for individual professional groups. Women lawyers created their own group within the Women's Agency; in 1937, it had 144 paying members.[79] The women physicians' association, likewise Nazified in 1933 and dissolved in 1936, did not have a distinct profile in the Women's Agency, but from 1937 Ursula Kuhlo worked for the political representation of women doctors inside the National Socialist German Physicians' League.[80] In fall 1938, she persuaded the Reich Physicians' Leader to set up a women doctors' section.[81] This body was to "attend to questions relating to the professional status of women doctors and represent their particular interests as women within the professional organization." Kuhlo, its director, saw the crucial tasks as being "to promote training opportunities for young women doctors in state and municipal hospitals" and "revise" the restrictions on female doctors introduced at the start of the Nazi dictatorship, which affected licensing and the authorization to carry out work for the statutory health insurance.[82] When the new Reich-level section was established, women members were appointed to all the 22 regional (*Gau*) medical chambers.[83] As for the women teachers organized in the National Socialist Teachers' League, in 1936 the German Women's Agency signed an agreement on close cooperation that was designed to optimize their implementation of Nazi women's policy.[84]

A similar organizational concentration can be identified in the case of research and universities—the domain to which the Weimar-era DAB had been particularly committed and that the Nazis were most eager to close to women. Gertrud Scholtz-Klink set the process in motion with her speech to the women's rally at the 1937 Party convention, dedicated to the theme of labor. She called for recognition that practical efforts for women's concerns must be underpinned by science and scholarship,[85] and argued that women academics therefore held "a great responsibility within work for women."[86] In summer 1937, Inge Wolff of the women students' office in the Nazi student leadership asked women to join the new University Community of German Women (Hochschulgemeinschaft deutscher Frauen), a female grouping within the "student combat aid" organization. It was intended to bring together all women "who take an interest in the German university, in higher studies for women, and in women's graduate occupations," irrespective of their educational and professional status. A 40 percent share of the hefty minimum dues of 2 reichsmarks per month was devoted to political training for women students, while 60 percent would help fund the establishment and maintenance of "women students' clubhouses, dormitories, trips abroad [to countries considered friendly to Germany], and support for individual, gifted women students."[87] As the junior lawyer Anna Kottenhoff, "commissioner for science and specialized education in the women students' office of the Reich student leadership," explained in early 1938, the economic and psychological support for women students offered by the new University Community of German Women was designed not only to counter the decline in women students' numbers, but also to pursue traditions that the DAB had nurtured before 1933. The University Community would offer opportunities to "female scholars and especially lecturers" to make contact with the women students following in their footsteps—for the "bearers of present and future female university work need to support each other, so that the former know why they carry on working, the latter why they carry on striving." Behind this renewal of intergenerational networking was the long-cherished hope that women lecturers, "as leaders in the universities," would work together with women students "for women to feel at home in the university, so that education and research, just as other areas of cultural life, are borne jointly by men and women in the highest intellectual seats of learning."[88] However, Kottenhoff drew a rigorous line between the new grouping and the "individualist elitism" of Weimar. The University Community was by no means a "matter of professional association"; its openness to any "racially German" woman proved it was calling on German womanhood to defend a "general interest." Enabling talented women to access university education and scholarship was not a mere "act of selection" that acknowledged and protected individual rights, but something that laid claim to the "care and attention" of all women because the results would benefit "everyone." An important goal of the University Community was to convey this fact to "as large as possible a circle of German women," in order to eliminate the "ill-fated isolation" hitherto experienced by female academics.[89]

In order to represent the concerns of women academics themselves and bind them more closely into the regime's "work of construction," in September 1937 a further body was created: a "science" section ("Sachgebiet wissenschaftliche Arbeit") within the German Women's Agency. Its brief was to link up with all the relevant Nazi professional organizations, Party offices, government departments, and the young generation of academics with the aim of "giving woman the opportunity to fulfill her duty in science and scholarship in all those fields that are in keeping with her nature."[90] Lawyer Dr. Ilse Eben-Servaes was made acting head of the new section. Eben-Servaes had been trying ever since the Women's Agency's establishment to get the concerns of women lawyers heard within the Nazi women's organizations.

The German Women's Agency science section was, as Eben-Servaes explained in a 1938 Agency bulletin, to be a "homestead for the academically trained woman."[91] Its most important task would be to "build bridges between science and practical work," thus making relevant research available for National Socialist work on women's issues and providing a scientific basis for handling problems that arose from practical campaigns. In concrete terms, this meant finding women lecturers and practitioners to offer lectures and promoting them through the relevant Nazi women's periodicals. But the section was also intended—like the DAB and RDA before it—to nurture connections between graduate women working in the practical and academic fields, covering both university teachers and "the practical physician, lawyer, schoolteacher, librarian, economist, chemist, engineer, and pharmacist"; the networking was to come about through working groups at the *Gau*, district, and local level.

On the Reich level, Eben-Servaes defined the section's task as being above all to "guide" the graduate professions: it was "a matter of urgency to steer the professional efforts of women engaged in intellectual labor so that it fulfills the *Volksgemeinschaft*'s need for woman's assistance in the wider endeavors of the *Volk*." Close cooperation with the Party women students' office and the University Community of German Women was therefore highly desirable, continued Eben-Servaes. The Women's Agency science section perceived its duty less in "training an ideologically reliable and professionally capable new generation" than in "integrating the woman engaged in intellectual work into efforts for women's issues." International work was another important domain for the section, as "the scientifically trained woman" was currently "the focus of interest from women's organizations abroad." Although responsibility for promoting international relationships generally lay with the Women's Agency "borderlands and foreign" section, the science section hoped to collaborate closely with it.

In institutional terms, then, by 1938 structures had been created on various levels that continued the DAB's networking strategies, staking and defending distinct claims by graduate women to an "organic integration in the service of the *Volk* as a whole" inside the Nazi hierarchy—and this despite the frequently repeated insistence that the separate representation of female academic interests was a feature of the obsolete bourgeois women's movement, for which there was neither justification nor necessity in the new Nazi state.[92]

The sparse surviving evidence suggests that the period between approximately 1935 and 1938 was one of intensified activity and a sense of new beginnings. This mood cannot be explained solely by the efforts to create new networking structures in the Nazi system. It is also reflected in reactions by graduate women when official Nazi publications—responding to a looming scarcity of qualified labor—began to encourage women to study and train again. In 1935, the Munich psychologist and university lecturer Maria Schorn observed exultantly that the "advancing development of the National Socialist state form" had now clearly refuted "all the fears that women would be pushed out of the professions and public life."[93] The situation also altered the attitude of Nazi women officials toward the previous women's movement and its educational and professional ideals. Helene Lange was now honored again, and tribute was paid to the history of women's university studies and women's professional achievements during World War I. The change embraced even those academic women who had dissociated themselves from the "new era" at the start of the Nazi revolution. Gertrud Bäumer noted with gratification that Nazi women's publications had now "completely transformed" their judgment of bourgeois women's achievements.[94] Even Marie-Elisabeth Lüders cherished hopes that academic women would soon be able to take up a role as a crucial societal "intermediary" between state leadership and *Volksgemeinschaft*, just as they had during World War I.[95]

It is difficult to say precisely how, and how successfully, the new networking initiatives came into their own in the period that followed. The Women's Agency science section held its first session with women representing various graduate professions on August 30, 1937.[96] From January 18–21, 1938, it joined with the women students' office to organize a women students' camp in the Reich Mothers' School in Berlin, where 32 active members of the National Socialist women students' working group came together "to consult with leading women academics on deploying the academically trained woman in the life of the *Volk* and to identify the most pressing areas of work."[97]

But the propaganda climax of the Nazi-led female academic networking offensive was a conference of women university lecturers organized by the science section in Berlin on January 2–6, 1938. The event was attended by 22 of the 36 female lecturers then still working in the German Reich at universities and technical colleges or on the new training programs for women teachers. In her welcome address, Scholtz-Klink told the assembled lecturers why she believed the time had come "to include the intellectual woman in the National Socialist work of construction." The first years of such work, she said, had been dedicated to building "foundations" for "practical work for women." Facing accusations of indifference and incomprehension toward the problems of women working in academia had not been easy. Yet it had been vital to wait "until the necessity of including the intellectual woman could be clearly justified on the basis of National Socialist work for women."[98] Now that the foundations had been securely laid, the need had emerged for "women's theoretical work" to "underpin what we have achieved so far."[99] The meeting would, Scholtz-Klink continued, provide initial contacts and a focused exchange of views between Women's Agency officials and women academics, in order to explore the possibilities of "making lecturers' specialized knowledge fruitful for work on women's concerns."[100] The conference was also intended to tackle the more general issue of integration into state structures, and the lecturers received a detailed briefing on the responsibilities of the various new departments. In turn, they were granted the opportunity to explain how important women's research and teaching in higher education was for "service to the *Volk* as a whole."

The majority of the 36 female university and college lecturers in Germany were receptive to the German Women's Agency's claim that the hour had now come when "woman as a researcher and educator in the universities can be organically and fruitfully integrated into work for women."[101] Responding to the conference invitation, mineralogist Doris Korn from Aachen said she had been filled with "an indistinctly happy feeling" that "we are not being asked to attend just to inform us that in our profession we are nothing but dispensable parts of the whole."[102] The economist Gertrud Savelsberg, also from Aachen, publicly avowed a passionate concern for female academic networking within the *Volksgemeinschaft*. Choosing her words carefully to distance herself from the liberal demands of the "old" bourgeois women's movement, Savelsberg said she had felt fulfilled by stepping out of "individualism" and meeting female university lecturers "who are willing to hold that position not as a 'woman's right to work' but as a task with which we are entrusted and that places us under an obligation."[103]

To be sure, statements like Savelsberg's reveal little about the personal motivations that led individual women to attend the conference. However, their professional biographies indicate that at least some delegates welcomed the Agency's networking initiative as an opportunity to improve their own position in their universities. Doris Korn, who graced a 1938 cover of the journal *Frauenkultur im Deutschen Frauenwerk* as a female symbol of hope for National Socialist promotion of young scientists, attended the conference partly in the expectation that building her links with the Women's Agency might help to remedy her marginalized and precarious position at Aachen's Technical University. Apart from a scholarship in 1936, Korn appears to have been without secure employment at the university between her habilitation in 1933 and a teaching contract in 1939, and in 1938 considered accepting a teaching post at Bryn Mawr in Pennsylvania.[104] Her colleague Savelsberg, an authorized university lecturer in economics since 1930, probably found herself in a similar situation. Between Savelsberg's senior assistantship, which ended in 1936, and her permanent appointment to a senior research post in 1939 there is a gap in her résumé that suggests she had no regular income at the time of the conference.[105] The urgency of personal economic circumstances and lack of perspectives can also be assumed for Germany's first female theologian to receive a habilitation, Hanna Jursch of the University of Jena. Jursch qualified in 1934, and until 1939 she experienced massive harassment from Hans Ebert, the leader of the local National Socialist Lecturers' League, who repeatedly attempted to have her status as an authorized university teacher revoked.[106] Another participant probably also attended the conference in the hope of consolidating her own position: geneticist Paula Hertwig, a part-time professor extraordinarius at Berlin University since 1927 and head of the zoology department at the University's Institute for Research on Genetics and Breeding. Hertwig stood for election on a democratic nationalist ticket for the Deutsche Staatspartei in 1933 and later belonged to a group of biologists trying to free genetics from the slogans of Nazi racial ideology. She joined the Nazi Lecturers' League, but came under considerable pressure at the University of Berlin due to her "insufficient political commitment." After an initial refusal, in 1934 Hertwig agreed to comply with the Dean's request to offer lectures on human heredity at the University's race research department. In 1939, there were questions over whether Hertwig should be allowed to keep her position as professor extraordinarius.[107]

Not all of those attending the lecturers' conference were faced with such professional difficulties that they would welcome the German Women's Agency invitation purely out of a sense of insecurity. The hydrographer Charlotte Möller, for example, had joined the NSDAP in April 1933 and been elected to the DAB board the following May. Although her position at the Institute for Oceanography in Berlin was initially at risk,[108] in 1934 the Prussian Ministry of Education and Culture allocated the institute a curator post, and Möller, who had supervised the collections from research trips for many years, now obtained a permanent job there. One year later, she was appointed a part-time professor extraordinarius, making her as well established in the academic world as it was possible for a woman to be at the time. She could continue her research work, though not directly in oceanography—she was not permitted to participate in expeditions at sea, which were often carried out on warships. Instead, under her custodianship the institute substantially expanded its research on inland waters. Möller brought modern oceanographic methods to hydrology, and initiated large-scale, application-oriented research projects in cooperation with local waterways administrations. In 1935, commissioned by Königsberg's hydraulic engineering department, she directed the hydrographic inspection of the Curonian Lagoon as the basis for planned drainage works. In 1936 and 1938, Möller was able to attend international conferences in her field in Helsinki and Paris. She never achieved a full professorship, but in 1940 she was elected to the German Academy of Sciences Leopoldina and from 1942 onward she directed the new department for continental hydrography as part of the Nazi "Generalplan Ost," the project to colonize and "Aryanize" central and eastern Europe.[109]

The statistician Charlotte Lorenz had joined the NSDAP in March 1933.[110] She was appointed part-time professor extraordinarius by the University of Berlin's political science department on April 20, 1937, after a spell as acting professor of applied statistics since the full professor's death in 1932. Lorenz had submitted her PhD dissertation on "The Woman Question in the Ottoman Empire" in 1919, then joined the research commission led by Max Sering in the Prussian Ministry of War to study women's employment during the World War I. From the mid-1920s, she worked for the Reich Statistical Office alongside her teaching duties at the University. The information available on her research during the 1930s indicates an interest in statistical and economic aspects of the "woman question" that continued unbroken by the upheavals of 1933. But Lorenz now adapted to the new ideological conditions, working in cooperation with the Nazi government and its institutions. An expert on state economic planning and "social questions," at the Statistical Office she carried out surveys of the "distribution of expenditure in the households of working *Volk* comrades." In liaison with the Women's Agency department "home economics—national economics," her research later contributed to making households, and housewives' work, visible as an important component of the national economy.[111]

At the time of the conference, Ilse Esdorn was a researcher and lecturer in applied botany at the University of Hamburg and the chief pharmacist of the League of German Girls, though she only joined the NSDAP in 1937.[112] She too pursued her career goals in close collaboration with the Nazi organizations, and her decision to focus on medicinal plants placed her in a domain that accrued great ideological capital through Germany's aspirations to autarky.[113] The German pharmacists' prize awarded to Esdorn for her work in 1937 indicates how successfully she had positioned herself within the Nazi science system.[114]

The Reich women's leadership used the meeting as an opportunity to report on its new institutions, and published illustrated profiles of the assembled female academic elite in its periodicals. The reports also included numerous photographs of the conference itself. The Nazi women were inserting their initiatives and delegates into a female tradition of scientific work—one that differed from the pre-1933 tradition chiefly by completely blotting out the pioneering work of Jewish women.[115] From 1938 until well into the war years, the science section in the German Women's Agency consistently pointed to female careers in research and university teaching when asked to portray the successes of German women's work.[116] Several women lecturers, in turn, indicated through their attendance at Agency events or their own publications in Agency journals that they expected to gain significant advantage from being co-opted as "poster girls" for Nazi policy.[117]

The continuities that can be identified throughout the Nazification of the DAB, its reorganization as the RDA, the initiatives of Nazi female academic networking policy, and the biographies of individual activist members make one point abundantly clear: the attempt to form a female educational elite, arising in the Weimar Republic on an Anglo-American model, did not vanish from the scene along with Weimar's democratic structures. Rather, National Socialist women's policy reinterpreted it as an important element of service to the *Volksgemeinschaft*, and continued to pursue it with determination after the previous association structures were dissolved in 1936.

In emphasizing these continuities, I do not wish to distract from the fact that for many academic women—even those who were not persecuted as Jews or on the grounds of democratic or socialist nonconformity—the Third Reich brought painful biographical ruptures, the end of professional careers, economic and political duress, and personal isolation.[118] The DAB membership presents a spectrum of such experiences. Agnes von Zahn-Harnack withdrew completely from public life and sat down at her desk at home to write a biography of her father, the influential theologian and founder of the Kaiser Wilhelm Society Adolf von Harnack, for which she was awarded an honorary doctorate by the University of Marburg's theological faculty in 1949.[119] Like

Zahn-Harnack, Dorothee von Velsen, who told Gertrud Bäumer in a 1941 letter that the "*Lebensraum* slogan" made her feel "physically ill,"[120] could afford to live from her family's resources in seclusion, without paid employment. The Marburg scholar Luise Berthold, a specialist in Old German language and literature, joined the resistant Confessing Church, shunned Women's League meetings, and in 1934 turned down the chance of a professorial appointment in order to avoid political exposure. Her memoirs describe the breakdown of many friendships and a restriction to a small circle of social contacts, although she herself remained unmolested in her research and teaching.[121] For the trainee lawyer Elisabeth Schwarzhaupt, 1933 brought the collapse of both professional and personal plans for the future. She decided against following her Jewish fiancé into Swiss emigration; a doctor, his permission for public practice had been revoked on April 1, 1933. The couple remained in contact until 1936, but Schwarzhaupt did not want to leave Germany without any professional prospects—though her hopes for better opportunities as a lawyer in the Reich proved illusory. For years she corresponded with the Higher Regional Court in Hamm about the resumption of her internship, which had not been extended after May 15, 1933. Through friends of her parents, she finally found a badly paid post in April 1934 as a legal adviser to the Nazified League of German Retirees in Berlin, and in 1936 moved to the legal section in the administration of the Lutheran Church, which by this time was dominated by pro-Nazi "German Christians."[122] The later professor of statistics Asta Hampe, who earned her PhD in physics in 1931 with a study of cathode ray tubes, was promoted in 1933 to run the laboratory of Hamburg's Barmbeck Hospital, taking the place of a Jewish physicist removed from his job. She, in turn, was dismissed just a year later at the instigation of Hamburg's Minister of Health, on the grounds that physics was an unsuitable occupation for women. Until 1943, her career played out in the Hamburg import and export trade, the experimental laboratories of Krupp and Phillips, and the Navy in Kiel.[123]

For the internationally renowned biologist and crop researcher Elisabeth Schiemann, differences of "worldview" and scientific theory led to a rift with her doctoral adviser Erwin Baur in the early 1930s and ended her employment at the Kaiser Wilhelm Institute for Plant Breeding Research in Müncheberg, Brandenburg. From that point on, Schiemann worked at the Botanical Museum in Berlin, sometimes supported by research grants from the Kaiser Wilhelm Institute for Biology, and taught at the University of Berlin. Schiemann's openly hostile attitude to Nazism left her in a perilous position, and her permission to teach was revoked in 1940. It was only during the expansion of plant breeding research in the final years of the war that Schiemann's situation improved: in 1943, she was appointed to the newly founded Kaiser Wilhelm Institute for Cultivated Plant Research in Tuttenhof near Vienna.[124]

At the end of 1934, Marie-Elisabeth Lüders was banned from publishing and public speaking, and—like Agnes von Zahn-Harnack (and incidentally also Gertrud Bäumer)—spent the years from 1933 to 1936 on historical research. In this period, she wrote *Das unbekannte Heer* ("The Unknown Army") on the achievements of women's work in the World War I. Published in 1936, it received positive reviews even in the publications of the German Women's Agency.[125] After a four-month Gestapo imprisonment in 1937, she spent some time in various sanatoria for fear of further persecution. Lüders's subsequent fortunes under the Nazis show that the personal ties between former DAB activists survived despite their deep-seated political differences: Lüders ultimately found professional sanctuary with her one-time DAB colleague, the oceanographer Charlotte Möller. She spent 1938 to 1944 at Möller's institute, carrying out politically unobjectionable research on trends in international fisheries management.[126]

Science, Womanhood, and *Volksgemeinschaft*

In her evaluation of the development of women's university studies and graduate professions in the Third Reich, Claudia Huerkamp concluded that National Socialist antifeminism, in its ambivalent repression and promotion of female ambitions, proved highly "adaptable to the constraints of a modern economy, and especially of preparations for war."[127] The politics of female academic networking under Nazism makes it particularly obvious that this flexibility did not arise from structural features alone. If a relatively large number of women retained a presence in graduate occupations and professional networks during the dictatorship's early years, despite the freeze on women's university enrollment and all the many other obstacles, that was principally due to the great strategic adroitness of the women themselves, who skillfully turned the regime's ambivalence to their own advantage. This aspect of academic women's preservation of their existing gains has long been ignored. As a rule, particular actors are only singled out in order to reconstruct the hindrances or persecution they encountered under Nazism. One reason for this limited view may be that most historical studies depend on autobiographical sources written after 1945, without reflecting sufficiently on the constraints of the postwar perspective. The result is the reiteration of assessments that sidestep such uncomfortable factors as women's professional success, self-assertion, and targeted career strategies under Nazi rule. In the final section of this chapter, in contrast, I use mainly contemporary sources to retrace the strategies and successes of academically trained women from the DAB milieu.

Among the women involved in the graduate professional associations who continued to make public statements after Nazification was completed, the strategy was to accept the division of society into masculine and feminine domains so noisily proclaimed by Nazi ideologues in the early 1930s, while at the same time insisting that their professional skills were indispensable to the feminine segment of the *Volksgemeinschaft*. Johanna Bleker, Eva Brinkschulte, and Sabine Schleiermacher have shown how successfully, and with what momentous consequences, women physicians deployed this strategy to safeguard their position as experts catering to the needs of women's and "racial" health care within the national community. It is surely no coincidence that physicians stood out—not only in their own association but also in the DAB milieu as a whole—as vociferous advocates of the new era, or at least successfully cultivated their careers under its wings.

A good example is that of Auguste Hoffmann. Born in 1902, Hoffmann took up the still relatively new specialty of sports medicine in the mid-1920s and was employed at the German College of Physical Education (Deutsche Hochschule für Leibesübungen) in Berlin on October 1, 1930. Much suggests that Hoffmann was one of those university-trained women who welcomed the Nazi dictatorship with optimism and confidence. She remained a member of the Nazified women doctors' association BDÄ despite finding its new National Socialist president Thimm "dreadful,"[128] and took on several high-profile responsibilities, also joining the National Socialist Teachers' League.[129] From May 1933 Hoffmann helped to build up the League of German Girls (BDM), for which she organized and directed medical care in Berlin, a role that eased her way to an appointment as BDM medical expert in the Reich Youth Leadership in August 1934.[130] For Hoffmann, volunteering with the BDM offered both great personal satisfaction and attractive opportunities for influence in medicine. She worked to persuade other women doctors to increase their cooperation with the girls' mass organization, also offering centralized training courses for prospective "Aryan" sports physicians.[131] Hoffmann also argued for BDM doctors across the Reich to carry out free mass medical screening, collecting medical histories in as much detail as possible, and to have the results recorded centrally in a standardized format. As well as contributing to care for individual girls, Hoffmann expected this plan to provide an important statistical basis for research on the health of large segments of Germany's female youth.[132] She regarded close collaboration with the BDM as promising distinct strategic advantages for women physicians as a group and their efforts to assert themselves professionally. When Hoffmann campaigned among her Berlin colleagues for more active commitment, she was also anticipating "that the BDM leaders will always stand up for the necessity of a woman doctor and thus support her in the struggle for recognition of her work."[133] With regard to the leadership of the BDM, this calculation initially paid off. When Reich girls' leader Trude Mohr met with the BDM physicians in late 1934, she publicly declared that "the female doctor" was "absolutely indispensable and can certainly not be replaced by a male colleague."[134]

However, although the principle of female doctors was uncontested inside the BDM leadership, from early 1935 it began to attract more negative attention from both male doctors and the political leadership of the Nazi youth organizations as a whole. On the one hand, male physicians successfully applied pressure from many different sides to prevent the institutionalization of independent female healthcare provision in the BDM. Despite vehement protest by the women physicians' association, the mass screening of BDM girls that Hoffmann had helped to initiate was halted on April 1, 1935, and replaced by individual examinations carried out by the Labor Front's Office for Popular Health.[135] The women physicians' panel in the Reich youth leadership, headed by Hoffmann, was dissolved in June 1935 "because the girls... were allegedly being tempted away from their family doctor to the female doctor."[136]

Dismissed from her roles in the BDM and the Reich youth leadership,[137] Hoffmann shifted her focus to the university setting. She researched and taught at the College of Physical Education (now named the Hochschulinstitut für Leibesübungen), worked on her postdoctoral dissertation on the influence of physical training on the skeletal muscles, and redirected her volunteering energies into the creation and management of training courses for women doctors in the association of sports physicians. At the 1936 Berlin Olympics, she contributed to the medical supervision of the 400 international women competitors.[138] In 1938, Hoffmann received her habilitation at the University of Berlin, and in October 1940 was appointed to teach there. She remained a lecturer with civil service status at the University until the end of the war.

Auguste Hoffmann was not the only academic in DAB circles who managed to secure a position at the University of Berlin during the Nazi period. The statistician Charlotte Lorenz, mentioned above, made the leap into full-time teaching at the University with a 1939 appointment as professor extraordinarius,[139] allowing her to give up her work at the Reich Statistical Office. During the appointment process in 1939, Lorenz listed as a research project her plan to establish a student working group that, "in concert with the German Women's Agency," would publish "a large-scale anthology on 'The German Woman in the *Volk* and Economic Life of the Present Day: A Statistical Reader and Reference Work.'"[140] The Hamburg botanist and crop researcher Ilse Esdorn, too, evidently tried to fulfill her career goals in the University with the help of NSDAP offices. This caused a serious breach with her departmental director in 1940. Esdorn weathered the storm and was transferred to Reinbek near Hamburg as a head of department in the Reich Institute for Foreign and Colonial Forestry. In January 1941, the University of Hamburg appointed her as a professor extraordinarius.[141] As for the Aachen economist Gertrud Savelsberg, who reinforced her ties to the Nazi system by joining the Party in 1938, she received a permanent research post in the library of the Kiel Institute for the World Economy in 1939; in 1944, she was additionally appointed professor extraordinarius for economics and political science at the University of Kiel.[142]

These academics specializing in medicine, botany, economics, or statistics were not the only ones to foster strategic or ideological proximity to Nazi institutions, or to adopt the Nazi segregation of society into male and female spheres and reinterpret it to their own ends. In the field of law, too, some women tried to counter their exclusion from Nazi jurisdiction by eagerly demanding responsibility for women's concerns in the Nazi state as female "preservers of justice."[143]

In 1937, the freshly qualified lawyer Wiltraud von Brünneck based her appeal for women to be granted a firm place in the legal system on the "total turnaround that the National Socialist revolution has entailed for all areas of life."[144] The "idea of the revolution," if "truly grasped," meant a "complete transformation of previous views and concepts" not only in jurisdiction but also in legal science. Instead of an abstract system of norms, "the idea of the

Volk" had "come to the fore of all scientific thinking with unprecedented rigor." The basis of jurisprudence, therefore, must no longer be a closed system of abstractions, but the concrete "reality of the *Volk*"; a positive example was the redefinition of property in the new Hereditary Farm Law. For the National Socialist reshaping of jurisprudence, Brünneck argued, women's participation was "not merely desirable, but quite indispensable," because the "changed stance of today's discipline" gave special value to "a way of thinking that is closely bound to the *Volk* reality." Women's "natural closeness to life, their strong ties with practical events" made them predestined for an active role and crucial to the success of a new, National Socialist legal system.[145] Brünneck passed her final legal exams summa cum laude in 1936 and 1941, and secured employment as an assistant professor in the University of Berlin's department of labor law. When a bombing raid destroyed the papers for her PhD project, she moved to the Reich Ministry of Justice, where she held a senior position in the land registry and inheritance law section until 1945.

Particularly interesting in the notion of the indispensability of female participation in science and the professions is that it significantly modified a key paradigm of the women's movement in its drive to achieve access to academia and professional careers: the gender-neutral ethos of science and of the graduate professions. In the late nineteenth century, women in Germany had fought their way into the universities using the argument that science, just like human reason, was not itself gender-specific but obeyed universal principles. Acquiring knowledge of fundamental scientific principles meant the same thing for men and women, and only when those principles had been successfully learned could one begin to speak of women putting them into practice differently than men due to the differing natures of the sexes. It was in the application of universal scientific principles, this argument ran, that female participation in higher education could, as a kind of complementarity, enrich and alter science—that the gender distinction could truly come into its own.

To be sure, during the 1920s the pressure of the labor market and competition with aspiring male professionals had already eaten away at this position. "Feminine" graduate professions were associated so closely with a "feminine nature" that the postulate of the underlying equal ability of the sexes became blurred.[146] After 1933, however, the "new women" distanced themselves far more rigorously from the "liberalistic" ideals of equality propounded by the earlier women's movement. Lawyer Martha Unger, who had worked closely with Marie-Elisabeth Lüders in the DAB in the early 1930s but then dedicated herself fully to the "new era," summarized this departure: "Of course we thank the old fighters for what they, as daughters of their era, won for women, but we do not follow them.... The general break with liberalism also holds true for us German women."[147]

This same dictum applied to the question of scientific and scholarly potential. The new protagonists of the politics of female professionalization deployed old gender-specific stereotypes to propound the special aptitude of women for work as gynecologists, "preservers of justice for women," and not least as scientists and scholars in a gender-segregated racial community. This allowed them to argue that women were the ideal guarantors of a holistic, National Socialist science, because practical and application-oriented principles and methods corresponded to their nature and because their sex protected them from losing themselves in "abstraction and individualism"; in every area of knowledge they "instinctively" found the fields for which this special characteristic best suited them.[148] Even if the DAB ceased to exist as an organization in 1936, therefore, three lines of continuity endured throughout the Nazi era: the strategies of structural networking, the demand for academically educated women to play a part in society's leadership, and women's assertion of their responsibility in the academic world.

6
Networks in Action: Assistance to Refugees

Even after the German DAB's enforced alignment with the Nazi regime, its official dissociation from the IFUW, and its subsequent dissolution in early 1936, many of the bonds formed in the Weimar era through its international networks remained intact. This applied not only to the personal contacts and friendships that had emerged from congresses and stays at the clubhouses of London, Paris, and Washington, but also to institutional connections, especially those associated with the fellowships program of the IFUW and its affiliates. The program acquired a new significance after 1933. With the beginning of Nazi rule, the IFUW immediately took on the function of a transnational network to assist academic women in their flight from persecution. Surprisingly, the existence of this network has hitherto received little attention, let alone systematic investigation, in either the history of science or exile studies.[1]

This chapter presents the development, structure, and practical implementation of the IFUW's assistance to academic women refugees between 1933 and 1945. I identify the motives, methods, and outcomes of the transnational network and its national affiliates outside Germany. Carefully reading rarely used source material, I trace their efforts to assist persecuted colleagues from nations under Nazi control. My analysis of the network's endeavors to aid its former members proceeds in four steps, based on the chronology and geographical locus of IFUW engagement. From 1933 until the outbreak of war on September 1, 1939, the focus of emergency work lay in Britain. Initially, between 1933 and 1938, British support took the form of research promotion, but as the stream of refugees began to swell in summer 1938 a second phase commenced: humanitarian forms of assistance came to the fore and were now extended to those women graduates not professionally active in the university sphere, from Germany, Austria, and Czechoslovakia. The third phase began in fall 1939, when the British borders were closed. British women's support for refugees now concentrated on those already in the country, while the work to aid emigration relocated to the United States and non-European countries, and covered persecuted academic women from all the areas under

German occupation. The United States' entry into World War II in December 1941 marks the beginning of the fourth and final phase. Thereafter, escape from Germany and the countries Nazi Germany occupied became practically impossible. From December 1941 until the end of the war, the work of the IFUW and its member associations—to the extent that these were still capable of functioning at all—was directed at helping refugee women to integrate socially and professionally into their host society and, in some cases, offering financial and moral support to colleagues trapped within German-dominated Europe. Thus, the first two sections of this chapter center on Britain and the BFUW for the period up to September 1939; the third section addresses the policy and global initiatives of the IFUW after 1939, and the fourth the United States and the AAUW.

Promoting Research, 1933–38

The political upheavals in Germany, with their life-threatening consequences for the community of academic women, had an immediate impact on the IFUW's research promotion program. The number of funding applications to the BFUW, AAUW, and IFUW rose dramatically in 1933. Whereas the summer of 1932 had seen 29 women apply to the IFUW and BFUW for the popular one-year fellowships, more than 50 project proposals a year arrived in London alone in the period that followed.[2] In addition, the BFUW received "a large number of applications for extra fellowships and for grants…by members of foreign Federations, who, for political reasons, were unable to continue working in their own countries."[3] In October 1933, the Dutch botanist Johanna Westerdijk, who had commenced her presidency of the IFUW at the 1932 Edinburgh conference, launched an appeal to all member associations to support German academic women in every way they possibly could. These women's deplorable situation, she said, offered the IFUW "a chance now of putting into practice our principle and of helping our friends in need."[4]

The BFUW was the first IFUW affiliate to react to the Nazi accession to power and its disastrous impact on so many members of the German federation, the DAB.[5] British women were already offering assistance and hospitality by the summer of 1933. Numerous academic women who had been barred from working in Germany asked for, and received, a temporary home at Crosby Hall. The most compassion was roused by those guests who had lost their posts due to having one Jewish grandparent. "They have never thought of themselves as anything but German, they have had no connections to the Jewish community, they and their parents before them have been Christians. They are utterly bewildered that they should have been cast out by Church and State."[6] All the more impressive was the "noble dignity" with which the German colleagues faced their sudden exclusion, commented the BFUW: the women referred to conditions in Germany only with the greatest reticence, and their unbroken sense of connection with their country of origin was admirable. Most had, it was noted in 1933, now returned to their homeland.

Because the leadership of the BFUW initially believed—probably due not least to information provided by former DAB members[7]—that the National Socialist revolution in Germany was nothing more than a passing phenomenon, they considered the most effective strategy to lie in enabling their colleagues to overcome what appeared to be a temporary professional crisis. The BFUW's officers thus resolved to use the federation's modest resources to enable a small number of outstanding women scholars to continue their research in Britain until they were able to return to Germany. This BFUW policy followed the lead of the Academic Assistance Council (AAC, later renamed the Society for the Protection of Science and Learning, SPSL), which had been formed in Britain in May 1933 in response to the dismissal of German university lecturers and boasted numerous prominent advocates within the universities and the British government.[8] Soon after the establishment of the AAC, the BFUW executive decided to do its utmost to support the new Council's work. This implicitly meant flanking the AAC's funding policy, which was clearly oriented toward world-class male academics, with measures to support their appropriately qualified female counterparts. In line with this principle, the BFUW concentrated its efforts on funding longer stays in Britain for established women scholars. The annual meeting in Cambridge in fall 1933 resolved to collect additional donations to the tune of £300, with the goal of offering a year at Crosby Hall to, initially, three "distinguished German academic women deprived of their posts," taking care of their accommodation and living expenses and helping to secure their academic progress.[9]

In seeking to understand why the British association reacted to the IFUW's appeal in this particular way and with such alacrity, one must consider the extraordinarily high social and political status that the funding of persecuted German academics in Britain attained after January 1933. As early as April 1933, the British government had relaxed its restrictive immigration regulations for those well-known university scholars, artists, and business people with private assets who had been forced out of their jobs in Germany—in the expectation of turning the Nazi policy of exclusion to Britain's own advantage. The eagerness of British professors to allow colleagues so shamefully treated by German universities to continue their academic work was rapidly declared by the government to be a matter of prime importance for society as a whole, provided, of course, that the financial needs of the learned asylum-seekers were fully covered by private donations.[10] In light of the fact that assistance for academic refugees had been proclaimed a national task, the BFUW was in no doubt that it must demonstrate its own engagement as the female counterpart to the AAC. After all, women university lecturers had played a key part in shaping BFUW policy ever since the federation's establishment in 1907, and its officers shared with the AAC a resolute antipathy to National Socialism. The role of Crosby Hall as a nexus of international contacts meant that Britain was particularly rich in personal ties with academic women in Germany; the German-speaking lecturers who had contributed to IFUW committee work also enjoyed high regard within the BFUW.

As the funding of all the BFUW's fellowships was based on collecting donations, the decision to raise additional money for the support of German academic women in distress entailed a substantial extra burden for the organization and its almost 3,000 members. A 1933 appeal to all members to offer special support to the annual Crosby Hall Christmas bazaar, dedicated that year to the German colleagues, met with an unexpectedly warm response. "Visitors positively seemed to *like* spending their money to show their real sympathy with the needs of those distinguished German women who have lost their posts," reported the BFUW journal: "money seemed to pour in at the door, the stalls, the side-shows."[11] Most local groups also organized "bring-and-buy sales" to raise money for the refugees, and were able to pass on considerable sums to the London office. As early as January 1934, record takings of £538 meant that three extra 12-month Residential Fellowships at Crosby Hall and numerous shorter grants were available for German academic women in addition to the seven existing one-year fellowships.[12]

From 1934 to 1938, the BFUW initially supported women who were already staying in Britain or who were personally known to the organization. Thus, the first year's award went to the microbiologist Emmy Klieneberger, the holder of a postdoctoral "habilitation" qualification in addition to her PhD. In August 1933, Klieneberger had been forcibly retired from her post at the Hygiene Institute in Frankfurt. Robbed of her pension, she left for London to pursue new career options.[13] Klieneberger's case illustrates the complementary nature of the assistance provided by the AAC and by the BFUW. With the AAC's help, Klieneberger quickly found a place at the Lister Institute of Preventive Medicine, where she was offered laboratory space in October 1933. A testimonial from the IFUW's Agnes von Zahn-Harnack paved her way to accommodation in Crosby Hall, for which the BFUW charged her the greatly reduced rate of 2½ guineas for the room and a daily meal. For the first few months, Klieneberger drew upon her German assets to meet her living expenses, as the position at the Lister Institute remained unremunerated. When, in early 1934, it became impossible for Klieneberger to transfer her own money out of Germany, the BFUW awarded her one of the three new German Scholar Residential Fellowships so that she could continue her research at the Lister Institute.

Another recipient of the 1934 supplementary residential fellowships was Betty Heimann, a professor of Indology from Halle. Heimann had spent a year in India in 1931–32 with the help of an IFUW Senior Fellowship, and in September 1933 was lecturing at English universities and BFUW events about her research trip.[14] The news that her permission to teach had been revoked reached Heimann during her time in England. She decided not to return to Germany,[15] and, like Klieneberger, took a room at Crosby Hall. From the beginning of 1934 Heimann received the second of the German Scholar Crosby Hall fellowships and taught Indian philosophy on a freelance basis at the University of London's School of Oriental Studies.[16]

The third grant recipient of 1934 was the art historian Helen Rosenau. Rosenau had come to Britain in fall 1933, after her research grant from the

Emergency Association of German Science was not extended and the habilitation thesis that would have qualified her to teach at university level was rejected by the University of Münster.[17] Rosenau, too, was known among BFUW members. During her year-long German Scholar Residential Fellowship at Crosby Hall, the BFUW supported her research on the comparative history of religious architecture by funding a visit of several months to Egypt and Palestine. Members of the local academic women's organizations in Jerusalem and Cairo opened their homes to her for the duration of her work at the American Schools of Oriental Research.[18]

All the women who had received one award had their funding extended—if possible for as long as it took for them to become reestablished in their field of expertise.[19] Emmy Klieneberger was granted a second year-long award for 1935, this time funded by a regular AAUW fellowship.[20] As a result, Klieneberger had two years' access to free accommodation in the heart of London and a small income for her activities at the Lister Institute of Preventive Medicine, within walking distance of Crosby Hall. Indologist Betty Heimann received a series of short-term awards from the BFUW once her one-year grant had expired, enabling her to live rent-free or at a reduced rate in the Crosby Hall residence and to continue to teach at the School of Oriental Studies. For periods of time, the BFUW also contributed to the payment of a modest lecturer's fee for Heimann when the School of Oriental Studies proved unable to pay her for her labors.[21] The art historian Helen Rosenau received a further year of support and accommodation in the shape of a regular Crosby Hall residential scholarship for 1935–36. Rosenau used the two years to publish in English the habilitation study that had been rejected in Germany, and to work on a second doctoral thesis, at the Courtauld Institute in London, on the architectural history of the synagogue.[22]

The photochemist Gertrud Kornfeld from Berlin and the art historian Adelheid Heimann from Hamburg, a specialist in medieval iconography, also received funding over several years. Kornfeld, who had initially taught at University College Nottingham with BFUW support, was awarded the federation's German Scholar Residential Fellowship in 1935, allowing her to pursue research at Imperial College in South Kensington. In 1936, Kornfeld was able to visit Vienna for a year with the help of an AAUW International Fellowship. Adelheid Heimann, who had taught at the Sorbonne from 1933 to 1935 and had been in Britain since 1936, received the Crosby Hall Residential Scholarship for 1937–38; the year after, she was awarded the Canadian Federation of University Women's Aurelia Reinhardt International Fellowship, which she used to continue her research in France.[23]

Outside London, too, BFUW members stepped in to help renowned women scholars from Germany. In the case of the respected archaeologist Margarete Bieber, a concerted campaign was launched in Oxford that brought together women's colleges and organized women graduates. The women's college Somerville awarded Bieber, who had been dismissed from the University of Giessen just before she was due to be appointed a full professor, an "honorary fellowship" for the academic year 1934–35; other colleges funded a lecture series for Bieber to help her earn extra money.[24] In Cambridge, BFUW members covered the expenses of Bieber's housekeeper and her six-year-old adopted daughter Irene, whom Bieber had taken in while still in Germany in 1932, at a time when she expected to be earning a regular professorial salary.[25] For Helena von Reybekiel from Hamburg, who specialized in psychology and Slavic philology,[26] BFUW members in Birmingham collected sufficient donations to fund a guest lectureship at the Birmingham and Midland Institute.[27] Reybekiel's salary as a lecturer in Polish was additionally funded by profits from Crosby Hall bazaars.

Between 1933 and 1938, a total of ten university lecturers from Germany received longer-term support through the coveted one-year fellowships awarded by the BFUW, IFUW, and AAUW.[28] However, the funding for German women scholars—made possible by the unexpectedly generous response to appeals among the BFUW and its milieu in the first year of the Nazi dictatorship—soon reached a limit. After two years, the initial willingness to support German colleagues through fellowships was replaced by a noticeably more sober attitude. In 1936 the limits of the BFUW's financial means became apparent. Like their male counterparts, the women of the BFUW had come to fear that German applicants could present undue competition for the next generation of British scholars.[29] As early as 1934, concerns were voiced that most of the women scholars from Germany did not belong to the target group previously envisaged for funding. They were, it was complained, too eminent, unfairly restricting the opportunities for younger women. In order to sustain support for the younger generation of female scholars, especially British ones, the BFUW resolved in 1934 "to devote a portion of the proceeds to assist young British graduates who may have otherwise suffered from the more intensive competition for Fellowships resulting from the candidature of German women with higher academic standing."[30] The offer of a dedicated young scholars' grant for British university graduates appeared all the more important in that the BFUW was facing general problems in attracting a new generation of members.[31]

In addition, once the German DAB officially withdrew from the IFUW in 1936 it became obvious that the situation of women scholars dismissed from their posts in Germany was not a temporary setback but likely to persist for the foreseeable future. Perhaps not surprisingly, in 1936 the IFUW also raised the issue of grant requests put forward by "overqualified" German applicants. Since the Nazis' accession to power, the IFUW noted, the proportion of applications from Germany had risen to a fifth of all grant applications, mostly coming from academically established university lecturers.[32] Yet the original purpose of the promotion program had been "to give women who have shown ability to carry on original work the opportunity of continuing to do so, thus fitting them to compete for higher academic posts."[33] At the same time, more and more graduates without an academic specialization were appealing for assistance, so that the grants committee of the IFUW recommended the creation of a specific assistance fund addressing the émigré members of formerly affiliated organizations.[34] The 1936 conference in the Polish city of Kraków adopted a resolution to this effect, agreeing to the establishment of a special budget for the concerns of refugees.[35] In its first year, this IFUW "Emergency Fund" disposed over the modest sum of £200; for 1937 and 1938 the annual allocation rose to £300.[36]

The growing tendency to promote women scholars from Germany not through regular or additional fellowships, but through resources from a dedicated refugee fund, impacted negatively on the university lecturers emigrating from Germany. In 1936, just one candidate from Germany, the behavioral scientist Dora Ilse from Berlin, received a full one-year fellowship.[37] Gertrud Kornfeld's application for the AAUW's Senior International Fellowship in Science was shortlisted but ultimately rejected.[38] Of the six German university lecturers supported by BFUW funds in the year after that, only Adelheid Heimann received one of the adequately funded, 12-month Crosby Hall grants. The support for all the other five women together totaled just £100, a sum that in 1934 had been defined as the minimum for a single one-year fellowship.[39]

Furthermore, the overall amount of donated money that the BFUW earmarked for women scholars from Germany fell from £368 in 1934 to only £100 in 1937, although energetic appeals in 1938 brought the sum back up to £240.[40] We can only speculate on the reason for these fluctuations. It may be assumed that the first surge of wholehearted support for the victimized German scholars dissipated in the mid-1930s, regaining momentum when shocking news began to arrive of the worsening persecution in Germany. To this extent, the figures reflect the dynamics of the policy of persecution within Nazi Germany and perceptions of that policy in western Europe. After the onslaught of discrimination and dismissals at the start of the dictatorship, the period up to the 1936 Berlin Olympics saw a "settling down" of public perceptions of the conditions in Germany. In 1938, the British public's desire to donate was reactivated in parallel with the bellicose National Socialist policy of annexation and the escalation of state violence.

Bound for Britain, 1938–39

In 1938, the BFUW's renewed efforts to keep alive its members' commitment to women scholars from Germany were overtaken by events in Germany and within Britain itself: as Nazi annexation and persecution gathered pace, immigration regulations for refugees to Britain became increasingly restrictive. The first women to emigrate had faced comparatively few difficulties in either leaving Germany or entering Britain. Initially, the recipients of the German Scholar Crosby Hall fellowships were also able to reenter Germany safely on trips to order their affairs or visit relatives. Emmy Klieneberger, for example, left Germany without impediment on the night train to London via Ostend on September 15, 1933; in spring 1934, she briefly returned to Frankfurt am Main to visit her mother and sister and to spend her remaining savings—which she could now no longer transfer out of Germany—at the famous Zeiss optical works in the city of Jena. She bought a "first-class research microscope" and a Leica camera with full darkroom equipment, which she brought with her on her return trip to England. Klieneberger's resourcefulness put her in possession of what she considered the very best available equipment for her research, something that the Lister Institute could not provide. Both German and British customs and border authorities allowed Klieneberger to pass unhindered carrying her valuable apparatus.[41]

In summer 1938, however, maneuvers of this kind were practically impossible. The logistical problems of emigration now took center stage for all the persecuted scholars, as well as for the organizations offering assistance from beyond the German border. For the BFUW and IFUW, too, the steadily rising number of refugees meant that correspondence with women seeking help, the authorities, and members became increasingly onerous. At the same time, almost all countries were introducing more restrictive immigration regulations. In May 1938, the British government had begun to demand visas for German and Austrian passport-holders, and although since November 1938, in response to the anti-Jewish pogrom of November 9, the government had allowed charitable organizations (including the IFUW) to draw up lists of names for "block visas," the issue of a "hospitality permit" authorizing entry to Britain in these cases depended on the provision of a six-month guarantee of financial support for each applicant.[42] This made the search for host families much more difficult and time-consuming, since each family's financial circumstances had to be individually investigated. All too often, the result was that potential hosts shied away from offering their "hospitality" at all.

The situation had thus already deteriorated considerably when, in 1938, Austria's annexation by Nazi Germany meant hundreds more academic women were suddenly faced with an imminent end to their professional lives, alongside a severe reduction in the options for escaping Nazi harassment via emigration. In the wake of *Anschluss*, the practical face of refugee assistance in the BFUW and IFUW underwent a fundamental change. In May 1938, the BFUW executive set up an ad hoc committee on refugee issues so as to be able to respond adequately to the questions and calls for help that were coming especially from Austrian colleagues.[43] But by summer 1938, it had already become impossible to cope with the burgeoning correspondence through volunteer commitment alone. In fall 1938, the BFUW decided to employ one person to staff the secretariat of the Committee on Refugees, someone who could be called upon for help more or less around the clock. The woman entrusted with running the refugee office was the 40-year-old Erna Hollitscher from Vienna. Hollitscher, the holder of a doctorate in English literature and Romance studies and a former employee of the Anglo-Austrian Bank in Vienna, had herself emigrated to Britain in June 1938 with the IFUW's help, and had spent the summer staying with a BFUW member in London.[44] Hollitscher took up her salaried post in September 1938, and at the same time moved into a room at Crosby Hall.[45]

Erna Hollitscher very quickly became both the heart and the brains of the BFUW's refugee work.[46] From the large volume of correspondence that survives in the BFUW archives, Hollitscher emerges as a person who shouldered her difficult task not only with commitment, reliability, and professionalism, but also with a great degree of personal maturity. The many messages of thanks indicate how happy the BFUW was with its choice for this office. Many of the refugees, too, maintained contact with Hollitscher for years.[47]

Historical analysis of the BFUW's refugee work is made possible by the great trouble the federation took to account for its activities in this area, always aware of the need to nurture the membership's willingness to donate. Despite her immense and emotionally grueling daily workload, Hollitscher fulfilled the task of reporting, like her other roles, in an exemplary manner. Hollitscher not only filed all correspondence with émigré women and kept a careful record of the office's activities, but also drew up systematic overviews of the refugee assistance work of the local BFUW branches. Hollitscher's documentation makes it possible to reconstruct the network of assistance in detail, enabling us to draw interesting conclusions about the émigrés' integration into society.

Because the financial focus of refugee work for academic women lay with the BFUW until September 1939,[48] Erna Hollitscher was faced with an stampede of requests from the very first day of her employment. Between September 1938 and April 1939 alone, her office received applications for help from 226 university women. Each request involved bureaucratic formalities and negotiations between the authorities, BFUW members, and the refugees themselves—usually a highly complex matter because, in parallel with the worsening circumstances in Continental Europe, British immigration policy had once again been dramatically tightened. From late summer 1938, each refugee had to present a personal guarantee of unlimited financial support or else proof of employment.[49] For female refugees, such employment was normally domestic work, for which a "domestic permit" was granted.[50]

The BFUW was not short of offers of assistance. On the contrary, after the temporary dip in 1937, members' willingness to support the refugees was now higher than it had ever been. Annual donations reflected this: in 1938–39, the sum of £799 was collected, more than double the amounts achieved in earlier years.[51] An additional burden of communication for the secretariat arose from many members' urgent desire to offer their help as guarantors or by offering employment to a domestic servant. Frequently, such offers included precise specifications concerning the potential guests or employees, requirements that had to be balanced with the needs of the refugees themselves. The task demanded considerable tact.[52] In addition, the various offers of short-term help had to be pieced together into "hospitalities," each of at least six months' duration. These composite hospitalities were fragile works of art that could be destroyed in a moment by, for example, the belated arrival of the "guest."[53] The tense situation in Britain heightened the need for particular care in the selection of the refugees to be granted assistance. In several cases, IFUW members in the women's countries of origin assisted by interviewing the applicants, trying to find out what social skills they possessed and how seriously to take their written declaration that they were "willing to take any work."[54] In London, the office's daily routine included counseling academic émigrés who needed support and practical advice on rescuing their parents, siblings, friends, or relatives.[55] Hollitscher and her assistants could generally do very little; the harrowing stories they heard probably served primarily to keep them informed about the likely difficulties of their own families and of the university women applying to the BFUW from the Continent. "A total of 1450 letters written in seven months," ran the BFUW's summary of the refugee work for its members in April 1939; "interviews every day and sometimes all day, constant telephoning, committee work and visits to the chief Refugee Committee Headquarters give some indication of the work involved."[56]

Virtually all the women who turned to the BFUW with a request for assistance in leaving their home countries were university graduates or else had been prevented from graduating when they were expelled from the universities of Nazi Germany's so-called Third Reich. Almost all disciplines were represented. Physicians and dentists headed the list, followed closely by natural scientists in the fields of zoology, botany, chemistry, physics, and mathematics. Slightly behind these in terms of numbers were philologists. Further significant, though numerically much smaller, groups were jurists, economists, and sociologists, along with historians, art historians, and archaeologists, and psychologists or psychoanalysts. A summary drawn up by the BFUW in October 1938 gives a precise picture of the distribution of professions among the women asking for help. Among 46 applicants, there were ten high school teachers and nine practicing physicians; seven applicants (including some with a medical training) had worked and researched in technical laboratories. Six women had acquired professional experience in state or private business administration. Two each came from the fields of librarianship, art, psychology, and law. One women had been a journalist. Four had worked as university-based natural scientists.[57]

In terms of age and marital status, the applicants' profiles corresponded with the membership structure of the university women's associations to which they had previously belonged. The 46 women who turned to the BFUW before October 1938 included only six younger than 30. More than half were in the 40–50 age-group, and five were over 50.[58] If one includes the correspondence of the AAUW, then the average age of the applicants for assistance shifts even further toward the older group. Of the 84 applicants whose age can be inferred from the American documentation, again more than one-third were between 40 and 50 years old, but in comparison to Britain the number of requests from 50–60-year-old women was noticeably higher, at 21. The great majority of those asking for help were unmarried, though this did not mean they had only to care for themselves. Some had children, while several aimed to emigrate together with their parents.

Only a few of the women were well-known scholars. True, the correspondence includes names like Lise Meitner, Marie Jahoda, Hedwig Hintze, Charlotte Leubuscher, Charlotte Bühler, Alice Salomon, and Helene Stöcker; a very few politicians, such as the former Social Democrat Reichstag delegate Adele Schreiber and the Czechoslovak social democrat Fanny Blatny, also applied to the network of academic women. But the great majority of the women who wrote to the BFUW are nowhere to be found in the reference works on emigration—even though, as their curricula vitae show, they had enjoyed successful and impressive careers in their home countries.[59]

Up to 1938, by far the majority of requests came from members of the DAB or its Austrian sister association, the Verband der Akademikerinnen Österreichs (VAÖ). After Germany's annexation of Austria in March 1938 and the pogrom that followed in November the same year, the office received increasing numbers of letters from academic women who had not themselves been IFUW members but who knew former members or other refugees who had received assistance. Their pleas indicate the rapidly worsening plight of persecuted women and men in Nazi-occupied Europe. The new circumstances gave rise to a de facto change in the IFUW, BFUW, and AAUW's refugee assistance that was never made official: whereas aid was offered first and foremost to members of the dissolved IFUW affiliates up to summer 1938, after that point it was no longer particularly important whether the applicants for assistance had been members or not. It was now sufficient for women to fulfill the general criteria for federation membership—that is, to have attained a degree or to have been prevented from doing so—in order for the assistance committees in London and Washington to regard them as belonging to the clientele eligible for allocations from the organizations' scarce resources.

When it came to rescuing respected and established members, the national associations in Britain and the United States and their refugee secretariats spared no effort, either before or after 1938, even when such cases involved special financial burdens and feats of organizational daring. In 1940, the AAUW arranged a highly complicated escape route through Portugal and Cuba for the former president of the Czech organization's German section, Käthe Spiegel—a venture which ultimately failed, as will be discussed in chapter 7.[60] The largest single allocation of funding from the British federation went to Johanna Hinrichsen, the DAB's last secretary, who had made contact with the BFUW through Marie-Elisabeth Lüders. Hinrichsen had come to Britain with her school-age son Peter, and the Croydon branch dedicated almost its entire refugee budget, £520, to pay Peter's boarding school fees. The longest-term support by the BFUW was for Helene Turnau of Frankfurt, a former DAB member of many years' standing. Turnau's plans to emigrate onward from Britain to the United States were frustrated in 1941. The BFUW granted her a weekly loan of £1 well into the 1950s after she was forced to give up her work as a teacher due to ill health.[61] Edith Mahler, an Austrian dentist and a member of the VAÖ's Vienna branch, owed her new professional start in England to the BFUW, which granted her £420.[62]

Another key factor was personal acquaintance. In this respect, the opportunities the IFUW had created for international networking through research funding, as well as the Crosby Hall clubhouse and the large, transnational membership conferences, had woven a web of connections that came into their own after 1933 as extremely valuable social capital. Photochemist Gertrud Kornfeld, for example, who received BFUW and AAUW funding for several years, owed her immigration into the United States to the director of the AAUW's International Relations Office, Esther Brunauer. The two women had met at the 1936 IFUW conference in Kraków, and Brunauer subsequently provided an affidavit guaranteeing Kornfeld's upkeep.[63] The archaeologist Margarete Bieber had been close friends with Virginia Gildersleeve, the founder and former president of the IFUW and an enthusiastic student of ancient Greece, since 1931. The women met when Bieber received that year's AAUW International Fellowship, and had maintained close contact from then on. In 1934, Gildersleeve arranged for Bieber to be invited as a visiting professor to Barnard College, where Gildersleeve was dean. Gildersleeve later helped her friend to secure a permanent appointment at Columbia University.[64]

In London, many connections and offers of help arose by way of Erna Hollitscher's old networks in Vienna. Hollitscher was in close touch with Hedwig Kuranda, who had been responsible for relations with the IFUW since the founding of the VAÖ (and had belonged to the association's democratic wing). It was Kuranda's direct contact with Hollitscher that enabled several Viennese women to escape to Britain. Before commencing her professional life in business, Hollitscher had been part of the small circle that clustered around the Romance philologist Elise Richter.[65] Her fellow student and best friend from those days, Leonie Spitzer, also received Hollitscher's help to enter Britain quickly in 1939.[66] Helene Adolf, another of Elise Richter's disciples, did not, in the end, need to make use of the "domestic permit" Hollitscher had obtained for her. She was able to emigrate sooner than expected to the United States, where she found accommodation with her sister Mona Spitzer in Philadelphia. A biochemist, Mona Spitzer had already emigrated in 1934 and was by that time a professor at Temple University.[67]

Between October 1938 and May 1939, 32 academic women from Germany, Austria, Czechoslovakia, Italy, and the Soviet Union succeeded in emigrating to Britain with the help of the BFUW; 75 more reached the country under their own steam but asked for help upon arrival. By April 1939, Hollitscher had found employment for 24 women, although a placement in "unpaid scientific work" was successful in only two cases. In the main, domestic or secretarial work was offered.[68]

In the period from summer 1938 to fall 1939, Crosby Hall took on a new role as well. Until 1938, the circle of émigré academics who were living at the residence remained small, with "hospitality" at reduced rates mainly offered for longer periods of some months at a time. But with the surge of refugees in fall 1938, the venerable building increasingly came to resemble a short-term reception center. During the fall, winter, and spring of 1938–39, 19 refugees were staying for around one week each at Crosby Hall, on the grounds that "every refugee on arrival needs a few days in London to register, make contacts, and try to get help for relations and friends." Because high rental prices in London made it extremely difficult to find private hospitality in the city, even a short stay at the residence was an invaluable help, especially as the Hall's staff could be relied on to provide sympathetic and competent advice.[69]

In addition to the BFUW's financial support for a relatively small number of women, the federation attached outstanding importance to efforts to assist refugees, otherwise cast completely on their own resources, in integrating into British society. Of prime relevance in this respect was Crosby Hall and the Hospitality Committee dedicated to looking after the Hall's Residential Fellows. As in the past, the committee was tasked with inviting the Crosby Hall guests to luncheons, teas, or dinners; presenting them to as many people as possible; buying them tickets to concerts and plays; and introducing them to the directors of schools and colleges, hospitals, libraries, and similar institutions.[70] Committee members also set up weekend "hospitalities" outside London so as to give the Fellows the chance to experience a "real English home."[71] For the women staying at Crosby Hall, the BFUW's clubhouse thus offered an excellent base—in both intellectual and social terms—from which to start their new life. This applied particularly to those women whose decision to leave Germany had already been made in the early 1930s. Emmy Klieneberger, for example, reminisced that she had experienced her dismissal in Germany and the resulting enforced emigration as a shock, but that "immigration itself, the achieving of a foothold in a new...country, the learning of a new language, the adaptation to another culture" had been a "great and satisfying enrichment" for her.[72] Her reception at Crosby Hall, Klieneberger stressed, had contributed crucially to this positive experience. It had been especially appealing to be able, right from the start, to live "in completely British surroundings," which had also benefited her progress in learning the language. During her time at Crosby Hall she made several close friendships, which played an important role in her daily London life when she moved into her own apartment in Chelsea in 1936.[73]

Certainly, Klieneberger was a particularly happy exception to the general rule in that, as well as enjoying the benefits of living at Crosby Hall, she managed to continue her scientific work almost seamlessly. However, gratitude also abounds in many letters penned by those arriving in Britain later, in 1938–39. For Erna Hollitscher, coming to Crosby Hall was a revelation:

> I cannot describe what it meant to me and other refugees when we were allowed to stay there, after the persecution and hatred we had undergone in "Greater Germany." In Crosby Hall we were not only tolerated, but welcomed, and we found an atmosphere of kindness and understanding which assured us that there was another world outside Nazi Germany in which we might be allowed to live freely, and perhaps happily. I feel sure that everyone who stayed in Crosby Hall felt that atmosphere, from whichever part of the world she came.[74]

For émigrés who did not live at Crosby Hall, the many events held in its clubrooms offered a popular opportunity for social intercourse. At the weekly "Friday Teas," women gathered in the relaxed atmosphere of the Hall's cozy salon, and the Hospitality Committee also arranged frequent receptions with concerts and lectures, often on topics related to Germany and the situation of the refugees, especially during 1933 and 1934.[75] Dinners, luncheons, and teas were organized for specific groups such as the German, Czech, or Polish refugees. Over the period from fall 1933 to spring 1934, around 150 British and more than 100 foreign guests attended the events at Crosby Hall. The guest lists show the regular participation of a firm circle of German academic women.[76]

Another much-valued Hospitality Committee institution was the "at home" offered by London members. "At homes," receptions or teas offered at regular intervals in the private homes of better-off members, were less formal than the meetings at Crosby Hall. Occasionally, a small musicale or a literary evening was organized. The elderly Alys Russell, whose determined efforts had played such an important part in the establishment of Crosby Hall, did not allow infirmity to prevent her from introducing the émigrés to the art of British conversation: she held court playfully from her sickbed.[77] "At homes" of this kind were an experience of both conviviality and social integration, offering émigré women the chance to connect with one another and BFUW members or to approach Erna Hollitscher personally for assistance.[78] In London, the BFUW succeeded in creating a network of social connections—one that, indeed, dissuaded more than a few emigrants from seeking work or accommodation outside the capital.[79]

That said, the BFUW succeeded in providing both assistance and hospitality outside of London as well. For example, the Nottingham branch kept up close contact with the Berlin physicist Gertrud Kornfeld, while Birmingham members awarded honorary membership to the 54-year-old Slavic literature specialist Helena von Reybekiel from Hamburg.[80] Even before the main influx of refugees began, several branches reported having benefited very immediately from their commitment to working for the émigrés. The campaigns to assist German academic women had considerably enlivened the branches' work at local level, they noted, and in some cases had also attracted new members to the groups.[81] The distinguished female scholars enriched the intellectual aspect of many clubs' activities by giving talks on their research. Dora Ilse, Betty Heimann, and Helen Rosenau, in particular, were popular and conscientious lecturers.[82]

With the increase in the number of refugees after fall 1938, almost every local BFUW group—after consulting with the London Committee on Refugees—offered "sponsorship," adopting one, two or, infrequently, up to ten refugees and sometimes their family members, from Germany, Austria, Czechoslovakia, and Poland. Branches in Birmingham, Bristol, Edinburgh, and Leicester organized local "at homes" for émigré women, collected money to fund small-scale support payments, looked for work or training places, assisted with onward emigration to the United States, and invited the refugees to social events. In Manchester in 1940, all the refugee academics were invited to become honorary members of the branch; in Croydon, the local BFUW Hospitality Committee, in cooperation with the town's refugee office, maintained a refugee hostel and organized language courses and concerts.[83] Members of the East Sussex, Oxford, and Cambridge branches offered private hospitality to a large number of refugee women, so that social contacts arose "of their own accord" both with BFUW members and among the émigrés themselves.[84] Above and beyond such branch activities, many members took on individual and personal commitments to look after refugees. High school teacher Gertrud Schlesinger, for example, mentioned "a charming afternoon in Croydon, for Miss Thompson drove us to the woods in her car and invited us to tea."[85] Several women who were waiting in Britain for a US visa and were taken in by BFUW members during this period retained grateful memories of their "sponsorships" in England. This was especially true of the women who subsequently faced far greater loneliness as immigrants to the United States.[86]

It is also worth noting that the BFUW women took care not only of their academic colleagues but also, to a considerable extent, of the women's families, a commitment not to be found in comparable form in other segments of British assistance for academic refugees. There appears to have been an unspoken agreement that women—even those who were active scholars or professionals—were more deeply tied to their family responsibilities than men. In the case of Dr. Gertrud Schüchterer, a chemist who found a position at the University of Birmingham, the local BFUW branch's offer of hospitality extended also to her mother, sister, brother-in-law, and four children.[87] As already noted, the BFUW in Croydon paid the costs of boarding school for Peter Hinrichsen, the son of the Weimar Republic's last DAB secretary, for many years; the Bradford branch supported the 12-year-old son of Else Hölzl in the same way.[88]

The outbreak of war in September 1939 brought about dramatic changes in assistance programs. An abrupt halt was called to all the BFUW's previously successful endeavors to help academic women leave the Continent for British shores. In all, between August 1938 and September 1939, the BFUW office had arranged for 53 academic women and 17 children to escape "Greater Germany." For 70 others, however, the outbreak of war thwarted all the British women's efforts to aid emigration.[89] The very last woman to receive the BFUW's assistance to emigrate was Hedwig Kuranda of Vienna. During 1938, Kuranda, the former president of the VAÖ's International Relations Committee,[90] had handled the foreign correspondence on visa matters for the Vienna Jewish Community. After that institution was dissolved, Kuranda's recommendations to the BFUW had still managed to help several colleagues escape to Britain.[91] Kuranda herself arrived in Oxford with her mother on September 2, 1939.[92]

Only a few weeks after the borders were closed, the Crosby Hall clubhouse had to be relinquished: the number of paying guests dropped so dramatically that the building's operating costs could no longer be covered. The BFUW offices relocated to Reading, while the IFUW continued its work from Chelsea. Even more than for the BFUW members themselves, the closure of Crosby Hall was a bitter loss for refugee assistance work and for the émigrés in London. "Many letters were received from university women exiles," notes the BFUW's 1940 annual report, "saying how happy they had been in its congenial atmosphere and how kind had been the Warden and sympathetic residents. They felt as if they had lost a second home and hoped for a re-opening at the earliest possible moment."[93]

From this point until the end of the war, the BFUW was compelled to do without its social and intellectual heart in London, even if in February 1942 the IFUW was able to return temporarily to Crosby Hall and to open the clubrooms once again.[94] Regular "at homes" were offered by London members as a substitute for the clubhouse hospitality.[95] For the women working at the BFUW office in Reading, wartime conditions brought added challenges. The evacuation of several members who had previously offered hospitality now put a temporary halt to the practice of free accommodation for refugees in BFUW members' homes—an arrangement that had not been easy even at the best of times. The internment of "enemy aliens" on the Isle of Man destroyed what had been promising new professional beginnings for many women, and necessitated continually new efforts to find work and housing as well as intensive personal reassurance for the beleaguered émigrés. Until the end of 1941, bureaucratic complications dogged efforts to assist women waiting for US visas. Only in December 1942 would conditions become more amenable as the war effort created an increased demand for labor. "Things are going quite well here as far as the Federation is concerned," Hollitscher was then able to write to Emma Reich, a Viennese physician with whom she had become friends in the course of Reich's emigration via England to the United States. "Nearly all our refugees have found more or less adequate work, rather more than less, only a few very difficult cases have not. All the doctors and dentists have been snapped up, and even our social workers have succeeded in finding jobs, so work here in the office is mostly quite cheering."[96] Until the end of the war, the focus of the BFUW's refugee work shifted to distributing large shipments of clothing sent by the Canadian federation, intervening in particular emergencies for the women for whom the branches or the refugee office held official responsibility as sponsors, and—increasingly—providing support for young emigrant women, whose studies were, in several cases, co-funded by the BFUW.[97]

Assistance across the World, 1939–45

With the closure of Britain's borders in fall 1939, the IFUW's activities took on a crucial role in supporting emigration from Germany, Austria, Poland, and Czechoslovakia, and later from France, the Soviet Union, Italy, and the Benelux countries. Shortly before the beginning of World War II, in August 1939, the IFUW held its eighth conference in Stockholm; aside from a 1940 regional meeting of the North and South American associations in Cuba, this was the last meeting of all IFUW delegates until 1946.[98] The central topic of the convention was supposed to have been career counseling and the employment market—but in view of the looming war, speeches and debates on the political situation predominated, culminating in a decision to concentrate on the IFUW's principle to unite "all the intellectual forces of women in forging links of understanding and goodwill."[99] The conference also agreed to vote on the resolution set out by the Council meeting in Budapest, the general debate on which had been postponed so many times. This resolution had been drafted by the Council in 1934 after talks with the new, National Socialist president of the DAB. It proposed that the IFUW's membership requirements be amended so that no national association could be admitted to, or remain part of, the international umbrella organization if it debarred women from membership in its own country on the grounds of "race," political opinion, or religion.[100] Several countries, including Norway, the Netherlands, Hungary, and the United States, had repeatedly voted against the introduction of a requirement of this nature or else had voted for the decision to be deferred. It was more pressing now, they argued, to preserve institutional links with members who were living in dictatorships and their spheres of influence.[101]

In Stockholm in the late summer of 1939, the majority of the delegates brushed aside such objections, showing both determination and a realistic evaluation of the situation. The president of the Polish federation observed that maintaining official links with Germany, Austria, Czechoslovakia, Italy, Spain, and Portugal had long since become impossible even without the new bylaw; the French delegate added that there was now an urgent need for the federation to stand up for its principles and stop dragging its feet. With the vigorous support of delegates from Britain, Sweden, Ireland, Yugoslavia, Belgium, and Switzerland, the incorporation of the Budapest resolution into the IFUW constitution was passed by a majority of 55 to 15.[102]

The election of the Polish medical scientist and university instructor Stanislawa Adamovicz as the new president of the IFUW was another clear signal from the 540 delegates, representing 25 countries, to demonstrate their solidarity with threatened Poland. The IFUW's choice meant that the federation lost contact with its president only two weeks after the Stockholm conference. Adamovicz had returned to Warsaw in August 1939, and although she survived the German invasion and occupation of Poland unharmed, she was unable to transmit messages across the border.[103] The remaining posts on the board were given to women from neutral nations: the economist and statistician Karin Kock from Sweden, the biologist and president of the Belgian federation Germaine Hannevart, and the Swiss chemist Jeanne Eder from Zurich.[104] Marguerite Bowie, from Britain, was reappointed as treasurer. This distribution of offices would soon prove to have been far-sighted: despite the myriad obstacles to international communication and the dissolution of most of the Continental European associations in the wake of German invasions, the IFUW's leadership remained capable of functioning via London, Stockholm, Zurich, and—through its former president Virginia Gildersleeve—New York.[105] This was of vital importance for refugee assistance throughout the war years.

Until the beginning of the war, the IFUW had delegated practical refugee work to the BFUW. Although an IFUW "Emergency Fund" had existed since 1936, it was only in July 1938, two months after the BFUW's Committee on Refugees first convened, that the IFUW set up a Committee for the Emergency Assistance of University Women to distribute the Emergency Fund's rather modest resources. From fall 1938 onward, a division of labor between the IFUW and BFUW committees gradually emerged: the IFUW attended to requests for help in emigrating to all non-European countries, whereas the

BFUW dealt primarily with women wishing to enter Britain, as well as coordinating the care of refugees who had already arrived. Responsibility for the initial selection of applications by German academic women for research grants was transferred from the BFUW to the IFUW's Emergency Committee.[106]

Up to this point, the IFUW itself had intervened only in exceptional cases and at the specific request of the British office. One such case was that of the learned sisters Elise and Helene Richter from Vienna. Elise Richter, a 73-year-old Romance philologist, and her elder sister Helene found themselves in a dire situation when Nazi Germany annexed Austria. Three years earlier, in 1935, the University of Vienna had already refused to grant Elise Richter, a former president of the VAÖ and a highly respected university teacher, the title of full professor—and thus a financially secure retirement—when she reached emeritus status. Immediately upon annexation, in March 1938, her permission to teach at the university was withdrawn, as was an ill-paid teaching contract she had held since 1923.[107] Because Elise Richter was an esteemed member of the IFUW, her treatment by the Germans soon became common knowledge. The Netherlands federation offered to take the sisters in, but Elise declined, making the much-cited comment: "Old trees can't be transplanted."[108] For some time, the Dutch women transferred monthly payments of 125 reichsmarks to help the sisters remain in their house in Vienna. In February 1939, the IFUW took over responsibility for the pension to the Richters.[109] The last payment was wired in February 1941; a year and a half later, the sisters were deported to Theresienstadt. Helene Richter was killed there a few weeks after her arrival on November 8, 1942, and Elise fell victim to the camp's appalling conditions on June 21, 1943.[110]

As well as practical help for particular individuals, the IFUW office worked with the Australian and the New Zealand associations to locate suitable candidates for emigration to Australasia, obtaining the necessary papers and paying the passage. Of 16 applicants to the Pacific, six succeeded in fulfilling the strict requirements of the Australian immigration authorities; another six managed to emigrate to New Zealand.[111] The IFUW forwarded a one-off payment to its members in the French city of Toulouse, as a way of supporting their spontaneous efforts to house 250 academic women and their families who had fled from Spain.[112]

An important component of refugee work in the IFUW's global network was the contribution of the Swiss association of university women SVA. The chemist Jeanne Eder, from Zurich, and the physician Mariette Schaetzel, from Geneva, were the chief protagonists of assistance for emigration and for refugees in Switzerland. They also operated in France. Among Eder and Schaetzel's projects in 1933 was their intervention to help the German experimental physicist Hildegard Stücklen, who had taken up an assistantship at the University of Zurich in 1925 and earned her qualification as a university teacher in 1931. In October 1933, Stücklen was threatened with being forced to return to Germany, whereupon Eder, via the dean, successfully applied to the immigration police for a reprieve, then recommended Stücklen for a fellowship from the Rockefeller Fund. Holding this fellowship meant that Stücklen was able to emigrate to the United States in 1934. Until 1939, she taught at Mount Holyoke women's college, where the president, Mary Woolley, served as president of the AAUW. In 1943, Stücklen was appointed professor at the women's college Sweet Briar, where she taught physics until her retirement in 1956.[113]

Jeanne Eder, who had grown up in New York, was president of the Swiss association. From 1939 to 1946, she held the office of third vice president of the IFUW. During World War II, she maintained communication with London and the United States. With the help of the president of the French association, Marie-Louise Puech, Eder's colleague Mariette Schaetzel rescued the Freiburg physician Else Liefmann, who was a founding member of the German association of medical women and had been part of the circle of active DAB members.[114] In 1941, together with Puech and locally active Quakers, Schaetzel managed to free Liefmann and her sister from Camp Gurs in the South of France and enabled them to enter Switzerland.[115] It was also through the efforts of Schaetzel and Puech that the 55-year-old medical scientist Eudoxie Bachrach received a $1,000 IFUW fellowship allowing her to remain in Geneva from 1944. Bachrach, an award-winning scientist and *maître de recherche*, had been expelled from her university position in Paris by the Germans. Since then she had received a small pension from the French association, which, however, was insufficient to cover the expense of living in Switzerland.[116] Similarly, the 72-year-old Helene Stöcker, now living in New York, was awarded a monthly supplement to her pension by the IFUW after pressure by the British-based American Alys Russell.[117]

Relying on America, 1936–45

The AAUW, too, made sustained efforts to aid displaced women scholars. During the war years, the AAUW became the financial backbone of international refugee work, relied upon by all the IFUW associations. Its activities were driven to a great extent by the director of the AAUW's International Relations Office, Esther Brunauer. As the story of Brunauer's endeavors reveals, it was often pure coincidence that determined how deeply the networks of women's academic internationalism outside Europe would engage in assistance for refugees. In 1925, Brunauer completed her doctorate at Stanford University on the German peace proposals during World War I, and in 1932–33 she took a sabbatical (funded by the Carl Schurz Foundation) from her position at the AAUW in order to revise her thesis in Berlin. As the guest of the university women of Berlin, Brunauer experienced the National Socialist takeover first-hand, and sent home reports of thought-provoking meetings in the apartment of physicist Lise Meitner, with whom she had stayed for some time as a guest. Brunauer's reports gave special emphasis to the date of February 27, 1933: she had spent that day consulting documents in the archives of the German Reichstag and the evening with other women scholars at her hostess's home. It was during this gathering that the news arrived first of Gertrud Bäumer's dismissal from the Prussian civil service and then, a few hours later, that the Reichstag had gone up in flames.[118] The impact of these events prompted Brunauer to change the topic of her research. The documents Brunauer had planned to work on were now no longer available, and the dramatic political upheavals encouraged her to begin following the events currently unfolding in Germany. She spent the rest of her sabbatical examining the structure and strategies of the Nazi Party and interviewing leading figures, culminating in a short conversation with Hitler himself.[119]

Brunauer's considerable knowledge of Germany and the many personal contacts she had established during her year in Berlin were key factors in ensuring that AAUW members remained well informed on the details and context of the persecution of Jewish university women. Brunauer's experiences were recorded in numerous articles for the association's journal and in information brochures distributed to the AAUW's local branches.[120] The interest of many association members in learning about events in Germany from an eyewitness was so great that, on her return to the United States, Brunauer was soon overwhelmed with lecture invitations.[121] Just as important as her policy of disseminating information among members was Brunauer's ability to impress vividly upon the AAUW's leadership the alarming situation of the persecuted German academic women, many of whom she knew personally.[122] Brunauer dedicated great energy to the difficult task of establishing an effective program of assistance that could be carried out by the AAUW's International Relations Office. Her project was not made any easier by the American public's widespread reluctance to countenance a more generous refugee policy. Added hindrances included the State Department's refusal to accept the seriousness of the Nazi campaign of persecution and destruction, together with the increasingly restrictive attitude of the immigration authorities toward refugees from Europe.[123]

Until 1940, the AAUW's International Relations Office operated, under Brunauer's leadership, with no budget for refugee assistance. Even so, the office managed—by passing on contacts and affidavits and corresponding with members, public officials, colleges, and a wide range of professional bodies—to clear the path for the association to aid persecuted women scholars in reaching the United States and starting a new life there. Only very rarely did efforts to find the refugee scholars positions at American women's colleges succeed: the many placement negotiations in which the AAUW played a part yielded a mere five appointments.[124] Still, in the United States just as in Britain, individual members stepped into the breach, offering their private assets when a limited sum was needed to surmount an immediate crisis. The wealthy historian Shirley Farr, who, as editor of the *American Historical Review*, had spent time in the AAUW's Washington residence in the early 1920s and later became AAUW vice president, forwarded several substantial remittances from Vermont to Washington.[125] In the case of the Hungarian chemist Elizabeth Roboz, for example, Farr paid for a train ticket from New York to Stockton, California, where Roboz had found a job in the laboratory of a small factory. Farr also loaned Roboz money to cover her living expenses when the start of her employment was delayed.[126]

From 1938 to 1944, refugee work laid claim to the lion's share of the International Relations Office's time. Parallel to the BFUW's efforts, the office's extensive correspondence offers insights into the many difficulties confronting women who were trying to emigrate or had just arrived in the United States. In 1940, the increasingly desperate refugee situation, along with the IFUW's urgent request from London that the focus of organized refugee work be relocated to the United States,[127] convinced the AAUW's national convention to adopt a resolution committing it to "provide all aid possible to resist the totalitarian aggression."[128] On this basis, the American women set up a War Relief Committee that shouldered responsibility for refugee assistance. At the Committee's disposal was a newly established fund, which had attracted more than $30,000 in member donations by the time the United States entered the war in December 1941. Only a fraction was spent on refugee assistance within the United States. In the subsequent five years, $12,000 went to Britain, $5,000 to Palestine, $2,000 to Switzerland, $1,000 each to Canada and Sweden, and smaller sums to other important stages in the European women's escape routes: parts of France, Lisbon, and Shanghai. A total of $8,000 in small grants and interest-free loans was awarded to refugees in the United States itself.[129]

When set against the disastrous plight of so many colleagues in Germany and occupied Europe, the impact of the financial assistance provided by the organizations of university women as a whole remained limited. Yet the importance of the female academic network cannot be judged solely in financial terms—the efficiency of its communications structures must also be taken into account, as I would now like to demonstrate through the case of the dramatic rescue from Wrocław (then Breslau) of Hedwig Kohn, the only fully qualified university physicist left in Germany in 1939.[130]

The initiator and driving force behind the effort to help Kohn emigrate was Rudolf Ladenburg, Kohn's friend from Wrocław and her colleague of many years' standing. Ladenburg had been appointed to a professorship at Princeton in 1931. An active contributor to organized academic assistance for refugees, Ladenburg traveled personally to Washington in early January 1939 to meet Brunauer and outline the desperate situation in which Kohn found herself. Robbed of all means of supporting herself, and without any relatives abroad, the physicist was, Ladenburg reported, entirely dependent on assistance from colleagues and academic organizations. Because Kohn was not well known in international specialist circles, Ladenburg considered it unlikely that the large academic assistance organizations and universities would take up her cause. Her research on illumination and radar technology was, however, potentially useful to industry. Ladenburg stressed that Kohn would be able to exploit her research to the benefit of the United States, either at a good college or in business. He himself was willing to provide an affidavit and cover Kohn's living expenses, but he could not hold out hopes of employment at Princeton or of the visa that depended on such a job offer—at that time, Princeton appointed only men to academic positions.

Directly after their conversation, Brunauer contacted Erna Hollitscher and Erica Holmes, the IFUW's secretary in London. Both women immediately took up Kohn's case. Through a BFUW member, Hollitscher managed to obtain a grant for Kohn and an assurance by a Scottish physicist that he would employ her as a research associate in Aberdeen for as long as it took for her US visa to be issued. The funds necessary for this purpose were supplied by Dutch university women, the IFUW, and the German Scientist Relief Fund that Ladenburg had helped set up.[131]

In mid-August 1939, Kohn's emigration via Britain seemed assured, but her carefully laid plans, like those of so many other would-be émigrés, were shattered by the outbreak of World War II. What ensued between September 1939 and summer 1940 was a convoluted and hectic search for a means to enable the 52-year-old physicist to flee some other way. All too aware of the highly complicated and constantly changing immigration regulations, Kohn's helpers agreed on a route via Sweden. In May 1940, the economist Karin Kock, IFUW vice president and lecturer at the University of Stockholm, together with Lise Meitner, who had emigrated to Stockholm in 1938 (like Kohn, Meitner was a physicist, and she had been an active member of the DAB with a place on the IFUW's Fellowships Committee), succeeded in obtaining a temporary Swedish visa for Kohn. It remained more than doubtful, however, whether Kohn would be able to escape Germany by this means. The Swedish temporary visa only permitted entry to the country if it could be demonstrated that the migrant's living expenses were fully covered right up to the day of departure for his or her ultimate destination—unfortunately, the funds available were only enough to support Kohn in Sweden for less than one year. Because it was certain to take longer than that to obtain a visa for the United States via regular channels, she would need a specific invitation by a US college or university, allowing her to enter the United States as a teacher, outside the normal quota arrangements, and thus to leave Sweden within a year.

Procuring an academic position for women in advance of their arrival had proved virtually impossible for the AAUW even before 1938. Accordingly, Brunauer took a gloomy view of Kohn's chances of obtaining one of the sought-after "non-quota" visas. But when, in late May 1940, Ladenburg telegraphed the news that Kohn had been threatened with deportation from Wrocław, Brunauer pulled out all the stops. She sent out an urgent appeal to the solidarity of the Seven Sisters colleges, supported by a copy of Ladenburg's telegram. The colleges' reaction shows the level of resources on which the AAUW, through its members, was able to draw in an emergency. Two colleges agreed, saving Kohn's life. Meta Glass, the president of Sweet Briar College, answered Brunauer's call for help by return of post, sending an invitation for Kohn to teach at the college for the academic year 1941–42. Glass had been president of the AAUW from 1933 to 1937, had attended the dramatic IFUW meetings in Budapest, Kraków, and London, and also knew Brunauer personally. By deciding to offer Kohn a position, Glass was circumventing the college's decision-making bodies and disregarding its own priorities. Sweet Briar did not need a second physics teacher, she told Brunauer, and Kohn would certainly be better off at a research-based university. She herself was, however, willing to do everything to assist in fulfilling the immigration formalities, and guaranteed to cover Kohn's living expenses for the duration of one year should no more suitable situation be found for her.[132] A few days later, Mildred McAfee, the president of Wellesley College and another active AAUW member, agreed to help as long as she could be assured that Kohn was a "bona fide refugee" and not a spy.[133] The fact that Kohn finally found herself neither at Sweet Briar nor at Wellesley was due to the efforts of fellow physicist Hertha Sponer, who had obtained a professorship at Duke University in North Carolina in 1936, thanks not least to AAUW support. Prompted by Ladenburg, Sponer argued doggedly, and ultimately successfully, for Kohn to be given a one-year contract for 1940–41 at the women's college of the University of North Carolina in Greensboro. On the basis of these three invitations—the fruit of well over a hundred letters and telegrams between Ladenburg, Brunauer, Hollitscher, Kock, Meitner, Sponer, Glass, and others on both sides of the Atlantic—the persecuted physicist was able to leave Germany in July 1940 and enter Sweden. After a long and arduous journey across Eurasia, and subsequently a serious illness, Kohn took up her post at Greensboro College in January 1941.

The commitment to refugee assistance by the associations of university women was not without consequences for attitudes and policies at home. For American women, in particular, working for the international community of academic women resulted in a changed political stance. As early as 1938, the AAUW became the first of the American "internationalist" organizations, and the first large-scale women's organization, to take a stand against the United States' unconditional neutrality. Starting in 1940, the AAUW called for military intervention in Europe.[134] If in 1919 the American women had hoped to make a lasting contribution to the preservation of peace through the ideal of female academic internationalism, circumstances now caused them to take leave of the idea of national neutrality. Their advocacy for US entry became more and more determined as events showed ever more clearly that political principles were ineffective weapons in the fight against Nazi expansion. As they knew first hand, the increasingly arduous work to assist refugees was powerless to counter the growing misery of academic women in German-occupied Europe—not least because of the many obstacles raised by their own government.

7
Marked by Persecution

Thanks to the rich archival material in London and Washington, it is possible not only to reconstruct the IFUW network's assistance to refugees, as I did in chapter 6, but also to trace the individual destinies of the university women who faced exclusion, dismissal, and exposure to persecution after the German university women's organization DAB aligned itself with the Nazi regime. Without the continuity of the international organization outside Germany, the research I present here would have been impossible—because the DAB's Jewish members disappeared almost without trace from the German sources immediately after its Nazification. Fortunately, the extensive correspondence of the BFUW, AAUW, and IFUW contains a wealth of information on how the marginalized and persecuted former DAB members in Germany and occupied Europe responded to the traumatic loss of their rights.

Between 1933 and 1945, around 500 persecuted university women approached the BFUW and IFUW for help; a further 140 contacted the AAUW's War Relief Committee in Washington. In this chapter, I ask how they assessed their own situation, what solutions they sought, how they rebuilt their lives in exile, what hopes they placed in their colleagues abroad, and how far those colleagues were willing and able to fulfill such hopes.

These questions allow a fresh discussion of some key hypotheses of exile studies. For example, a focus on married women and mothers has often led to sweeping assumptions that all female émigrés found it easier than men to adapt to their forced emigration from Germany, because they had less to lose in professional and social terms.[1] It was women's alleged family orientation, rather than their professional prospects, that motivated them either to leave Germany or to stay as long as possible despite the rapidly worsening situation; for similar reasons, the argument continues, women in exile found it less difficult than their husbands or male colleagues to handle the move to badly paid physical labor. In the still rare studies of émigré graduate women of various professions and academic fields, one hypothesis is that emigration opened up new professional opportunities for younger women that they would not have enjoyed to the same extent in Germany.[2] Older women, already better established in their profession or academic domain, found the new beginning considerably more difficult than did men: firstly because their employment opportunities were more limited, and secondly because their academic careers were often neglected "out of consideration for their family ties and responsibilities, or at least were far more strongly shaped by their family background" than was the case for men. For women, "professional and family identity" were "evidently much more closely linked."[3]

I will address these issues in four parts. The first section looks at the processes of deciding whether and when to initiate emigration. In the second, I examine the initial years—usually the most difficult ones—of the new beginning abroad, asking how women academic refugees coped with loss of status, material insecurity, and hard physical labor and how they managed the process of social and cultural assimilation. The third section considers how far and why certain émigré women succeeded in picking up their previous careers over the long term; the fourth is dedicated to those members of the international network who, despite all efforts, did not manage to escape from Germany and its sphere of power in Continental Europe.

To Emigrate or Not to Emigrate?

A few Jewish graduate women began to think about moving their professional or academic activities abroad even before the Nazis expelled them from their jobs. Such deliberations were not a purely private affair: they were also discussed within the DAB. In her memoirs, microbiologist Emmy Klieneberger recounts having considered emigration as early as March 1933, before the wave of dismissals of Jewish professors began: "In the first weeks of the Hitler regime I would gladly have left the country at once. My work was the greatest factor in my life and I was just 40 years old; and I knew there were no further possibilities for me in Germany." It seems that the DAB leadership, under Marie-Elisabeth Lüders until her resignation in May 1933, kept up a dialogue on this question with the Jewish members who were threatened by dismissal and exclusion. Klieneberger, at least, took time off in spring 1933 to travel from Frankfurt to Berlin for a week "for consultation with older members" of the federation.[4] The DAB women advised her "not to leave before I was dismissed lest people abroad should not believe that I could no longer practise my profession in Germany."[5] This cautiously strategic counsel is typical of how the last democratically elected DAB executive tried to deal with the Nazi regime; it also shows how carefully the options open to Jewish members were weighed and explored. Klieneberger followed her DAB colleagues' advice. She waited for her dismissal in September 1933 and then, as described in chapter 6, went directly to London, where she immediately found a home in Crosby Hall and fresh opportunities for work—albeit unpaid at first.

It is difficult to know how many other Jewish members sought advice from the DAB executive in this way. Marie-Elisabeth Lüders and Agnes von Zahn-Harnack, certainly, remained in close contact with colleagues affected by dismissal, exclusion, and persecution well beyond the dictatorship's early days. They reported on the activities of the IFUW, BFUW, and AAUW, and passed on contact addresses. It was on Agnes von Zahn-Harnack's advice, for example, that Susanne Engelmann turned to Esther Brunauer in Washington in fall 1934.[6] Engelmann, just 47 years old and the principal of Berlin's first girls' high school, the Viktoria-Oberlyzeum, had been forced into "retirement" one year previously. She had spent 1912/13 as a German Scholar at Bryn Mawr College, and hoped she could use her connections from that stay to ease her move to the United States. Since being driven out of the state education service, Engelmann had been giving private lessons in literature, psychology, and educational theory in Berlin, but her 1934 letter to Brunauer told of her increasingly distressing sense "of being cut off from my work twenty years earlier than might have been expected in a regular course of life." She added that emigration was only an option for her "if a position as a professor, a lecturer or an instructor at a college could be found for me."[7] However, even in the early 1930s, only Germany's topmost scientific elite could expect to obtain such work without being able to apply in person, and Engelmann was not prepared to emigrate "on spec." In Germany, she was still entitled to a small pension, and unlike Klieneberger, whose unmarried sister remained in Germany with their mother, Engelmann herself lived with her elderly mother and did not want to leave her alone.

For many academic women, responsibility for elderly parents was a compelling reason to defer emigration or else seek a way to emigrate together—thus making it more difficult, and often impossible, to escape. There are numerous examples of this.[8] One of them is Käthe Spiegel, a historian and librarian at the national and university library in Prague. She put off applying for a US visa until 1939, aged 42, when the doctors told her that her ailing mother would not live for much longer. The physician Johanna Maas, a long-time member of the DAB's Karlsruhe group, wrote the BFUW in March 1939 that she was seeking a post as a doctor in an English retirement home: her plan was to continue her work while staying close to her mother, whom she did not want to leave alone in Germany.[9] Lucie Adelsberger, who had worked at the Robert Koch Institute and been an active member of the German women physicians' association until 1933, was offered a post as lecturer in bacteriology at Harvard University during a short visit to the United States in November 1938. Although she could easily have remained in the United States, concern for her aged mother took her home even after the vicious "Kristallnacht" pogrom in Germany that month. Back in Berlin, she tried to organize official entry to the United States for the two of them.[10]

Solicitude for their parents, and especially for widowed mothers, was central to the lives of German academic women, the majority of whom were unmarried. Ultimately, though, the realization that they no longer had a professional future in Germany prompted the decision to emigrate. Faced with the irreversible collapse of all their professional prospects, they began arranging their emigration—if possible accompanied by their elderly parents, but after the November 1938 pogrom if necessary without them, at least for the time being. Käthe Spiegel from Prague contacted the AAUW for help with the practicalities of a move to the United States in 1940, when her mother had died and she had lost her job once and for all.[11] Although Johanna Maas from

Karlsruhe was still permitted to practice as a doctor as long as all her patients were Jewish, she anticipated that this livelihood would not be secure much longer in view of the rapidly advancing destruction of the Jewish community. This was her paramount reason for emigrating.[12] Lucie Adelsberger, too, still had a modest professional and economic base in Berlin when she returned from the United States in November 1938: like Maas, she was among the Jewish doctors who were still permitted to treat a restricted circle of patients under the new anti-Semitic regulations.[13]

As these cases indicate, the Jewish "subculture" produced by social exclusion in Germany at first created opportunities for women to remain active in their profession beyond 1933. This was true not only for women doctors, but also and especially for former university lecturers and schoolteachers. As Jewish students were expelled from mainstream education, a separate Jewish schooling and training system mushroomed, and working in this system presented an acceptable alternative to emigration for both younger and older women until 1939. In fact, the demand for female teaching staff probably even rose as male teachers and lecturers emigrated. Women moved up into positions left empty by their male colleagues.

Alice Apt, for example, managed to complete her PhD on Caroline Schlegel and the society of early Romanticism in 1936, aged 26, in Königsberg, then found a teaching post at the Jewish adult education institute in Dresden. She escaped to Britain in summer 1939 with a domestic permit.[14] Ruth Ehrmann was employed at a respected Jewish private school, the Waldschule Kaliski in Berlin's Grunewald suburb, after earning her PhD in English literature at the University of Basel in 1933. She taught in Grunewald until 1938 with great success: graduates of the Jewish teacher training institute were shown her lessons as a model to be followed. Ehrmann had been considering emigration for a long time, and, from 1934, she took private Spanish lessons to prepare for a possible future outside the English-speaking world. But she only began to organize the move in June 1938, when the opportunity arose to get her mother safely to Paris with her brother and sister-in-law and to take up a school-teaching position in Britain. Ehrmann entered Britain on a domestic permit with the help of the BFUW, and taught refugees from Germany at a girls' boarding school in Bristol.[15]

Those women who firmly defined themselves as "non-Aryan Germans," and accordingly kept aloof from Jewish educational work, had more difficulty in continuing to teach in Germany after 1933. The high school principal Susanne Engelmann was a staunch Protestant and a member of Martin Niemöller's Dahlem congregation of the Confessing Church. She had to depend on private tutoring from 1933 until 1935, when she was appointed to manage adult education at the Paulusbund (St. Paul's League) in Berlin. This "union of non-Aryan Christians" expressly distanced itself from "Jewish" cultural work, offering classes for "non-Aryan" members of the Christian churches.[16] When the Paulusbund expelled all those of its members defined by the Nuremberg race laws as "full Jews," Engelmann too lost her job.[17] She began to offer preparatory courses for the Cambridge University language examinations and private English classes for the people euphemistically designated "willing to emigrate."[18] In November 1938, she applied to the AAUW again, this time with great urgency, to assist her in emigrating and finding a teaching position in the United States.[19] "I am in very great need of help now," she wrote, "as I am, though a Protestant, considered and treated as a Jew in this country, and am about to lose my and my old mother's home in the near future, besides being deprived of the possibility of scientific or educational work."[20] With the help of her brother, who had been working in the Ministry of Economics in Turkey since 1936, Engelmann managed to escape with her mother to Istanbul in February 1940. There, she offered courses in educational psychology at the YWCA Social Service Center and tutoring in German literature for the children of exiled professors at Istanbul University. After her mother, Martha Engelmann, died in Istanbul in June 1940,[21] Susanne Engelmann obtained a US visa. Traveling via Russia, Siberia, Manchuria, and Japan, she reached Palo Alto, California, in winter 1941.[22]

In terms of the persecuted women's professional identities, it is revealing to note how few of them moved into a different field as a way of preparing for a nonacademic life in exile. Only for art historian Adelheid Heimann, who specialized in medieval iconography, is there evidence of this kind of reorientation. Heimann earned her PhD in Hamburg in 1930 and worked there without pay with the art historian Erwin Panofsky. After a period teaching art history at the Sorbonne in Paris from 1933 to 1935, she returned to Germany temporarily to train as a photographer in Berlin, in preparation for her emigration to Britain in 1936.[23]

For as long as a choice of destinations was still available, the majority of women preferred to remain in Europe. Almost all of them, unlike the majority of their male colleagues, spoke at least one foreign language—often not just English, but also French and in many cases Italian. Yet outside Europe, they favored the English-speaking world, including Australia, New Zealand, and North America. The United States was the first choice of those women who already knew the country, such as the former Bryn Mawr scholar Engelmann or the Prague historian Spiegel, a specialist in US history who had spent one year in Washington, DC, as a Rockefeller Fellow in the late 1920s and had met many AAUW members personally. Family connections also played an important role, as in the case of the psychologist and economist Erna Barschak, whose sisters had already emigrated to the United States. In some cases—those in which emigration was regarded as a professional opportunity—intellectual curiosity about the United States was also a motivation. Thirty-seven-year-old Auguste Jellinek from Vienna, for example, was a trained opera singer but also a scientist, who had specialized in children's hearing and language disorders in Rome at the University's otology clinic after 1933.[24] Forced to emigrate again when Italy introduced its own anti-Semitic legislation, in September 1938 she told the AAUW: "Being a Jew, I cannot return to Vienna and no other European state will allow me to work there. Here in Italy it is absolutely impossible to get in touch with the big international organisations created for Jewish emigrants. But since many years I wanted to go to America and I am glad that it comes to it now."[25]

The New Beginning

Auguste Jellinek reached the United States within a few weeks of this letter, thanks to Brunauer's intervention with the US consulate in Naples, and managed to pick up her scientific career where she had left off. By December 1938, she was already working in St. Louis at a well-established and pioneering institution in her field, the Central Institute for the Deaf. Its founder, Max Aaron Goldstein, had studied in Vienna with Jellinek's own teachers. A year later, the international networks of otology also came to the aid of Jellinek's mentor Emil Froeschels, who initially emigrated from Vienna to the Netherlands: at her instigation, he arrived at the Central Institute in 1939.[26]

As a rule, university women who left Germany and its growing sphere of power in 1938 or later faced far less promising career prospects than earlier émigrés. Correspondence with the BFUW, IFUW, and AAUW refugee offices in Britain and the United States shows how difficult these women found it to overcome the loss of their profession, as both an intellectual activity and a social environment. Despite having already been deprived of their status in Nazi Germany, many had not anticipated how hard the complete collapse of social privilege would hit them in exile. Even in early 1939, just before arriving in Britain, Johanna Hinrichsen—a PhD in economics and the last secretary of the Weimar DAB executive—told the BFUW that she would bring with her not only the complete furnishings of a small Jewish children's home she had been running since 1933, but also her own furniture, including some valuable Empire and Baroque items, a Persian rug, a good Rembrandt copy, and more. This plan met with a scathing response. The BFUW's Croydon branch accepted only a few bunk beds for the Croydon children's refugee home that Hinrichsen was to direct. There was, wrote her sponsor, absolutely no space for such bourgeois luxuries as elegant furniture. Hinrichsen would have to make do with one room and at most be able to hang her pictures there. "She will find life very hard if she has been used to such a grand home. We can only rightly give our refugees the scale of living that we give our own distressed people—working men's standard, not middle class. I hope she is preparing her mind for this."[27]

Alongside the bitter experience of reducing their personal possessions to the contents of a few suitcases, another serious problem in Britain was the employment restriction to domestic labor that the immigration regulations imposed on women refugees. The women writing to the BFUW and AAUW carefully stressed not only their professional qualifications but also their domestic skills—yet actually having to work in a private home and satisfy the demands of British or American employers thrust many into deep depression. Most of them were probably accustomed to delegating their own household responsibilities to domestic staff. Gertrud Schlesinger found a place in a Surrey home with BFUW assistance after leaving Berlin, but she was deeply mortified at having to serve tea to her "mistress" and the Viennese physician Susanne Jahoda—staying in the same house—while she herself had to sit in the kitchen with the charwoman. After only three weeks, she confessed to Hollitscher that she was "quite desperate, because I see no prospect at all of doing anything other than this soul-destroying housework; it's wretched, and even purely physically I will not be able to stand it in the long term."[28] Not every émigré woman will have dared to turn to the BFUW with such concerns, but there can be no doubt that the restriction to domestic labor imposed by immigration rules (especially in Britain) and the resulting frequent job changes amounted to a stressful rupture in most of the graduate women's lives. As other letters confirm, it posed challenges they certainly did not find easy to overcome.[29]

Despite the limited funds at their disposal, the BFUW and AAUW tried hard to offer at least occasional assistance, especially in the difficult initial period. Often, that help was crucially important for the émigré women, both in the short term and for their future professional careers. In some cases, the BFUW gave its protégées the chance to improve their housekeeping skills, thus keeping down the costs of the sponsor's own duty of maintenance. It paid for the young philologist Hildegard Rosemann to attend a course in domestic science at the sanatorium of Homerton College in Cambridge after her miserable failure in her first housekeeping job.[30] The Manchester branch sprang to the aid of 26-year-old Viennese doctor Nina Bleiberg, rescuing her from virtual enslavement as a nanny. The women offered her free accommodation and paid her costs for a course in midwifery.[31] Probably less than happy with this solution, Bleiberg—a specialist in speech disorders—dropped out of the course as soon as she received her US visa, but she continued to enjoy the Manchester BFUW's hospitality until she set sail in summer 1940.[32]

The initiation phase for refugees differed between the United States and Britain, as becomes clear in the cases of women who moved to America after staying in Britain for some time. The physician Emma Reich from Vienna viewed the imminent move with some trepidation. She and her husband had been cared for by their British sponsor with an enthusiastic attentiveness that had sometimes felt suffocating; they hoped the United States would give them the chance to live independently again in both professional and personal terms. Nevertheless, America seemed a little "suspect" in terms of its treatment of refugees, of its "charity and friendliness."[33] Several colleagues reported success in finding their way back into their careers, and said they felt far less alien in the United States than they had in Britain. However, although these optimistic messages were reassuring, they did not always reflect reality.

The difficulties, sometimes overwhelming ones, that academic women faced when they arrived in the United States are illustrated by a collection of pioneering case studies on women lawyers, psychologists, and social workers and on women doctors who had been involved in abortion reform during the Weimar Republic.[34] Lawyers and highly specialized physicians faced the greatest adversity in reestablishing their careers, but these studies show that the majority of the academically trained émigré women had to watch their hopes of reattaining their professional status fade away, obscured by the inexorable need to secure their immediate survival through housework, cleaning, or heavy manual labor. For many, the grueling first years in the United States were marked by physical exhaustion, the fear of total impoverishment, self-doubt and worries about their relatives, feelings of loneliness and alienation, and the painful realization that they would have to completely reinvent their lives.[35]

In contrast to Britain, in the United States it was rare for close and cordial contact to arise between AAUW members and the refugees. The BFUW gave advance warning of its protégées' arrival in America, and the New York City branch's Hospitality Committee, especially, always took care to welcome the newcomers, invite them to tea, and introduce them to refugees already living in New York. If they had experienced and appreciated the care of the BFUW branches in Britain, the new arrivals themselves immediately sought contact with their local AAUW branch. Yet, with a few exceptions, no American city offered an openness and integration comparable to that in London and other large British cities. For many of the new refugees, the only reliable, considerate, and welcoming contact person was Esther Brunauer, the secretary of the AAUW International Relations Committee. Attempts to network more widely often petered out or never got off the ground, as in the case of Friederike Fleischer in Boston and Elisabeth Anrod in Washington, DC.[36] Katharina Flesch from Vienna, a PhD in classics and an experienced social worker, told Brunauer in 1941 that she had not maintained her contacts with the AAUW branch in New York City after her initial enthusiasm because, as a packager in a factory, she had felt out of place among the American women, as had her husband. This one-time successful businessman was now a leather worker, having failed in his attempt to retrain as a butler. The university women of New York, wrote Flesch, had shown little appreciation of the couple's situation, though one branch member had invited them home and kept up close contact since then. Flesch regarded this acquaintance as a happy exception in an otherwise bitterly disappointing picture: "I think every refugee who comes to this country without any means whatsoever is delighted to meet at least a few people who judge by his background and by his attitude if he is 'commensable.'"[37]

Flesch, then, interpreted the American women's difficulties in dealing with their refugee sisters as a problem of social acceptability in a cultural or material sense. The Americans themselves did not share this view. Instead, the new arrivals were measured in terms of their capacity to shield their hosts from the desperate difficulties of their situation. This becomes obvious in the worry that the Dutch doctoral candidate C. R. Meibergen raised at one of the AAUW's New York afternoon teas in 1942. Meibergen wanted to complete her doctoral dissertation, begun in the Netherlands, in New York. The unaccustomed overheating in New York rooms was preventing her from writing, and she asked if there would be any chance of temporary hospitality in New England. The AAUW Committee on Refugee Aid unanimously adjudged Meibergen's request a clear case of self-pity, a "typical refugee attitude." Shirley Farr, the committee's chair, put the American position in a nutshell: "It certainly is unfortunate that the refugees do assume the attitude of self-pity.... I wish somehow we could convey to them as a part of their initial American training, that Americans would probably be sorrier for them if they did not do it for themselves."[38]

The incisive, self-deprecating, and witty study by psychologist Erna Barschak of her own assimilation in the United States confirms this kind of expectation and adds further facets. In her 1945 book *My American Adventure*, Barschak explains with perspicuity why many academic women from Europe found it so difficult to make social contacts in the United States outside of émigré circles.[39] She describes her own acclimatization as an odyssey through the New York offices of the refugee organizations and job agencies. Her attempts to follow well-meaning advice on standing out in job interviews—as a "charming personality" with freshly set hair, varnished nails, and her skirt taken up an inch or two—give striking insights into the world of American white-collar employees and the very different forms of femininity familiar to German academic women, who were only partially able to adapt.[40] They were not used to a "pleasant appearance" playing a role in a professional setting, nor (with some exceptions) did they have the experience or energy to match the American women's adroit, convivial demeanor at sociable luncheons or teas.

Barschak's book addresses another troublesome gap in expectations, one that is also mirrored in the AAUW correspondence and sometimes had a substantial negative impact on the refugees' employment opportunities. Whereas the European women were anxious to recommence their previous careers, their American hosts insisted on a high degree of pragmatism. This applied to immigrants in general, but especially to refugees from the Nazi regime—and especially during wartime. The highest precept for a new beginning was held to be the newcomers' determination to stand on their own two feet as soon as possible, even if that meant relinquishing the idea of working in their own domain or a similarly qualified one. Barschak's memoirs reflect this stipulation in the shape of harrowing doubts: "Had I already forgotten the persecutions, the humiliations, the anxieties in Nazi Germany...?...What claim had I to ask for a 'better job,' a job in my own line? Only the idea that I could do better in an intellectual job?"[41]

In her book, Barschak proposes a solution to this dilemma that reads like a sure formula for American crisis management. She identifies the turning point in her refugee fortunes as an interview with the Oberlaender Trust, one of the organizations that cooperated with the Institute of International Education's Emergency Committee in Aid of Displaced Scholars in New York to support émigré academics through fellowships, subsidies for publishing fees, or—ideally—by finding them positions in a college or university. Asked by the director how she could best be helped, Barschak skillfully replied: "I need to learn your ways of doing things, your behavior. I need, in other words, American experience. I need a chance to learn! And this chance I want. No fellowship, no research grant!"[42] Won over by Barschak's enthusiastic manifesto, her interviewer made the redemptive phone call that opened her way into a university teaching job.[43]

For the AAUW, the prime objective of financial assistance to the refugee women was to help them find their economic feet fast. Suitable means might be a loan to open a medical practice or a grant toward a practically oriented college degree. At times, this pragmatic approach had the effect of further cementing women's anyway difficult situation on the academic employment market. This is illustrated by the case of a Berlin specialist in sexual medicine, the Vienna-born doctor Sidonie Fürst. Among other things, Fürst had built a reputation in male contraception research at Magnus Hirschfeld's Institute of Sexology in Berlin. When she arrived in New York in 1940, she complained in a letter to the AAUW that the National Refugee Service would fund only her husband to prepare for the US medical licensing examination; the Service argued that it was enough if the husband in a family was enabled to become the breadwinner. She herself had to work in Margaret Sanger's birth control clinic by day and as a nurse by night, leaving her little time to prepare for her own examinations. She failed them, and when she applied for support to help her try again without such an exhausting workload, the AAUW refused, citing her age and the failure of her first attempt. Fürst was sent a small donation from Vermont, but it was much too small to be of any real assistance.[44] A similarly unforthcoming response faced women who made a second application after receiving assistance in the past.

In Britain, in contrast, women who had already received BFUW help, whether established academics or not, could expect further assistance if special hardship arose. There was also more sympathy for the refugees' desire to continue their previous academic and professional careers if at all possible. As in Adelheid Heimann's case, short-term grants were awarded to enable a woman to seek a position in her own field instead of driving herself to the edge of exhaustion with domestic labor. This different emphasis in the BFUW's refugee care, especially toward scholars in the humanities, is indicated by the support given to the Viennese philosopher Rose Rand. Rand had entered Britain in 1938 with IFUW help, and her nursing job at a London psychiatric clinic had taken her to the point of physical collapse. She placed all her hopes in gaining a fellowship from Harvard University. Erna Hollitscher and Erica Holmes, the secretary at the IFUW office, considered Rand "quite unfit for practical life and only interested in her philosophy researches." It was thus necessary to cover her living costs until Harvard came to a decision, rather than expecting her to carry on with heavy physical labor. She was awarded a small fellowship on the grounds that "Dr. Rand might not receive her Harvard fellowship and, if so, might be compelled to do domestic work." The modest grant was made on the slightly tongue-in-cheek condition that the philosopher spend the money "only on food, not for books."[45]

There is no evidence of similar decisions in the US setting. On the contrary, within the AAUW it was generally considered more or less pointless to support research by humanities scholars; instead, their energy was to be directed into more practical channels. A successful example is the archaeologist Alice Mühsam, who applied for an AAUW fellowship in December 1940, aged 51. The widow of journalist Kurt Mühsam, who died in 1931, she began studying archaeology in Berlin in 1929 at the age of 40, and, in 1936, completed her PhD dissertation on Attic grave reliefs of the Roman era. Helped by her eldest daughter, Ruth, who had emigrated to Hollywood in 1937, Alice Mühsam reached New York in 1940 and there made her application to the AAUW. Her objective was to recommence the research she had been forced to abandon in Berlin. But the AAUW refused Mühsam's request for a one-year, part-time fellowship that would gradually allow her to give up the physically strenuous cleaning work that barely paid her bills in New York. Instead, Esther Brunauer arranged for Mühsam to get advice from her fellow archaeologist Gisela Richter, a curator at the Metropolitan Museum of Art. Richter suggested that instead of pursuing her ideal of research work, Mühsam should retrain in the restoration of ancient artifacts. Brunauer and Richter joined forces to persuade the head of restoration at Brooklyn Museum to train Mühsam at his workshop for two years. The AAUW's War Relief Committee approved an annual $250 grant to support her in this project.[46]

It is striking that the academic refugees in the United States who immediately felt at home there and had good experiences with the AAUW were either those able to return to their own profession immediately, or those with exceptional communication skills, a light and charming manner, and great optimism that withstood even the greatest difficulties. One of them was Auguste Jellinek, who not only quickly found her feet professionally in St. Louis, but also felt she had been received with particular warmth by the local AAUW branch. She reported to Brunauer with effusive gratitude that she had joined the branch and was enjoying the "spirit of high sociability and companionship which makes life so pleasant for us newcomers."[47] The archaeologist Margarete Bieber came to Barnard College in New York in 1934 at the invitation of Dean Virginia Gildersleeve, initially for a one-year visiting professorship, and, like Jellinek, she had every reason to fall in love with America. In Britain she had felt the full impact of her refugee status, lacking not only money but also, and most critically, accommodation that she could have shared with her daughter and her German housekeeper. At Barnard, she was allocated a whole suite, and later, after she moved to the archaeology department of Columbia University, an apartment of her own.[48] In addition, her housekeeper—joining her directly from Germany—managed to bring Bieber's entire library and her furniture, rugs, and pictures to America. Bieber was introduced to archaeological circles and the highly select Archaeological Club by the curator Gisela Richter, a longstanding acquaintance of hers. She was well provided with both intellectual stimulation and opportunities to supplement her salary with lectures. There is no doubt that her impression of a greater warmth and kindness in the United States than in Britain was related to the exceptionally favorable material and academic conditions in New York.[49]

Susanne Engelmann, who left Germany for Turkey in 1940, will have benefited from her excellent language skills and knowledge of American life when she arrived in California in early 1941. We do not know whether she approached the local AAUW branches independently or had recommendations; what is certain is that the 54-year-old former school principal made an immediate name for herself as a speaker at the YMCA, AAUW, and various church congregations. Her experience at Bryn Mawr in 1912–13 most likely contributed to this quick progress. In the spring term of 1941, Engelmann attended the Stanford University School of Education course that would qualify her to teach in Californian high schools. The energy and success with which Engelmann's branch worked to get her funding from the AAUW refugee aid budget attests to a particularly warm personal relationship. The branch enthusiastically described Engelmann as not only "a brilliant, splendidly trained woman" but also "a gracious person, of great charm and integrity."[50] Everyone who met her was impressed "by her courage and determination with which she has worked to begin a new life."[51] In August 1941, Engelmann received $500 to sit her examinations.[52] She also worked hard to make contacts within the Hoover Library and Mills College, Oakland; it was at these institutions that Engelmann first began to earn her own income again, at first by translating and soon also by teaching adult education classes. When Engelmann's encouraging new start in California was halted by a tightened curfew on "alien residents" in March 1942, the YMCA and the AAUW came to her aid again.[53] With recommendations from the two organizations, she was invited as a "Refugee Scholar" to Wilson College in Chambersburg, Pennsylvania, a liberal arts college for women that had also opened the route to an academic teaching career for Erna Barschak.[54]

For those women not lucky enough to meet with such a welcome or to have previous experience of the United States, the initial period tested not so much their academic expertise as their physical endurance and their domestic, social, and communicative skills. To the considerable chagrin of German university graduates, their Austrian counterparts were generally considered more charming and pleasant-natured, easing their reception in Britain and, to an even greater extent, in America.[55] Thus, the socially resourceful Alice Friedmann, who had directed a boarding school for problem children in Vienna until 1938 and arrived in New York from Britain in December 1939, gave a glowing account of her encounter with the AAUW. This remedial teacher and psychologist—who, like Engelmann, tackled her new professional start with verve and courage—reported that she had been received very cordially by AAUW members in both New York and Baltimore, and had made important professional contacts with their assistance.[56]

The Viennese dentist Helene Erlach was another woman who matched up to the ideal of the brave immigrant remaining cheerful against all odds. She benefited from the efforts of the AAUW's Chicago branch, where her experience in international networking no doubt helped her to make and maintain good contacts. She had attended IFUW conferences for many years as the Austrian delegate, and she "talked easily on her feet."[57] Erlach came to Chicago in 1938 with her husband and two teenage children. According to the local branch, she made superhuman efforts to build a new home and livelihood for herself and her family there. With the AAUW's help, in 1939 Erlach found a job as an assistant in the dental practice of an elderly, widowed colleague. This dentist's death in 1942 created an opportunity for Erlach to take over her practice, and members of the Chicago branch worked hard to get her $400 from the refugee aid fund, as capital to buy the practice.[58] Their successful application included explicit reference to Erlach's personal qualities: she was a "cheerful, hearty sort of person. She seems to glow with warmth and friendliness and has an iron bound courage."[59]

Erlach and Jellinek were among the very few émigré graduate women to join the AAUW. The determining factors for refugees' integration into the US university women's organization were rapid professional success and the ability to adapt to an American style of communication—as is indicated not only by Erlach's and Jellinek's experience, but also by the case of Frieda Wunderlich, who was the only member of the AAUW New York branch apart from Margarete Bieber to have immigrated from Europe. Wunderlich very quickly found a position at the University in Exile, New York (today the New School for Social Research), and she taught political and social sciences there from 1933 until 1954.[60] In contrast, women who never managed to resume their careers in New York, or did so only after a harsh and long-drawn-out transition phase, found no new organizational home in the AAUW. The social capital they had acquired in Europe in the shape of personal and social networks was not easy to transfer to the American context. Only in combination with linguistic and communication skills earned in America itself did that social capital become a convertible currency and help the refugees to put down new professional and personal roots.[61]

The case in Britain was different. More émigré women joined the BFUW and sustained these bonds over a long period. This was probably due in large part to Crosby Hall's congenial atmosphere and the indefatigable assistance of Erna Hollitscher until she retired in 1954—Esther Brunauer, the crucial point of reference for émigré women graduates in the United States, left the AAUW as early as 1944 to become the US State Department's first senior woman official.[62] But the commitment of the local groups across Britain also contributed greatly to the refugee women's integration there. In 1943, physician Alice Blau answered Hollitscher's query about joining the Federation as follows:

> I should be only too pleased to become a member—actually the wording is, I think: I should consider it an honour and a privilege—but I mean it—to become a member of the Federation. I have not forgotten that a long time before our work was needed to a certain extent, at a time when the great majority of the English looked upon us with a mixture of pity and disgust, the Federation treated us like as their guests and tried to make things easier for us in a special sense, to make us feel like friends among friends, or, in one word, as equals.[63]

When the BFUW appealed to its members in 1960 for donations to build a second residential wing in Crosby Hall, Hollitscher and her friend Edith Mahler from Vienna (a cousin of Gustav Mahler's) raised the sum of £1,000 from one-time refugees. This contribution gave the donors the right to name one room in the new building after a personality of their choice. Out of gratitude for the help they had received before and during the war, said the women, they wanted to dedicate "their" room to Gustav Mahler, to honor the protagonists of refugee aid in Britain and keep alive the spirit of internationalism that was expressed in Mahler's music.[64]

Professional Connections

Considering how often it is assumed that women were much less likely than men to continue their previous careers after emigration, a surprisingly large number of the applicants to the AAUW, BFUW, and IFUW refugee committees managed to rejoin their field or successfully entered new professional domains. This was true of both younger women and those aged over 40 or 50.[65]

The first group to mention in this context are the approximately 40 female academics who were funded by the network, as far as their later careers can be reconstructed. Bacteriologist Emmy Klieneberger emigrated to Britain in fall 1933 and was the first beneficiary of the new Residential Scholarships for refugees at Crosby Hall; in 1935, she found a post at the Lister Institute of Preventive Medicine in London, where she made groundbreaking discoveries on mycoplasmas.[66] After a year in Oxford, from 1934 Margarete Bieber taught at Barnard College and the Graduate School of Fine Arts at Columbia University, New York, initially as a visiting lecturer and then, from 1937 to 1948, as associate professor. Once she reached the official retirement age of 60, she failed to get her appointment extended for another ten years, despite vigorous efforts.[67] From 1949, Bieber worked instead as "special lecturer" at Columbia University's School of General Studies, and from 1954 she also took up duties at Princeton and the New School for Social Research.[68]

The art historian and photographer Adelheid Heimann, based in Britain since 1936, found work with commercial publishers specializing in illustrated volumes after several years of funding by the BFUW. Her subsequent professional and academic career was closely tied to the Warburg Institute, a renowned photograph-based collection on Renaissance art that had relocated from Hamburg to London in 1933. Heimann published several articles in the Institute's journal, and in 1939 she was employed for a two-year period in its photographic studio, where she worked with an experienced photographer. Heimann then moved into photojournalism, contributing articles and photographs to the pioneering liberal news magazine *Picture Post* from 1941 to 1952. She spent the last ten years of her working life as deputy curator of the Warburg Institute's Photographic Collection—her first adequately paid position as an academic. In this period she also made a triumphant and highly productive scholarly comeback.[69]

Another Heimann, Betty Heimann (no relation to Adelheid), taught at the University of London in the Department of Indian Philosophy on half pay until 1944. The department was specially created for her. In 1946, she was granted a full position as senior lecturer at the University's School of Oriental and African Studies. Shortly afterward, Heimann moved to the University of Ceylon in Colombo, where she taught Sanskrit and Indian philosophy until her retirement in 1949.[70] Like Bieber, she was unable to extend her contract in Colombo beyond the official retirement age, and returned to Britain aged 61.[71]

Although photochemist Gertrud Kornfeld and biologist Dora Ilse did not return to academia, they found attractive positions in industrial research. Kornfeld emigrated to the United States in 1937 with an affidavit from Esther

Brunauer when her BFUW and AAUW funding ran out. She joined the Kodak research laboratories in Rochester, New York, and remained there for many years, carrying on her scientific work.[72] Dora Ilse, a behavioral scientist and the creator of popular butterfly films, initially made her living in Britain as a schoolteacher. In 1944, after a protracted thyroid illness, she was advised to cease teaching. She managed to break into industrial research with Reckitt & Colman, a well-established chemicals company in Norwich that had gained a new global market in 1933 with the antiseptic Dettol.[73] It is quite possible that Ilse brought her expertise in entomology to the company, because after 1945 Reckitt & Colman dominated the market for domestic pesticides. With an annual salary of £400, Ilse was probably—at least for a time—the highest earner among the graduate women refugees. It is not known whether she was able to remain with the company permanently.

The case of art historian Helen Rosenau again suggests that women who had worked in academia back in Germany may have been relatively well equipped for exile: they had developed the staying power required to carry on their own research for many years, even when there seemed to be virtually no hope of making it the economic basis of their lives. When Rosenau's BFUW funding expired in 1935, she had to earn a difficult and uncertain living from lectures in order to keep up her own scholarship. In 1938, she married the Jena physician and PhD Zwi Carmi, who had a small import-export business; this seems to have improved her economic situation at least for a time. When the borders were closed at the outset of World War II, however, the couple lost their livelihood at a blow. In the period of despair that followed, Rosenau canceled her BFUW membership because she could no longer afford the dues.[74] Her plans for emigration to America came to nothing. Even then, despite extreme economic distress, the art historian did not abandon her research, and in 1940, now aged 40, she completed her second doctoral dissertation, this time on the architectural history of the synagogue. From 1941, she worked with Karl Mannheim at the London School of Economics on the social status of women in artistic representations, publishing her study in 1944 as *Woman in Art: From Type to Personality*. It was only from 1947, as a University of London lecturer and the mother of a toddler, that she began to earn a more or less adequate income. She taught in Manchester between 1951 and 1968, then returned to London as a senior lecturer at the University and Leo Baeck College. By this time she was well known and respected as "one of Great Britain's leading art historians," belatedly receiving recognition and with it an affirmation of her life's choices.[75]

There is no doubt that the women who left Germany before 1938 had substantially better chances of resuming their previous occupations than all those who emigrated at a later point. Nevertheless, individual cases show that even late in the day it was quite possible for women to reestablish themselves in their own field of research, especially in the United States. Erna Barschak, who came to America from Britain in 1940 aged 52, spent a year working at Wilson College before being appointed professor of psychology at Miami University in Oxford, Ohio, around 30 miles northeast of Cincinnati.[76] Susanne Engelmann, who was 54 when she arrived in California in 1941, also managed to find a niche in research and teaching, even if she initially had to eke out a living with short-term contracts at various locations across the United States. Engelmann's visiting professorship at Wilson College in Pennsylvania in 1942 was followed by a year funded partly by an Institute of International Education research grant, partly by lecturing at the Institute of International Relations in Wichita, Kansas. From 1943, she taught for two years at Smith College in Massachusetts, one of the prestigious "Seven Sisters," and in summer 1944 also held courses at the Institute of International Relations in Des Moines, Iowa. For the academic years 1945 and 1946, she taught in Texas at Weatherford College, Dallas.[77] Finally, in 1946, she was appointed to an associate professorship at Mary Washington College, University of Virginia, where she remained until she retired in 1952. Engelmann took US citizenship in 1948.[78]

Hedwig Kohn, too, was able to continue teaching in the United States until her retirement. After an initial year at Greensboro College in North Carolina, the 55-year-old physicist took up the guest professorship at Wellesley College that had been arranged in 1939 as a precondition for her immigration. The post was regularly extended until 1952, when she retired and moved back to North Carolina to spend the last 12 years of her life on independent research at Herta Sponer's spectroscopy lab in the Duke University physics department.[79]

Finally, art historian Alice Mühsam also managed to find her way back into academia with the help of the practical retraining that the AAUW had arranged for her. Mühsam's peers soon came to value her work as a restorer of ancient Egyptian and Greek ceramics, but until 1945 she remained dependent on income from cleaning and babysitting. From then on, she paid her way by tutoring Columbia University art history students in the German language. The doctoral dissertation she had completed in 1936 appeared in book form in 1956, and ten years later she published a second monograph, on ancient Jewish coins.[80]

For the great majority of the academic women who emigrated in or after 1938 and were forced into housework or other demanding physical labor, this initiation remained a harsh but temporary transitional phase. In Britain, a general move back into professional employment can be observed from 1940 on, especially for women teachers, doctors, dentists, and social workers. The war encouraged this mobility in some ways, but in others interrupted and impeded it. Analyzing each case individually shows how much a successful integration process depended on the personal determination of women who defied all adversities to build themselves a new home in their own professional fields. This involved heavy workloads, frequent relocation, considerable material hardship, and often also the decision to study for a British university degree despite their age.

Gertrud Schlesinger's career in Britain—slow to start but ultimately successful—exemplifies this. It was only after five years of searching that the schoolteacher found an appropriate position, teaching science subjects at a girls' grammar school in London. She had originally hoped to work at her previous school, which the principal Leonore Goldschmidt had relocated from

Berlin to Folkestone, but this remained a brief and unpaid interlude between October 1939 and May 1940: before her teaching contract could be signed, the boarding school was closed due to the threat of bombing raids. Schlesinger only worked in Folkestone for her board and lodging as a "house mother." After a short spell of unemployment, she had to return to domestic work in Surrey until, in December 1940, she found a position at a small private school near Birmingham. Though very meagerly paid, the job did allow her to acquire a work permit as a teacher—but because Erna Hollitscher could not find her free or cheap accommodation with a BFUW member, she was still not earning enough to pay her way. Schlesinger therefore gratefully accepted a job that Hollitscher found for her as a "Turkish news typist" with the BBC, making use of the knowledge of Turkish and Turkey she had gained in exile there. The BBC agreed to arrange for Schlesinger to be released from the teaching service on a temporary basis. This formality was important to her, because she did not want her BBC job to stop her returning to the teaching profession later on.[81] Schlesinger continued working in broadcasting from 1941 through 1944, a period when she seems finally to have achieved some material, psychological, and emotional stability. She kept up regular contact with BFUW members, gave lectures on Turkey, and joined with other émigré women and their children for companionship on the Protestant holidays.[82] In summer 1944, eleven years after being dismissed from the education service in Germany, Gertrud Schlesinger—now aged 43—found a job at the East Ham Grammar School for Girls in London. In 1947, in her last letters to Hollitscher, Schlesinger reported that she was still teaching science there.[83] Even during her days as a domestic servant, she had made plans to improve her opportunities as a teacher by taking a further degree in English, and this project was completed gradually over several years. She probably began her studies while working for the BBC. In December 1947, she took some months off from teaching to stay at the newly reopened Crosby Hall and prepare for her examinations.[84]

Many émigré women in Britain struggled well into the postwar period to regain their professional careers, remaining in close contact with the BFUW. In the United States, close and lasting links with the AAUW were far less common, but the transition phase of hard physical labor was usually much shorter than in Britain. The Viennese social worker Katharina Flesch and her husband, for example, were able to abandon factory work relatively soon. When the couple moved from New York to Detroit in summer 1941, Flesch quickly found a job in the School of Public Affairs and Social Work at Wayne State University; her husband worked in the automobile industry.[85] The Reichs, who were friends of Erna Hollitscher's, both passed their US medical licensing examination at the first attempt, and Emma Reich found employment at a New Jersey hospital after only six months, while her husband opened his own practice in Manhattan. In 1948, the Reichs moved to Phoenix, Arizona, where Emma Reich gave up work after several years of exhausting shift duty.[86]

For all the women involved, building a new life and career in exile was a lengthy and extremely demanding process. In many cases, it took them years even to find an apartment of their own, and as time passed, their hopes waned of regaining the standard of living they had known in Germany or Austria. The success or failure of their new beginning depended on many external factors, including their year of arrival, marital status, professional field, knowledge of English, and age, and not least on sheer serendipity. But previously active members always found the IFUW a reliable network, a source of support in the absence of family, friends, and their own professional peers. The surprising extent of their activity and productivity, often well into old age, indicates that university-trained women did not simply obey economic constraints: they also, and importantly, drew on their professional identity to create continuity across the traumatic ruptures in their lives.

Only one of the women discussed here is known to have returned to Germany permanently, a fact that tells its own tale.[87] Former high school principal and professor Susanne Engelmann returned from the United States to what was now West Berlin aged 67, after retiring in 1952, and lived in the suburb of Dahlem until her death in 1963. She probably chose the district due to her links with the Lutheran congregation there, dating from the 1930s. Engelmann successfully claimed a pension from the compensation authorities in Berlin and assumed German citizenship again.

Engelmann's correspondence with the AAUW reveals that she had been thinking about a return to Germany, with clear goals in mind, even before 1945. Asked by the AAUW women whether she could collate bibliographical information for the education branch of the Control Council for Germany, she answered in August 1945 that it had been her "fondest hope through all these years…to be allowed actively to contribute to the reconstruction of post-war schools not only in Germany but in the world." To this end, she added, she had already put together a selection of school texts to be used in postwar Germany for the publisher Bermann Fischer in New York, and had drawn up a lecture series on "Education for World Citizenship."[88] Shortly after arriving in America and still based at Stanford, she had begun working on a book that appeared in New York in 1945 under the title *German Education and Re-education*.[89]

It is not known whether Engelmann reentered the DAB, where university-trained women—especially those of her age—joined forces again in 1949, but she certainly never again played an active role in the federation.

Vocation and Survival

The correspondences of the AAUW, BFUW, and IFUW refugee aid offices say much about the desperate plight of the university women fleeing Nazi rule, and about the network's energetic efforts to assist them itself or mobilize support from elsewhere. On balance, the record of success in the IFUW's assistance for escape and for the refugees in exile is not positive. The letters and committee minutes clearly show how difficult, complex, and costly it was to assist emigration and resettlement in each individual case. Just as in every other refugee organization, there was a large number of women for whom the network could do very little or nothing at all. The IFUW's efforts to get the cancer researcher Gertrud Wreschner out of Berlin were unavailing, and she was killed in Auschwitz; the escape of Viennese physicist Maria Anna Schirrmann foundered at the last moment, when she informed the Viennese authorities in August 1939 of a typing error in the visa she had just received.[90] We do not know exactly how many IFUW members still succeeded in leaving Germany and its growing sphere of power after 1939 by other means, but it can be assumed that many of them were unable to escape after the outbreak of war.[91] Still, as the following examples indicate, the international network could also play a supportive role for those women who had no choice but to remain in Germany or the occupied areas.

For the historian and librarian Käthe Spiegel, who had once headed the German section of the Czechoslovak university women's association, close contact with the AAUW up to late summer 1941 was a practical and psychological lifeline. The organization seems to have been Spiegel's only link to the outside world. She had lodged at the AAUW's international guesthouse during her Rockefeller year in the United States in 1927–28, researching the book on the American Revolution that she tried in vain until 1936 to submit as a habilitation thesis at the University of Prague.[92] In Washington, she had met Brunauer and other activists personally. The AAUW correspondence with Käthe Spiegel exudes the special sense of affinity that arose from these transatlantic encounters, and the American women evidently set a high priority on Spiegel's attempt to escape from Czechoslovakia. Brunauer kept all the AAUW members who had met Spiegel updated on her fortunes. Highly articulate, Spiegel was able to explain her plight convincingly. The high quota number allocated to her by the immigration authorities meant she would not be able to immigrate in the foreseeable future, and, in 1940, there was little the US members could do in practical terms. However, they regularly sent her encouraging letters and assured her they were both willing and able, whenever it became relevant, to get her an affidavit and the necessary cash to come to America.[93] For her part, Spiegel made no secret of placing all her hopes in the AAUW. Shortly after the deportations of Viennese Jews to occupied Poland began, she put her life in the hands of her American colleagues with a moving petition, evidently aware that this turn of events had made immediate emigration a matter of life and death. As Spiegel wrote to Washington in March 1941:

> I confess I am very miserable now, and it is really an SOS letter I am sending you and ask you to help me as much as you can.... May you clever American Women... arrange all things for me as you think is best and please let me know what you are planning about me and my future! It is now the question to be or not to be and I am asking now life or death, success or perish to complete for me and I am very eager to know which one of both will succeed to get me first![94]

Spiegel's urgent call for help touched a chord within the AAUW. Strategically circulated by Brunauer, it reached all the members assembled for the AAUW's

1941 National Convention in Cincinnati, where a spontaneous collection brought in enough money to start the complex and expensive rescue attempt. In late May 1941, sponsors for an affidavit had been found; passage with the Hamburg America Line was booked and paid in advance. The passage to Cuba cost almost $400, the Cuban visa $265, the "landing fee" $500, the deposit for the onward voyage to the United States $150. In addition, $150 had to be paid as an insurance deposit to the bank and the women had to guarantee a further $40 for every day that Spiegel would spend in Cuba.[95] The expenses amounted to around $2,000, by far the highest sum that the AAUW had raised to support one person.

For all their commitment, the American women were no longer able to help Käthe Spiegel. In late summer 1941, an AAUW member managed to visit her personally in Prague[96]—so she almost certainly received the news that all the necessary steps had been taken and paid to enable her emigration and thus survival. But it was too late. Spiegel's booking was canceled by the Hamburg America Line on December 8, 1941, the day the United States entered the war.[97] She was deported to Łódź on one of the first death transports, and died in 1944 in a concentration camp in occupied Poland; it is not known which one.[98]

In February 1939, the Austrian sisters Elise and Helene Richter, by then both over 70 years old, declined the IFUW's offer to help them emigrate to Britain and resolved to stay in their home near Vienna.[99] Romance philologist Elise Richter explained their decision in an autobiographical manuscript dated 1940:[100] the ailing sisters feared they would have to spend their last years as "mere objects of charity" abroad, in poverty and isolation. In view of their age, familiar surroundings and contact with close friends seemed more important than safety from escalating Nazi harassment. At the time of writing her autobiography, Elise Richter still believed the decision had been correct. A spell of serious illness soon after refusing the IFUW offer had made her appreciate "the whole blessing of being at home," where she was nursed by friends and could be treated by her own doctor with care and consideration.[101] Up to 1940, regular payments from the IFUW buffered the sisters' rapid impoverishment; they also received letters from former students who had emigrated with BFUW or AAUW help—until the attack on Pearl Harbor "put an end to the exchange of our thoughts and feelings," as the Romance philologist Helene Adolf, who had emigrated to the United States in 1939, wrote in her 1948 obituary of Elise Richter.[102]

Though unwilling to leave Austria themselves, the sisters were eager to ensure that their writings at least, would be rescued. Helene Richter asked for IFUW help to get her last large-scale study of English literature, on John Keats, taken to London in 1939. She wanted the manuscript, which no longer had any chance of appearing in Austria, to be housed and catalogued in the British Museum's Department of Books. To her distress, the BFUW could not fulfill this wish, but in the end her works went to the library of Royal Holloway College, University of London, through the efforts of the RHC principal Edith Batho, an important IFUW official.[103]

Deprived of all opportunities for research from 1940 on, 75-year-old Elise Richter channeled all her powers of resistance into writing her autobiography, which bluntly denounced the new era. She produced two copies ready for typesetting, and in spring 1941 gave the bundle to her friend Christine Rohr for safekeeping. The text survived the Richter sisters' forcible admission to a Jewish retirement home in spring 1942, their deportation in fall 1942, and their deaths soon after at the Theresienstadt camp.[104]

For five of the academic women who unsuccessfully appealed to the IFUW, BFUW, or AAUW, there is evidence that they survived deportation and the death camps.[105] Three of these were physicians: Olga Weiss from Vienna, born in 1885; Johanna Maas from Karlsruhe, one year younger; and the Berlin doctor and scientist Lucie Adelsberger, born in 1895. In 1938, Weiss had rejected the idea of going to Britain as a housekeeper, while Maas and Adelsberger did not want to abandon their mothers by emigrating before 1938. All three women practiced as physicians right up to their deportation, and continued to do so in the camps—Weiss and Maas in Theresienstadt,[106] Adelsberger in Auschwitz. Their vocation was a crucial factor enabling them to resist death there.

After liberation in 1945, Weiss and Adelsberger renewed their contact with the BFUW and IFUW. Their British colleagues issued invitations to Crosby Hall or the homes of members outside London to help them recover from the ordeals of their imprisonment and attend a "refresher course" in their field. Olga Weiss's sister Grete had established herself in Britain as a teacher and joined the BFUW, and she passed Olga news of the opportunity, but visa restrictions made it impossible to accept until 1947. Olga Weiss spent several months in London during 1947 and 1950, working in renowned clinics. The draconian immigration regulations put paid to her plans to join her sister in Britain for good.[107]

The university women's association in the Netherlands helped Lucie Adelsberger—a pediatrician, internist, and immunologist—to sign up for a period of convalescence at Crosby Hall in 1945, but in the end her trip fell through because of food shortages in Britain. Instead, the Dutch women supported her during a stay in Amsterdam.[108] From there, Adelsberger prepared for emigration to the United States, where her younger siblings had gone pre-1938. Her hopes of taking up the post at Harvard that had been promised in 1938 came to nothing; however, she managed (probably from the Netherlands) to obtain a contract with the Montefiore Hospital in New York.[109] It is not certain whether Adelsberger tried to contact the AAUW once she arrived, but we do know that no long-term bonds were forged. According to her sister, Lucie Adelsberger remained a lonely figure, deeply marked by the horror she had experienced. She spent the remainder of her life focusing on her medical practice and research.[110]

* * *

Bearing in mind the wide range of decision-making processes and experiences described in this chapter, it is not easy to draw general conclusions, but in terms of the factors that proved key to the graduate women and academics associated with the IFUW in emigrating and reestablishing their careers, the following points stand out. The women who asked the AAUW, BFUW, and IFUW for help in emigration were usually at least 40 years old and unmarried, and the majority of them were at the peak of their professional powers. When they deliberated whether and when to emigrate, the question of career opportunities in Germany and abroad was a decisive one for these women, no less than for their male colleagues. If there was a reasonable chance of finding satisfying work abroad, as was the case for natural scientists, certain women emigrated right at the beginning of the Nazi dictatorship. But in contrast to some of their male colleagues, far better situated in this respect, women did not receive job offers from abroad, and until 1938 many shrank from emigration "on spec," fearing they would find themselves even further down the career ladder than they were in Nazi Germany. Inevitably, women's stronger family orientation also played a role. This mainly took the shape of responsibility for elderly parents, especially because many of the unmarried women shared their homes. If emigration with their parents appeared difficult or impossible, most of these women decided to remain in Germany for the time being—provided that they could continue to work in their field after being dismissed from state positions, for example in the Jewish education system or in segregated healthcare. By the end of 1938, all the women involved saw emigration as unavoidable: their chances of working in Germany or the countries under Nazi rule had dwindled to almost nothing. At this late stage, as a rule women could only manage to escape Germany if they agreed to enter domestic servitude in Britain and (at least temporarily) abandon the hope of continuing in their own profession. Most women now turned out of necessity to this last remaining, gender-specific chink in the immigration regulations.

A key result of this study is that for university-educated women aged more than 40 to 50, the decision to emigrate was the single most difficult step toward a new career abroad. After 1938–39, many were no longer able to leave Germany. The correspondence of the IFUW and its networks documents the failure—despite considerable efforts—of many attempts to escape, and also many conscious decisions to stay put.

The later professional paths of the women who did leave suggest that much existing scholarship has tended to paint too bleak a portrait of highly qualified women's careers in exile, often based on assumptions that do not stand up to empirical examination. The great majority of the women who emigrated with IFUW help succeeded in reestablishing themselves, regardless of their age or field of work. Certainly, the initial stage of that process was harsh, as women forfeited their social prestige, working as domestic servants, and were excluded from their accustomed professional environment. Women certainly did not find the transition any easier to overcome than did men. Yet most kept sight of their goal of returning to their profession or research at all costs, even if the process dragged on for many years. Their persistence was underpinned by forms of cultural capital that were more readily available to women than to men. Unlike many of their male colleagues, graduate women generally had an adequate grasp of their host country's language. Added to that, they were highly motivated, having had to fight for their profession with considerable determination in Germany previously—again unlike most of their male colleagues. Despite the inevitably greater obstacles to emigration for academic women, especially older ones, a surprisingly large number in all disciplines and graduate professions managed to resume their previous careers. This appears to have been closely related to the special significance of a woman deciding to pursue an academic or professional career in the early twentieth century. Even before 1933, a woman's decision to follow her academic "vocation" often demanded great patience and a willingness to tolerate very uncertain prospects for the future. Anyone who set off on this mostly convoluted path had to be self-disciplined, purposeful, and resolute enough to assert herself in the male-dominated world of the academic professions or the university, to accept financial privation and social isolation, and often to forgo social recognition (as a married woman, for example). If so many of the women forced to emigrate kept up their previous professions, or broke into new ones, despite the social and economic tribulations of persecution and the growing burden of family obligations, this demonstrates their firm and crisis-tested professional identity: an identity perhaps in many cases stronger than that of their male colleagues.

8
Continuity, Memory, and the Cold War

The development of the atom bomb and its detonation in Hiroshima and Nagasaki changed both the landscape of science and the fabric of international organizations forever. When the United Nations was founded in summer 1945, a new era began for the IFUW, which was granted permanent observer status in all UN bodies and an advisory role in UNESCO.[1] This new international framework transformed the IFUW: previously limited mainly to the North Atlantic sphere, the women's network now assumed global dimensions. By 1968, the number of nations it represented had grown to 50, including states from all continents. Associations of university women formed in Bolivia and Chile; in Thailand, Japan, and Korea; in Pakistan, Indonesia, and Burma; in Egypt, Zimbabwe, Uganda, Kenya, and Nigeria; in Iran and Turkey. The federation's focus shifted to postcolonial problems, poverty, and human rights, with a high priority on promoting educational opportunities for girls. From the mid-1950s onward, less and less significance accrued to professional, and especially research, opportunities for female graduates in Europe.[2]

Nevertheless, in the immediate aftermath of the war the close links between the IFUW, Europe, and specifically the German university women endured—providing a continuity that was vital to the creation of new national and international networks among women in Germany (at least West Germany).[3] This chapter asks how German university women reorganized after 1945, and what role the international network played in their redefinition. The chapter addresses issues of institutional and biographical continuity and the interpretations of the past that were propounded in national and international settings.

Rebuilding Networks, 1945–49

Most of the IFUW's European member associations that had been closed down or brought to a standstill during Nazi occupation were able to resume work as soon as the hostilities ended, if not before. This rapid restoration was possible because within each organization a nucleus of members had defied official bans on political assembly to stay in contact throughout the occupation. Equally important was the endurance of personal and institutional links across national boundaries, an endeavor in which émigré women played a preeminent role. A group of Polish women in exile pursued their association's objectives first from Paris and then from London.[4] A Belgian member who had fled to London kept up her connections from there, and even continued paying IFUW dues so that the Belgian women's official membership remained active. The Danish organization took a similar approach, paying its IFUW dues through members in Swedish exile.[5]

When World War II came to an end in Europe, there was no immediate shift in the emphasis of political activity for the IFUW or its member organizations in Britain, the United States, Sweden, and Switzerland, the major players in academic women's wartime assistance for refugees. In April 1945, the IFUW Refugee Fund was transferred to a new committee set up in Zurich by the Swiss federation's president, Blanche Hegg-Hoffet: the Committee for the Relief of War Victims. The BFUW refugee office in London, headed by Erna Hollitscher, remained active, providing refugee and emergency assistance for distressed university women in the liberated regions. These efforts focused on distributing donations (mainly from Canada and the United States) of clothing, food,[6] and money for travel and books; arranging sanatorium treatment in the Swiss Alps for university women suffering from tuberculosis; and taking care of colleagues who had approached the organization as displaced persons or concentration camp survivors.[7] Crosby Hall funded free-of-charge stays in its rooms so that women could recover from the emotional and physical ordeals of imprisonment. British members secured places on professional refresher courses for the survivors of Nazi racism, to help them return to their careers.[8] In the United States, the AAUW arranged funding for students and doctoral candidates from the liberated countries "whose education was interrupted in varying degrees by the war." International Study Grants were introduced in 1946, and by 1946–47 they had already enabled 37 women students from previously Nazi-occupied areas to spend 12 months in the United States, studying and researching at a wide range of colleges and universities.[9] A year later that figure was 54, and in 1948–49 a further 56 women benefited from the program.[10]

The international community of academic women also stepped in to help when university women in Germany began to rebuild their networks after 1945. The first contact with former DAB members was made as early as summer 1945, as part of the occupying powers' efforts to draw reliable representatives of the German educational elite into their project of "re-education."[11] Major Mary S. Bell, based in Germany as a member of the United States Group Control Council's education branch, approached the AAUW in Washington, DC, in August 1945 requesting the names of former DAB members who might be suitable candidates for cooperation.[12] Bell was an AAUW member and the dean of women at a small Iowa college, which had released her for her military duties. The director of the AAUW International Relations Office, Helen

Dwight Reid, sent Bell the names and addresses of the DAB's last democratic board and of the three German delegates who had attended the IFUW's Edinburgh conference in 1932. Reid could only comment personally on one of these, the high school principal Anna Schönborn from Berlin, whom she had met in summer 1933. In Reid's opinion, Schönborn was at that time "definitely not a Nazi," though "strongly nationalist in her views."[13] The AAUW International Relations Office had not maintained links with university women in Germany, so Reid asked Bell to pass whatever information she found straight to the American association, which was eager to know of "any responsible university women with whom we might renew our contacts."[14]

This revitalized contact between university women in Germany and the Allied nations developed its own internal momentum in each of the Allied occupation zones, which had little interaction in the initial postwar period.[15] In the American zone, links initially arose primarily through the AAUW members who were stationed in Germany as part of the occupying authorities. Alongside Major Bell, another important figure was Dr. Ruth Woodsmall, who gave up her post as general secretary of the World YWCA in 1948 to work in Frankfurt between 1949 and 1952 as chief of the women's affairs section of the US High Commission for Occupied Germany (HICOG).[16] Woodsmall joined the Frankfurt DAB branch, reestablished in 1948, and supported the founding of DAB groups in Heidelberg and other cities in the US zone.[17]

Even more important for the reinstatement of an academic female network in Germany and its integration into the international umbrella organization were the relationships that arose between university women in Berlin's British sector and the BFUW, especially through Crosby Hall in London. Just as in Weimar days, after 1945 the London clubhouse became a nexus of personal contacts. This mode of rapprochement preceded the rebuilding of the DAB in Germany by several years. Friendships renewed with British women at Crosby Hall paved the way for the DAB to re-form in West Germany in 1949, and for the West German women's successful reintegration into the IFUW during the early 1950s.

The crucial figure on the British side was the Austrian emigrant and longstanding secretary of the BFUW refugee office Erna Hollitscher. Hollitscher assumed British citizenship in 1948, and did not return to Austria. But at the end of World War II, she worked to help university women in defeated Germany, and to promote their readmission to the international organization, no less energetically than she cared for those émigré women and survivors of Nazi occupation who appealed for her support. On the German side, the key protagonist was Agnes von Zahn-Harnack, cofounder and first president of the Weimar-era DAB, who tended and multiplied contacts with Britain one year after the war. An opponent of Nazism from the very start, Zahn-Harnack had retreated from public life from 1933 to 1945, but spent that period in close contact with friends in resistance circles.[18] As the editor of a celebrated bibliography on the "woman question" in Germany, she was well known on the international stage, and in 1945 could draw on a stock of high regard within the IFUW.

Zahn-Harnack, now aged 61, took the initiative to reestablish a women's organization directly after hostilities ended in Berlin. Thanks to her efforts, a German Women's Union (Deutscher Frauenbund) was constituted in the city on July 21, 1945. The strict Allied licensing of political organizations meant that this association initially adopted a more modest title, named after one of the city's British-occupied districts: Wilmersdorf Women's Union 1945 (Wilmersdorfer Frauenbund 1945). In terms of its claims and impact, however, from the beginning the Wilmersdorf Women's Union extended beyond Berlin's British sector, and from 1947, now called the Berlin Women's Union (Berliner Frauenbund), it opened up to women across the city. Until her death in May 1950, Agnes von Zahn-Harnack continued to embody the leadership, initiatives, and reach of Berlin's biggest women's organization, setting her signature on its policies and a "tradition-conscious new beginning."[19] Unlike the numerous women's committees that arose all over the city after the war, the Berlin Women's Union did not contribute directly to what was called "the practical work of survival." Instead, it aimed to provide an arena for a thoroughgoing process of rethinking women's issues, cultivating "the exchange of views by members of various occupational groups," and promoting reentry into women's international democratic organizations. The program of the Wilmersdorf, later Berlin, Women's Union was similar to that offered by the DAB in Weimar days. Lectures on social, cultural, and political topics attracted considerable interest, with average audiences of 150. This approach drew a disproportionate number of university-trained women into the organization, as can be inferred from 1947 accusations that Zahn-Harnack's Union was an "intellectual clique out of touch with the problems of the present day" and offered too little practical assistance.[20]

Despite the preponderance of academics in her new organization, Zahn-Harnack was initially wary of limiting membership to women university and college graduates, on the model of the Weimar Republic's DAB. Encouraged by Major Mary S. Bell to write something about the situation of German women and especially university women for foreign observers and the AAUW, in summer 1946 Zahn-Harnack presented a pessimistic report that explained why she considered it neither advisable nor possible to institutionalize a supraregional network of German academic women at that time.[21] For one thing, she argued, the occupation statutes ruled out any such supraregional organization; secondly, at present the general situation of university women in Germany did not permit it. The quality of higher schooling and training for young women had suffered severely under Nazi rule, so that the graduates of German high schools, universities, and teacher training institutes could hardly be described as a new generation of academic women. Zahn-Harnack particularly emphasized that these younger women did not possess the capacity to think in an abstract and logical manner. Although university women of the mid–age brackets, securely rooted in their professions, were currently at a clear advantage over their nonuniversity-trained contemporaries, the domain of practical women's social work that once employed so many female graduates had been completely discredited by the experience of

Nazism. As Zahn-Harnack had emphasized when the Wilmersdorf Women's Union was founded, women in this field had refused to understand "whom their work was really serving: the power politics and immoral purposes of a criminal administration."[22] With respect to women's professions open only to university or college graduates, Zahn-Harnack's report noted the rigorous policy of exclusion in the Third Reich that had kept women out of responsible positions and prevented them from acquiring professional experience. Of the handful of female academics who had not been forced out of the universities, even fewer had resisted the pressure to join the NSDAP, and as former Party members these women (along with their male colleagues) had now been suspended from their posts until further notice.[23] In Zahn-Harnack's judgment, the academic women who were older than fifty when the Nazis came to power were the ones most likely to have remained immune to Nazi ideology, as they were able to draw on a store of education and culture accumulated before 1914.

Zahn-Harnack's 1946 insistence on her own generation's moral integrity and the values of the Wilhelmine Empire reads like a déjà vu of the early Weimar years. As I showed in chapter 4, after World War I the pioneers of German women's university education dissociated themselves from the generation succeeding them in a very similar way—but in the end, influenced by the IFUW, they created an organization intended to build both cross-disciplinary and intergenerational solidarity among academic women and to establish a female academic tradition in Germany.[24] However, the generation gap in 1945 was deeper and more politicized than in Weimar days. Zahn-Harnack's image of the female educational elite reanimated a Weimar perception that had favored the pioneer female students of the Wilhelmine Empire as the ones capable of imprinting their "intellectual being with the loftiest means of the era."[25] Zahn-Harnack regarded the new generation not only as intellectually inferior to these women, but also as being, no less than the rest of the population, fully pervaded by the Nazification of society. Writing in 1946, she found it impossible to imagine simply resuming the Weimar project of a cross-disciplinary, cross-generational female academic tradition.

Zahn-Harnack's assessment changed surprisingly rapidly in response to a visit to Britain shortly afterwards. In winter 1946, just a few months after Zahn-Harnack wrote her gloomy report on the state of graduate women in Germany, the British occupying authorities officially invited her to London to represent the Berlin women's movement. During this short visit, leading BFUW members held a small soirée at Crosby Hall in Zahn-Harnack's honor. The welcoming committee included some of her closest IFUW colleagues from pre-1933 days. Among them were Winifred Cullis, a professor of medicine who had led the IFUW between 1929 and 1933, and the IFUW office's secretary Erica Holmes, another acquaintance of Zahn-Harnack's from her days as DAB president and IFUW delegate. Edith Batho, principal of Royal Holloway College, also attended, as did the émigré Erna Hollitscher, who had come from Vienna to London in 1938 with BFUW help and shortly afterwards began to manage the BFUW's refugee aid office.[26]

Zahn-Harnack was deeply touched by her reunion with these women and her warm and unprejudiced reception at Crosby Hall in December 1946. If Crosby Hall "had not given me such an overwhelming welcome last year," she wrote to a Scottish school principal in fall 1947, she would never have dared to initiate contact with "any of my former friends of the University Women," for the feeling of guilt arising from war, destruction, and persecution hampered the Germans' own attempts to make contact. "We cannot expect anybody abroad to forget what has happened," wrote Zahn-Harnack, "nor can we explain it. But we feel most thankfully that your great kindness overbridges the terrible abyss that has opened between our two people[s]."[27]

Zahn-Harnack's 1946 visit to Crosby Hall resulted in renewed friendships between her, Erna Hollitscher, and several other German university women, along with lively correspondence and a scholarly exchange of thoughts and ideas. At Zahn-Harnack's suggestion and with Hollitscher's help, the first German academic arrived at Crosby Hall for her six-month stay in spring 1947, years before the German women were permitted to rejoin the IFUW and thus became officially entitled to use the international network's facilities. Plant geneticist Elisabeth Schiemann, appointed professor extraordinarius at the University of Berlin in late 1946, was invited to London by the Commonwealth Bureau of Plant Breeding and Genetics. Schiemann spent March to September 1947 researching at the John Innes Horticultural Institution. At Crosby Hall, she was welcomed as kindly as Zahn-Harnack had been the year before, in part because she was introduced to the BFUW women as a "friend of Agnes von Zahn-Harnack and an absolutely reliable person."[28] Schiemann became a BFUW member for the duration of her residence at Crosby Hall. The stay enabled her to reconstruct her international research connections and gave her the opportunity to see her old friend, physicist Lise Meitner, again—nine years after Meitner had been forced to flee from Berlin to Stockholm.[29]

Schiemann worked hard to further the interests of German academic women by making as many contacts as possible and exploring options for reestablishing the DAB.[30] On May 1, 1947, she asked Hollitscher whether the BFUW "could give any help in persuading the British Control Authorities in Germany to grant permission for the establishment of an Association of University Women in the British zone of Germany."[31] Schiemann liaised with Agnes von Zahn-Harnack throughout her stay in London, and her approach to Hollitscher suggests that Zahn-Harnack had by then reversed her previous skeptical stance on refounding the DAB. There is no evidence that the BFUW presented the German women's case to the British occupation authorities. Still, in July 1947, shortly after Schiemann's sally in London, Zahn-Harnack took a further step toward the reestablishment of a German federation of university women by creating an "academic commission" within the Wilmersdorf Women's Union. It was chaired by the physicist Luise Holzapfel of the Dahlem-based Kaiser Wilhelm (from 1948 Max Planck) Institute for Silicate Research. Writing to the BFUW, Zahn-Harnack described the commission as a "little sapling" that might one day grow into a reconstituted DAB.[32] Like the old DAB, the academic commission in the Women's Union accepted only university and college graduates as members. The minutes of the commission's first session reveal that it intended to renew networks in response to "an urgent need to cultivate personal contact between the members of the various different disciplines and professions" and discuss "professional questions." The commission was also to "attend to those areas where women are already being disadvantaged vis-à-vis men in their professions."[33] As Zahn-Harnack told Hollitscher, this was a reaction to the reignited controversy over whether "dual earners" (in other words, married women in employment) should be allowed to take qualified jobs and over the massive enrollment restrictions that many western German universities were now once again imposing on women students.[34]

Apart from these points, with regard to "practical everyday issues" the academic commission largely applied the intergenerational approach developed in the 1920s. Older university women agreed to offer women students access to their private books and libraries, hoping in return for assistance in finding firewood and similar daily routines. Mutual practical help between seasoned academics and women students also included invitations to dinner and modest financial assistance for their university studies.[35]

Agnes von Zahn-Harnack kept Erna Hollitscher (and Elisabeth Schiemann during her Crosby Hall stay) informed of every stage in the evolution of the academic commission. The BFUW and the US and Canadian federations shored up Zahn-Harnack's efforts to reestablish university women's networks in Berlin by donating food and clothing. Starting in fall 1947, a series of packages was dispatched to Berlin via Crosby Hall; Erna Hollitscher was the driving organizational force for this campaign, as well. Despite the worsening food situation in Britain, high costs, and shipping problems, the Berlin women received a consignment of food, coats, clothing, and shoes monthly from late 1947 until 1950. Importantly, the flow of assistance remained unbroken even during the Soviet blockade of West Berlin in 1948–49. The academic commission used the goods to set up a clothing store, from which they issued garments to especially needy members or women students once a week. The distribution of the shipments in West Berlin was coordinated by Agnes von Zahn-Harnack, Luise Holzapfel, and Margarete Scherer, who taught English language and literature at the new Free University of Berlin. They viewed this material aid from their Allied colleagues as invaluable moral support for German women's reintegration into the international academic community.[36]

In summer 1949, the Wilmersdorf Women's Union academic commission opened a residence of its own for women students, picking up on a core pre-1933 DAB initiative to nurture a female academic tradition. Aided by the Free University of Berlin and the West Berlin city government, the commission rented and furnished a villa in the southwestern suburb of Zehlendorf, with a large garden and easy access to the Free University.[37] The building offered accommodation for 30 female students, while the academic commission could use the spacious clubroom for its meetings and conferences. It was here that the first meetings were held to prepare the refounding of the DAB and a Berlin local chapter.[38] The villa was also the venue for the new DAB's inauguration on

June 19, 1949, Agnes von Zahn-Harnack's sixty-fifth birthday. The minutes of the constitutive session itself have not survived, but it was certainly planned and carried out in close liaison with Crosby Hall. Almost all the executive committee members had by then spent time in Crosby Hall and were well known to the British women. Some of them, including Marie-Elisabeth Lüders, Emmy Beckmann, and Dorothee von Velsen, had also visited the United States.

Poor health prevented Agnes von Zahn-Harnack from taking on the presidency of the new DAB, and Emmy Beckmann, a senior official in the Hamburg education authority, was elected in her place. A year later, in August 1950, Beckmann and the Berlin librarian Luise von Schwartzkoppen attended the IFUW conference in Zurich as observers.[39] The question of readmitting the West German university women was discussed at this conference, although a decision was postponed to the next Council meeting, to be held in the Dutch town of Oosterbeek in 1951.[40]

As we have seen, the renewal of university women's networking in Germany was initiated by the IFUW via Britain and America. However, at local level it resulted in inevitable, and very specific, confrontations with the Nazi past. During the resumption of local and national networking between academically trained women in Germany's Western occupation zones, structures and dynamics arose that closely paralleled those undergone by the universities as they struggled to deal with the Nazi past: particular individuals whose political record, social status, and relationships with the occupying powers gave them special standing in the postwar context acted as "facilitators" in a process that Kai Arne Linnemann has called West Germany's "transition to a civil order." Linnemann's study of the Göttingen educationist Hermann Nohl vividly portrays this integrative mediation between actors with profoundly different experiences, careers, and complicities in the Nazi state.[41] For Linnemann, Nohl typifies the academic citizen of integrity who, like Zahn-Harnack, championed Wilhelmine values of education, bourgeois citizenship, and morality. Nohl became a societal facilitator because he dedicated his position and public reputation to the goal of helping his students find their feet in society again after the dictatorship, irrespective of whether they had been classified as "offenders" or "exonerated," or had recently returned from exile.[42]

Agnes von Zahn-Harnack took on this role of mediating between past and present as the West Berlin women's movement was rebuilt and female graduates began to organize again. At the constitutive assembly of the Wilmersdorf Women's Union in summer 1945, Zahn-Harnack had announced that she wanted to look forward, not backward. Every woman present, she said, knew of "our suffering and our culpability."[43] Each individual woman was to examine her conscience, a strategy that Zahn-Harnack considered imperative to the success of her project to rebuild a women's organization on democratic principles. Certainly, Zahn-Harnack's watchword—not passing public judgment but leaving it to individuals to face up to their own experiences and actions in the Third Reich—was one of the crucial preconditions for West German women in general, and university women in particular, to join forces again in the new organizations.[44]

University-trained women in other West German cities began recreating their networks in a similar way to the West Berlin case. The initiators, or in Linnemann's terms "facilitators," of this renewal were with few exceptions women of Agnes von Zahn-Harnack's generation who had been part of the DAB or other women's organizations prior to 1933. Almost all of them, furthermore, had personal connections with British or American women, acquired either directly or through other university women. In Freiburg, it was Johanna Kohlund, the *grande dame* of the local women's movement, who took up the networking role that Zahn-Harnack filled in Berlin.[45] Six years older than Zahn-Harnack, Kohlund was 69 when the Freiburg university women's organization was reestablished in 1947. She had been among the founding members of the first Freiburg women students' club in 1910, active in the Freiburg group of the women's education association "Frauenbildung—Frauenstudium" from 1913 and in its national executive from 1923. Kohlund also helped to establish the Freiburg local chapter of the DAB in 1929. Her membership of the German Democratic Party (DDP) and her pre-1933 political activities resulted in her expulsion from school teaching in 1934. Even so, Kohlund remained a member of the local DAB chapter, which continued to exist right up to the British bombing raid on Freiburg of November 27, 1944.[46] In 1947, she was elected as the first president of the newly founded Freiburg association of university women. Kohlund's facilitation of this new beginning included taking an active part in exonerating her predecessor, Maria Plum.

A highly respected lawyer in the region, Plum became president of the Freiburg DAB branch in 1931. She remained in office until 1944. Plum joined the NSDAP in 1933, and shortly afterwards was appointed to head the "district arbitration office" of the Nazi Women's League. Like all lawyers who had been Party members, in June 1945 Plum was initially suspended from her profession. Plum's denazification papers reveal how she defended her actions in the Nazi regime and interpreted her professional decisions from the vantage point of 1945. According to her statements, both her NSDAP membership and her work at the arbitration office were motivated exclusively by a commitment to the DAB.[47] It was to keep the Freiburg chapter alive, Plum claimed, that she had remained chair of the group even after its incorporation into the German Women's Agency. "All the events in those days," Plum emphasized in her denazification questionnaire, "were dominated by the struggle to continue the work of university women in Germany, of whom women lawyers were most severely affected." In Plum's account, the chapter's activities had not changed; it had only been required to invite a representative of the district Nazi Women's League to the individual meetings.[48]

Johanna Kohlund wrote one of the four notarized witness statements for Plum that were submitted to the denazification panel in 1946. She supported her colleague unequivocally, confirming that Plum had joined the NSDAP and the German Women's Agency in order to ensure that the Freiburg group was "left in peace." Kohlund emphasized that, "to our joy on all sides," Plum had managed to enable meetings of the previous members that included women classified by the Nuremberg race laws as "mixed-breed" (*Mischlinge*).[49]

In the postwar period, such references to having extended a helping hand to people of Jewish origin were part of the standard repertoire of justifications and exculpatory witness statements. Their purpose was to highlight the human decency of the accused person.[50] If Kohlund praised Plum's commitment to securing the participation of so-called *Mischlinge* in the Freiburg DAB's meetings, she remained silent on the exclusion of "full Jews" that took place under Plum's presidency. Neither the surviving documents of Plum's denazification proceedings nor the Freiburg group's historical self-portraits after 1945 make any mention of the Jewish lawyer Erica Sinauer, the Freiburg chapter's first president from 1929 to 1931.[51] It is no longer possible to reconstruct how and when Sinauer left the DAB. Like Plum, she had been a senior researcher at the University of Freiburg's department of legal history before opening her own legal practice. We know that Sinauer remained in Freiburg until 1940, then fled to France, where she was interned at Camp Gurs and deported to Auschwitz in 1942. She was killed there on an unknown date.[52]

Kohlund's intercession for Plum was welcomed by the Freiburg group. In 1949, two years after the group was officially reestablished, its members unanimously elected Plum as their president once again. She declined the position for the time being because her denazification process was still incomplete, and passed the presidency to the biologist Magda Staudinger. Plum nevertheless remained a key figure in the Freiburg association; the group's office was housed in her legal practice. In 1957, Plum resumed the office of president, holding it until her death in a car accident five years later.[53]

The reestablishment of the Karlsruhe DAB group and the election of its executive followed a similar pattern to that in Freiburg. The local chapter's reconstitution was initiated by members who had resigned in 1933, along with its former president, high school teacher Gertrud Carl, and several other women who had continued their membership after the federation's Nazification. Having been district adviser for research and girls' education within the German Women's Agency, Gertrud Carl was initially dismissed from public service by the denazification authorities, but appealed successfully against the verdict.[54] She managed to return to her old school, the Bismarck-Gymnasium in Karlsruhe, in March 1947. Like Plum's, Carl's denazification defense included the assertion that she had remained president of the Karlsruhe DAB after Nazification "in order to be able, wherever possible, to stand up for the interests of women's academic work with the relevant authorities."[55] In her address to the court, she claimed to have practiced resistance to Nazi schooling reforms.

However, it was not Carl herself but another high school teacher, Maria Roth, who carried out the reestablishment of the DAB's Karlsruhe group. Roth had campaigned alongside Carl in the DAB up to 1933, but stepped away when the Nazis came to power; during World War II, the Gestapo twice had her transferred for disciplinary reasons. After 1945, the Allies appointed her to chair a denazification panel.[56] In this role, Roth echoed Agnes von Zahn-Harnack's comment that nobody had the right to pass judgment on others for ideological reasons. She was fulfilling this "duty," she said, not in order to settle personal accounts, but to help build firm foundations for reconstruction.[57]

The official reconstitution of the Karlsruhe group in 1949 took place in Karlsruhe's first grammar school for girls, where Roth was now teaching again. Almost all the members of the Nazified DAB, including Gertrud Carl, were among the 20 university-trained women who gathered there. Maria Roth and her retired colleague Luise Pander ensured the smooth and democratic founding of the new Karlsruhe DAB. Pander took on the presidency until Carl was reelected in 1953.[58]

In Hamburg, Emmy Beckmann, a high-ranking education official dismissed in 1933, began to work for a reconstitution of her local DAB group in 1945, shortly after being reappointed to the city's education authority. In Bremen, Johanna Lürssen became the new president: responsible for higher schooling in Bremen until 1934, in July 1945 Lürssen resumed her civil-service career in education as Bremen's superintendent. In Marburg, it was linguist Luise Berthold who brought the city's university women back together, while in Heidelberg chemist Käthe von Kuenssberg played the key role. When her husband Eberhard von Kuenssberg, a renowned legal historian, died in 1941, Kuenssberg was left without the protection of her "Aryan" spouse and threatened with deportation; she had to go into hiding several times. In 1946, she emigrated to England, partly to join her five grown children, who had settled there in 1938, but decided to return to Heidelberg in fall 1948.[59] There, she was urged by the US occupation authorities to become the first president of a German-American "ladies' club," and then, in 1949, brought the city's DAB group back to life.[60] The renewal of the DAB in Frankfurt am Main was initiated by journalist Gabriele Strecker after returning from a visit to Crosby Hall in 1948.[61] A former physician, now in charge of the new women's radio station within the regional broadcasting service and later a Christian Democrat member of the West German Bundestag, Strecker was only 44 at the time, certainly Frankfurt's youngest, and probably also its best-known, female academic citizen. Her endeavors were backed by the high school teacher and later federal president of the DAB Marga Anders—once a protégée of the Mannheim professor Elisabeth Altmann-Gottheiner, who had played such an important role in founding the original DAB before her unexpected death in 1930.

In Göttingen, the driving force behind the revival of the local chapter was the principal of the city's high school for girls, Ida Hakemeyer. A student and friend of Hermann Nohl's, Hakemeyer had spent a year with members of the AAUW branch in the university town of Ann Arbor, Michigan.[62] In 1952, the Göttingen group counted 63 members, among them Lotte Möller and Charlotte Lorenz, both qualified university lecturers and both dismissed from the University of Berlin in 1945 due to their former NSDAP membership.[63] Lore Liebenam, once the Nazi president of the Halle local chapter, also joined the Göttingen group.[64]

In several smaller West German cities without a university, local groups were set up by former DAB members who were originally from Berlin but ended up in the Western occupation zones after 1945. A group was established in the Westphalian town of Detmold in 1948 by Anna Schönborn, who had directed the DAB's welfare for women students in Berlin until 1933 and represented the federation at the momentous IFUW Council session of 1934 along with her Nazi colleague Friederike Matthias. Schönborn had been stranded in Detmold two years earlier after her evacuation from Berlin.[65] In Celle, Else Wex, a PhD in economics, initiated a DAB chapter. Born in Munich, Wex had moved to Celle in 1906 and then to Berlin in 1922, where she contributed to the women's movement and liberal party politics as a member of the DDP and ADF (Allgemeiner Deutscher Frauenverein) executives. In 1945, aged 61, Else Wex joined the Celle city government for the SPD, and later became president of the local branches of the Social Democratic welfare association and the federation of women's organizations Deutscher Frauenring.[66] Dorothee von Velsen, a historian and one-time president of the ADF, helped to found a group in Munich.

Almost all of these facilitators were older women who looked back to the Wilhelmine period and aspired to recreate new organizations "in the old spirit," as Anna Schönborn put it when the Bremen group opened in 1948. Like Zahn-Harnack, the 68-year-old Schönborn had little faith in her juniors: she described young women as having been "imprinted by National Socialism for fifteen years," while women of the intermediate generation "lacked initiative" and fled from responsibility.[67]

Yet that intermediate generation—women who had embarked on their careers when the Third Reich began and pursued them with greater or lesser concessions to the regime—joined the post-1945 DAB in large numbers. To be sure, at first it was the older women who took the most active role both locally and nationally, describing themselves as untouched by Nazism. Thus, Emmy Beckmann (born 1880), a Weimar DAB representative and superintendent for girls' secondary education in Hamburg, was elected as the organization's first president in 1949 because, as Johanna Lürssen put it on the occasion of Beckmann's seventy-fifth birthday, she enjoyed "the confidence of large numbers of women in Germany" and had a good name within the international women's organizations.[68] But among the approximately 100 "leading women of the academies, universities, colleges, administration, and teaching"[69] who assembled in Stein, near Nuremberg, for the DAB's second annual meeting in 1951, there were also several representatives of the younger generation, such as the university lecturers Auguste Hoffmann, Ilse Esdorn, Charlotte Lorenz, and Gertrud Savelsberg.[70]

At the federation's third annual meeting, held in Göttingen in 1952, sports physician Auguste Hoffmann was elected as the new DAB president. Judging by the report on the meeting, it had proved hard to find a successor for Emmy Beckmann when she stepped down.[71] However, on October 6, 1952, the local newspaper reported that Hoffmann seemed to offer the DAB a "younger force who is capable of succeeding Emmy Beckmann for the future representation of the Federation's goals."[72] This was a genealogy that effectively obscured Hoffmann's nationalistic and *völkisch* stance during Weimar and the historical caesura of the Third Reich.

The obfuscation may be partially explained by Auguste Hoffmann's professional biography in the earliest days of the Cold War. From 1946–47 onward, the East-West conflict played an important role in advancing renewed solidarity and networking among university women in western Germany and especially in the western sectors of Berlin, deep in the Soviet occupation zone. Hoffmann was appointed a professor when the University of Berlin, located in the city's Soviet sector, reopened in January 1946, but she faced concerted political attacks there due to her former membership of the National Socialist Teachers' League, and finally, after five years, resigned on September 1, 1951.[73] At the time of her election as president of the DAB, Hoffmann was in the process of building a new career in the western sector of Berlin. In this delicate situation, the federation presidency promised to raise her profile and gain her access to networks throughout West Germany. During the election process, there was no mention of Hoffmann's activities as a sports physician in the Nazi state or of the political background to her difficulties in the Soviet zone, later East Germany, which were specifically due to her previous Nazi allegiances. Instead, what was known about Hoffmann in DAB circles was her experience of the East German universities' communist restructuring. Her inaugural speech in Göttingen, on the "forms of academia established on the pattern of the 'Soviet economy,'" made an "extremely strong impression" on her sisters.[74] Hoffmann was able to bridge the period of professional uncertainty between 1952 and 1955 as president of the DAB; after being appointed full professor of human biology at Berlin's College of Education in 1955,[75] she did not stand for reelection.

The sparse post-1945 statements by women university and college graduates about their experiences during Nazism bear traces of their era's political discourse on dealing with the past. Like other organizations of the time, the DAB managed its postwar reconstruction with scant references to the Nazi period. Those who had played an active role in the racialized system of *Volksgemeinschaft*—for whatever reason and with whatever private reservations—seemed convinced of the rightness of their actions; in contrast, there was rigorous self-criticism among those who had tried to evade the system. The post-1945 statements of Agnes von Zahn-Harnack, for example, several times comment on questions of personal responsibility during the Third Reich and stress that she did not exempt herself from a "bad conscience" due to her "sins of omission."[76] Linguist Luise Berthold, too, admitted her own political errors of judgment when she reflected on the dictatorship's final years.[77] And when Marie-Elisabeth Lüders condemned the recrudescence of anti-Semitism in West Germany in her famous 1960 speech as honorary president of the German Bundestag, she explained that she felt compelled to speak out by West Germany's almost complete failure to confront its recent past. An "explicable, but indescribable *inner* dread" was responsible for the silence on the Nazi period, "because we are afraid of looking at our own distorted mirror image of past years."[78]

Lüders's analysis pinned down the reasons why, when they reactivated their networks, German university women had so little interest in addressing the Nazi-era past of the DAB and its members. As Lise Meitner wrote to the sister of her close friend Elisabeth Schiemann from Swedish exile in 1947, "the past is over and done for them." Meitner had realized that after the war, friends and colleagues who had stayed in Germany were not—despite their opposition to the Nazi regime—willing or able to reflect on their own inevitable involvement in the dictatorship with a political analysis that went beyond questions of personal guilt.[79]

In view of Meitner's observation, DAB members' statements about their past in the Nazi state might be interpreted as an admission that the whole of society was involved in Nazism—albeit an admission that was indirect, repressed, and denied.[80] West German university women's resumption of networking on pre-1933 lines reflects a more general state of postwar society, a widespread psychological need for moral rehabilitation and reintegration. According to historian Norbert Frei, the extraordinary force of this need accounts both for the failure to publicly commemorate the victims of Nazi terror and for the fact that even those who were persecuted during the Third Reich supported the "course of inner pacification."[81] Frei's hypothesis is borne out by the detail of the DAB's rebirth. In all the minutes of the academic commission's meetings in Berlin during 1947, there is only one indication of the new DAB's members taking any interest whatsoever in the fortunes of their previous Jewish colleagues: the commission's chair, Luise Holzapfel, asked whether anyone knew what had become of her fellow physicist Marie Wreschner, who had been unable to emigrate from Germany in time. None of the women present had an answer.[82]

Contact with Jewish DAB members who had managed to escape after 1933 was, as in the case of Lise Meitner and Elisabeth Schliemann, restricted to individual meetings and correspondence. Meitner expected a public apology for the wrongs committed under Nazism, specifically from her fellow academics and from colleagues such as Schiemann and Otto Hahn, her best friends during her Berlin days; none was forthcoming. Although in 1958 the DAB executive won Meitner's promise to speak at the annual general meeting, to be held that year in Berlin, the official documentation shows no hint of an intention to welcome Meitner not merely as a renowned physicist, but also as a former DAB member and victim of the Nazi regime. In turn, Meitner's lecture manuscript includes no personal references.[83] When Meitner had to cancel at short notice for health reasons, her former colleague and Nobel Prize winner Otto Hahn was persuaded to replace her, and thus by sheer chance the annual meeting featured a personal appreciation of Meitner's research achievements and persecution by the Nazis, although Hahn touched only briefly on the circumstances of her escape via the Netherlands to Sweden in 1938. Hahn's speech failed to mention Meitner's crucial role in the discovery of nuclear fission, a discovery generally attributed above all to Hahn himself.[84]

The DAB as a whole did not initially publicly acknowledge the persecution and murder of its own members.[85] Neither did the women who had faced Nazi persecution, survived, and joined the local groups that formed immediately after the war see any reason to draw attention to their own experiences. Biologist Käthe von Kuenssberg, the founder and first postwar president of the Heidelberg group, dedicated just a few hazy sentences of her published autobiography to the three years she had spent hiding from deportation.[86] She reserved her descriptions of these experiences for her own children.[87] Only rumors indicated that Marga Anders, the longtime president of the Frankfurt group who was elected DAB president in 1958, had been unable to present an "Aryan certificate" because of her one Jewish grandmother, and had only managed to continue working as a teacher thanks to protection by her school's principal.[88] Erna Scheffler, later a judge at the Federal Constitutional Court, who joined the Karlsruhe DAB group in the early 1950s and became its president in the mid-1960s, likewise refused to discuss publicly how she had survived by hiding in a Berlin allotment garden. Notably, however, it was the DAB members who had been victims or resolute opponents of Nazism who showed the greatest interest in recreating the federation's international connections. They early on campaigned for readmission to the IFUW and sought links with university women outside Germany. By her own account, Käthe von Kuenssberg spent every free moment working for the International Federation of University Women.[89] Marga Anders and Erna Scheffler, too, set a high priority on reanchoring the DAB in its old international contexts. Even if these women did not say so explicitly, they seem to have been especially eager to position themselves as academic citizens in a transnational network, perhaps partly as a way of vouching for the "good" Germany.

If the DAB's activities during the 1950s nevertheless focused chiefly on the internal German context, and the lively academic internationalism of the late 1920s was never resuscitated, the reasons for this went beyond Germany alone. Although the Cold War accelerated the Federal Republic's integration into the West, for the transnational network of university women as a whole the East-West confrontation had a stultifying effect. Bloc formation meant that women in East Germany did not join the DAB when it was refounded in 1949; the associations of university women in Central and Eastern Europe also ceased to exist and abandoned their membership of the IFUW. In the United States, the AAUW's internationalist outlook attracted McCarthyite scrutiny from Congress. As a result, in the 1950s this biggest and wealthiest member of the IFUW had to devote extensive resources to defending the previous policies of its International Relations Committee against McCarthyite denunciation campaigns. For the AAUW, that defense involved dissociating itself from the interwar and wartime principles of internationalism and adapting to the anti-communist rhetoric of the day.[90] It also meant parrying personal smears from Senator McCarthy himself.[91] From summer 1948, McCarthyism threatened the professional career and personal future of political scientist Esther Caukin Brunauer, the woman who had constructed and managed the AAUW's assistance to refugees with great commitment up to 1944, when she left the organization for a post at the US State Department.[92] In 1951 she lost that position, baselessly accused of sympathizing with communism and having associated with communists in the course of her international work at the AAUW and later as the State Department's UNESCO liaison officer.[93] Until the end of the 1950s, the AAUW leadership was preoccupied with countering accusations that the "Pink Ladies of the AAUW" cherished pro-communist sympathies.

A bulging file with this label in the AAUW archives testifies to the destructive intensity of the denunciations.[94]

Alongside these political developments, which ate away at the AAUW leadership's previous international orientation, personnel changes reflected the underlying deterioration of women's academic career opportunities in postwar America. The generation of US women college presidents and professors who had left such a lasting mark on the IFUW's policies and external image right into the war remained largely without successors. As Margaret Rossiter has shown in detail, prosperity after 1945 resulted in the remodeling of many US universities into full-fledged research institutions, which was accompanied by a tacit, but rapid elimination of women from leading positions in higher education.[95] With the disappearance of college presidents and professors from the AAUW and IFUW leadership, accompanied by Cold War anticommunism in the West, scientific internationalism lost the centrality to the IFUW's agenda that had proved so attractive and unifying in interwar Europe. And despite the well-documented close relationship between the United States and West Germany in the 1950s, for the female academic network such links did not go beyond officially sponsored US trips by German university women.[96]

In the DAB's policy, the concerns of academic professionalization remained largely consistent from Weimar through the Nazi period and into the era of the Federal Republic, with the topic of women in academia now attaining renewed urgency.[97] The reason was that postwar West Germany, just like the United States, had signally failed to create new opportunities for women, and especially for university lecturers. Apart from Auguste Hoffmann, Doris Schachner, and Elisabeth Schiemann, who obtained full professorships in 1946 and 1949, none of the women with professorial qualifications who continued their research during Nazism received a full, civil-service professorship after 1945. Most of them were employed in positions inferior to their qualifications and achievements, forcing them to live in poverty.[98] Furthermore, when former NSDAP members were dismissed from the universities in 1945, the women among them faced a more difficult route back into science and scholarship than their male colleagues did—despite the fact that the men had in most cases been far more deeply involved in the Nazi system.[99]

However, whereas appeals to end discrimination against university and college teachers came from a small number of lecturers in the DAB, the more general call for legal and practical professional equality elicited an enthusiastic response throughout the organization. Campaigning for this demand was not only highly relevant to contemporary life, but also had advantages related to the politics of dealing with the past. For one thing, such activism offered DAB members a clear set of political coordinates within the new democratic setting. Among the advocates of equal rights in higher education and society, the most prominent were those DAB members who, during Nazism, had tried to continue representing the interests of a female academic elite under totalitarian conditions and within the setting of Nazi organizations. The objective of securing societal and professional participation for women in general, and graduate women or women academics in particular, had been crucial ever since the DAB's founding in the Weimar Republic, but during the Nazi period it had also served several members of the federation, with almost unbroken continuity, as a vehicle for asserting their place in the racialized Nazi community. After 1945, the call for equality gave these women a way to raise their voices again, this time in a democratic key.

Because of its historical continuity with the pre-1933 era, the demand for equal rights also enabled women in West Germany to downplay their own associations with Nazism. While it was accurate to call the Nazi regime a "male state" with a special hostility to highly qualified women, this frequently repeated description soon became little more than an empty formula. Reference to the "masculinity" of Nazism allowed every woman's career in the period to be narrated as a story of victimhood, even when the career in question had actually been marked by affinities with Nazi ideology or had been as successful as was possible under the circumstances.[100] Beyond all their political differences regarding Nazism, the members of the DAB shared an unspoken agreement that "as women" they had faced far more difficulties in the public and professional life of the Third Reich than their male colleagues, and were not willing to take a back seat again.

The Nazi Past and the Resumption of Tradition: Karlsruhe 1968

Under these circumstances, it was difficult for the German university women to achieve the DAB's one-time core objective of establishing a female academic tradition. This section will show just *how* difficult, by detailing the attempts of DAB members to reflect upon the history of their federation.

Shortly before her death in June 1950, Agnes von Zahn-Harnack wrote a history of the Deutscher Akademikerinnenbund from 1926 to 1933. Her text gives a comparatively detailed account of the process of Nazifying the DAB. Zahn-Harnack had resigned from the DAB executive on May 18, 1933, in protest at its alignment with the Nazi regime—yet her narration is remarkably silent on both events. For the initiated, her artfully wrought text probably did reveal the historical course of events but, in line with her approach to rebuilding the Berlin women's movement as discussed earlier in this chapter, Zahn-Harnack clearly did not want her history to be read as a personal settling of accounts. The text favors passive grammatical constructions that enable her to outline the Nazification process without actually naming the women involved and without indicating who brought about the May 18 decision not to disband the DAB but to mold it into conformity with the Nazi regime. In Zahn-Harnack's account, her own executive resigned at the crucial May 1933 meeting "because the influence of Nazi tendencies could no longer be halted."[101] The new president was not, Zahn-Harnack continues, a Nazi Party member; having "misjudged the possibility of holding the DAB back from further movement in this direction," she "therefore" resigned from office "shortly afterwards."[102] These remarks refer to Johanna Willich, a member of the nationalist-conservative DNVP, who was elected to the DAB executive on May 18 together with the Nazis Friederike Matthias and Lea Thimm. The eventful year between Willich's election as president in May 1933 and her resignation in late May 1934, ordered by the Reich Women's Leader Gertrud Scholtz-Klink, is collapsed here into a short period apparently free of major repercussions. A veil is drawn over Willich's explicit agreement to continue the federation on Nazi lines. Zahn-Harnack accords far more weight to the DAB's incorporation into the Nazi German Women's Agency: this "subordination," she writes, was carried out in early 1935 under Willich's Nazi successor at the head of the DAB executive. Zahn-Harnack thus defines the appointment of Friederike Matthias as DAB president in May 1934 as the essential rupture in the federation's post-1933 history. She fails to mention that the NSDAP activists Matthias and Thimm had been elected to the new executive alongside Willich in 1933. Instead, she underlines the emphatic claims by the Nazi president (that is, Matthias) to the IFUW Council in Budapest in September 1934 "that the DAB continued to condemn all discrimination against members for reasons of race, religion, or politics." Zahn-Harnack points out that the president "must already have known that this could no longer be correct." This carefully chosen wording unmasks Friederike Matthias's speech in Budapest as a tactical maneuver intended to deceive the IFUW Council about the DAB's anti-Semitic exclusionary policy—but on the other hand, the passage leaves open how the DAB actually stood on the "Aryanization" of its members after May 1933. In fact, at the time of the Budapest meeting the exclusion of Jewish university women from the DAB had long since been completed. It was Johanna Willich, the president elected in May 1933, who instructed all local chapters to expel their Jewish members, and she did so as early as summer 1933. The exclusion of the Jewish women was implemented by the very same Weimar officials and members who had backed the Nazification of the DAB in May 1933, long before a Nazi woman was installed to lead the federation. Yet in Zahn-Harnack's account, the implementation of the anti-Semitic policy appeared to be a purely Nazi affair. She wrote that in 1934—and thus apparently in response to the anti-Semitic measures—the DAB membership shrank first to around 200, then to a mere 40, because the members had refused "to join what was now a National Socialist body."

Through her particular weighting of the facts and her suggestion of interpretations that exonerated DAB actors, Zahn-Harnack contributed to a reading of history that would make it easier for the federation's Nazi past to be simultaneously remembered, reinterpreted, and repressed. But if Zahn-Harnack at least gave some sketch of the Nazification process, most of the DAB's public statements before and after made no reference whatsoever to the federation's turn to Nazism.[103] Thus, Emmy Beckmann's call for the reestablishment of the DAB in July 1949 merely noted: "The old federation of university women, founded in 1926, was forced to dissolve itself in 1935, like so many other associations."[104] Subsequent references to the past followed this pattern, even when they were clearly based on Zahn-Harnack's history. Abbreviated versions of the text reinforced the potential for exoneration that Zahn-Harnack had offered, while embroidered versions recast the DAB and its members as actual opponents of Nazi anti-Semitism. At a 1959 event commemorating the seventy-fifth anniversary of Zahn-Harnack's birth, DAB president Marga Anders said that in 1933 the DAB's executive had resigned and membership had fallen to just 40 women, at which point the federation had withdrawn from the IFUW on the grounds that "in the new Germany there was no longer any place for an organization grounded on intellectual or educational distinctions," but most importantly "because the DAB would not adopt an Aryan paragraph into its statutes as the National Socialist government demanded."[105]

It was not until the international IFUW conference of 1968 that the DAB's members were confronted with a more forthright view of the past. This was the first IFUW meeting ever to be held in Germany, the gathering planned for Berlin in 1936 having been rescheduled to Kraków. The German women were all the more gratified when the international executive accepted their invitation to Karlsruhe, hoping that the prestigious event would finally erase the remaining resentment toward the DAB, especially as it would coincide with the festivities to mark the fiftieth anniversary of the IFUW's founding in 1919. The first speaker on this historic occasion was the British member Marguerite Bowie, treasurer of the IFUW from 1936 to 1947 and then its vice president until 1950. To the audience's surprise, Bowie did not stick to courtesies. Instead, she devoted a substantial part of her speech to the dramatic period in the 1930s and 1940s when the IFUW had worked to assist refugees from Nazi-dominated Europe. In particular, Bowie's account of the 1934 Council meeting in Budapest did not spare her hostesses' feelings.[106] Her description of the Nazi DAB president's appearance at the meeting was met with incredulity by the bewildered German women. According to biologist Magda Staudinger from Freiburg, Bowie said that "a German woman came and offered the Federation money for fellowships if it would maintain the membership of the DAB, by then aligned with the regime."[107] In response, the IFUW Council had resolved that no member association would be allowed to refuse admission on racial, political, or religious grounds, and the German club was expelled.

The German university women felt humiliated by Bowie's speech. Especially for those attending an IFUW conference for the first time, it prompted indignation and embitterment. Internationally experienced officials like Staudinger, too, were staggered at being confronted—and by a British woman—with details of the federation's past that "are unknown to us and that, as regards the DAB's self-dissolution, were told us very differently by the women who were members at the time," most of them now dead.[108] Bowie had a reputation in the IFUW for occasionally forgetting diplomatic niceties and being a "bull in a china shop."[109] After her speech, she declined to enlighten her German colleagues about the background to her information.[110]

Some angry German delegates interpreted the speech as an atrocious "act of vengeance" concocted by an "English Jew,"[111] but there is no reason to believe that Bowie harbored any particular resentment against the Germans. In fact, after World War II, Bowie had played an important mediating role to their benefit. In the run-up to the negotiations over readmitting the DAB in 1951, she visited many local chapters in West Germany; on the basis of her impressions, she argued strongly in favor of the controversial admission of Germany into the IFUW.[112] However, nobody at the Karlsruhe event in 1968 remembered her visits.

The DAB executive felt pressured by Bowie's speech to "clarify the actual facts and circumstances." A high-ranking committee was hastily assembled, including the retired Constitutional Court judge Erna Scheffler, former Federal Minister of Health Elisabeth Schwarzhaupt, and Magda Staudinger. Its brief was to retrace the DAB's final months in the Third Reich, then publish a reply to Bowie in the IFUW newsletter,[113] aiming first and foremost to prevent the DAB losing even more face in the international organization. Staudinger had lobbied enthusiastically for the conference to come to Karlsruhe, and felt a special responsibility to shed light on the case for her younger DAB colleagues so as "to avoid damage to the tender shoots of international work, which anyway find it difficult to flourish in what is now a very provincial country."[114] She seems to have made the historical investigation a matter of personal pride. In 1933, Magda Staudinger and her husband Hermann, professor of biology in Freiburg, had openly protested the dismissal of Jewish colleagues from the university, an act that exposed them to considerable political duress.[115] She was "quite unnerved" to find that there could be inaccuracies in the version of history circulated by the DAB's Weimar members, according to which the federation had dissolved itself voluntarily without waiting for Nazification.[116]

The course of Staudinger's research indicates how profoundly knowledge of the events early in the Nazi dictatorship had been obliterated by the end of the 1960s. None of the German women with whom Staudinger now corresponded had "ever heard anything" about a Nazi president of the DAB or about conflicts with the IFUW between 1934 and 1936.[117] The most eminent members from those days were no longer alive, and it proved very challenging to locate reliable documentation or DAB papers from the period. In the end, pertinent information was obtained with the help of British and Swiss IFUW members. Staudinger was dissatisfied with this material, which did not fully illuminate the question of how the DAB was finally dissolved in 1936.[118] Nevertheless, the documents collected by the committee provided enough information to retrace the DAB's Nazification in May 1933, its altercations with the IFUW, and its incorporation into the German Women's Agency.

Staudinger recorded her findings in a two-page chronological outline that she circulated in 1969 along with the documents.[119] It is striking that Staudinger granted no significance to the DAB's crucial fall from grace on May 18, 1933. She made a fleeting reference to the "reelection of the executive," but neglected to mention that on the same occasion the board had decisively rejected Lüders's proposal either to temporize or to disband the federation, instead voting for Nazification and electing the NSDAP activists Friederike Matthias and Lea Thimm to the new executive. Staudinger preferred to emphasize that the DAB was given Matthias as its Nazi president in late May 1934 and that by September 1934 its membership had melted away to 200 women.

In line with this narrative, the final version of the justification concluded in very general terms (and without mentioning May 18, 1933) that when the Nazis came to power the DAB, like all other organizations in Germany, was forced to reorganize its executive and exclude its "Jewish" members. Matthias had been appointed as an acting leader in 1934 in order to implement these instructions—but, wrote Staudinger, even before Matthias took office most of the DAB's member organizations had dissolved, and afterwards the total membership shrank even more. In Budapest, therefore, Matthias represented a DAB that was "reduced in number and nearly non-existent."

Staudinger nowhere found any references supporting Bowie's statement that Matthias as a "representative of the DAB offered money for fellowships in order to make the IFUW retain the 'Aryanized' DAB as its member."[120] She thus advised her colleagues that the DAB's official vindication should not present any assumptions regarding the truth or falsity of Bowie's assertion, but simply ignore it. The text on which the committee finally agreed emphasized that Matthias had gone to Budapest only as an acting or provisional delegate; the remaining rump organization had been far removed from the "old" DAB. Before its official demise, the DAB had quietly wound itself up, with some local chapters continuing to meet in private. Shortly afterwards, the official seal had been set on its dissolution.

Despite its initial objective, then, the committee had once again written the federation's Nazification out of history. As a result, the narrative of the DAB's voluntary dissolution spread even more easily than before. Numerous statements in the 1970s and 1980s claimed that the DAB as a whole ceased to exist not in 1936, but in 1933, and disbanded on its own initiative. Thus, a 1978 article by the DAB president at the time, Dorothea Frandsen, asserted that the organization had disbanded under Marie-Elisabeth Lüders's leadership in 1933, "an act that took on particular significance after World War II when the time came to discuss readmission to the IFUW" in 1951.[121]

A glance at the minutes of that 1951 IFUW Council meeting shows how deeply rooted the myth of the DAB's voluntary dissolution had become by the mid-twentieth century. At the meeting in Oosterbeek in the Netherlands, the DAB was represented by Elisabeth Lürssen, Magda Staudinger, and Käthe von Kuenssberg, the president of the Heidelberg group. The official transcripts record an unusually controversial, politically and emotionally highly charged debate on whether the reestablished DAB should be readmitted into the IFUW. Especially energetic opposition was voiced by Sophia Berger Mohl, representing Israeli university women. Like Bowie, she had attended the Budapest Council meeting in 1934, and believed the current political climate in West Germany gave little reason to hope that the Germans had truly walked away from Nazism.[122] In remembrance of the victims of Nazi crimes, and out of respect for the suffering of the survivors, she therefore intended to vote against the German women's readmission.[123] Responding to this speech, it was Marguerite Bowie who found a compromise which, after a lengthy discussion, Mohl was able to accept: the IFUW should readmit the DAB precisely *because* it did not want the German crimes to be forgotten. The IFUW's task was to promote friendship and understanding, and only as members could the German university women be integrated into the international community of peace and bound to its goals. Admission would also be a way to support liberal tendencies within Germany. In the course of her visits, Bowie said, she had become convinced that the DAB was championing democratic interests, especially in terms of women's rights.[124] With the help of Bowie's advocacy and other personal pleas to the Israeli delegate, a unanimous resolution to readmit Germany was finally reached. Probably in order to show how effectively the IFUW applied the rules of international understanding, the report on the meeting stressed that Mohl was among the first to congratulate Elisabeth Lürssen when she took her Council seat for the Federal Republic.[125]

Looking back on the meeting in 1968, Kuenssberg wrote that, in Oosterbeek, Lürssen had "insisted that the German groups had dissolved themselves."[126] Staudinger, in contrast—who, after all, had helped to investigate the accusations raised at Karlsruhe in 1968—portrayed the crucial negotiations in Oosterbeek as a sentimental reconciliation scene. In 1978, she told Dorothea Frandsen that as far as she could remember, the resolution had been passed after some further questions to Lürssen, whereupon "a Jewish colleague came up to Frau Dr. Lürssen, embraced her and kissed her. This act of admission was nothing but a readmission, as the IFUW had never expelled the DAB, but only noted its self-dissolution and held its membership in abeyance."[127]

There is much to suggest that the growing embellishment of the DAB's past for its members paralleled a wider trend in 1980s West Germany, the emergence of an increasingly broad-based public discussion of the horrors of Nazi annihilation policy.[128] This would explain why the many texts published in that decade to commemorate the sixtieth anniversary of the DAB and its branches also date the federation's dissolution to May 1933 and extol it as a glorious chapter in the history of educated human decency.[129] A chronicle of the DAB written by the Frankfurt group in 1989 claimed: "When the Nazi government demanded that the 'Aryan paragraph' be adopted into its statutes, the DAB resolved unanimously to dissolve itself. The Jewish members abstained."[130] This may possibly have been what happened in Frankfurt. However, taken out of its specific, local context and extrapolated onto the DAB as a whole, the supposed self-dissolution became a master narrative of the federation's past, brought up to date for the sixtieth anniversary celebrations. A much-cited history of the DAB published in 1987 by Ella Barowski, president of the Berlin chapter for many years, acclaims the "DAB's self-dissolution" in 1933 an example of its "clear-sighted and steadfast" resistance to "the temptations of Nazism."[131]

Correspondence with a member of the Frankfurt group in 1999 indicates that the myth of the DAB's self-dissolution derives from a particular historical misattribution. To substantiate her assertion that the federation had dissolved itself rather than submitting to Nazification, this member sent me a copy of the minutes of a May 1933 board meeting. That meeting was, however, not the last session of the DAB board on May 18, 1933, but the historic final meeting of the BDF under Agnes von Zahn-Harnack's direction on May 15, 1933. These minutes had been regularly co-opted for the DAB's history, sight unseen, since the late 1970s—with the misleading addendum that the DAB had belonged to the BDF.[132]

The DAB's difficulties in dealing with its past are still evident in the twenty-first century. The Karlsruhe group possesses important and incriminating documents from the period of Nazification, but refuses to grant scholars access to the material.[133] Although the federal executive has now abandoned the heroic tenor of its 1980s statements, a visitor to the DAB website in 2013 will find a time line that yet again recycles fragments from Agnes von Zahn-Harnack's 1950 report: "1933: In May the executive resigns because Nazi tendencies are becoming increasingly dominant.... 1935: The German Federation of Academic Women is subordinated to the German Women's Agency. As a result, its members no longer enjoy racial, religious, and political freedom. This is followed by the definitive dissolution of the Federation."[134] By reconstructing the DAB's Nazification and unpicking the specific circumstances surrounding the writing of Zahn-Harnack's 1950 text, we can disentangle the ambivalence of such statements and reconcile them with the historical facts.

9
Conclusion

This book has investigated the four decades of networking history that lie between Caroline Spurgeon's momentous tour of the United States in fall 1918 and the IFUW's fiftieth anniversary, celebrated in the German university town of Karlsruhe in 1968. The IFUW emerged from an inter-Allied military and educational strategy developed in Britain and the United States at the end of World War I. Explicitly directed against the Central Powers, the project aimed to standardize the higher education systems of the Allied nations and intensify exchange between them; the ultimate aim was to quash Imperial Germany's scientific dominance and its influence on the United States. It was in the context of this educational offensive, backed by the governments of Great Britain, France, and the United States, that American and British women university lecturers stepped into action. With the war over, the League of Nations established, and women's suffrage finally achieved in key countries, many academically trained women felt called to transfer their political activism from the national to the international stage. To this end they founded an academic network, modeled on existing national organizations, that addressed university- and college-trained women across disciplinary and national boundaries. Their aim was to cultivate an elite of educated female global citizens. The dense web of contacts and friendships they wove was to make the new International Federation of University Women a pioneer of international understanding, at the same time furthering cross-border strategies for the advancement of women in science and society.

The drive to establish the IFUW was welcomed enthusiastically in much of Europe and North America. The organization burgeoned, drawing on the incipient international female educational elite that had been taking shape in the universities of the Western world since the late nineteenth century. Many women had spent time studying abroad—and thus building international contacts—in order to overcome barriers to higher education in their home countries. The innovative power of the IFUW's founders lay in their ability to bring together the first generation of female university graduates after World War I and to transform these women into a viable political force.

The IFUW distinguished itself from other international bodies of the day through the networking structures it fostered from the early 1920s onward, notwithstanding the difficult economic situation in Europe. IFUW fundraisers built international guesthouses in Washington, DC, Paris, and London; of these, the London clubhouse Crosby Hall became the federation's special intellectual and social hub. The "spirit of Crosby Hall," analogous to the "spirit of Geneva" that characterized enthusiasm for the new League of Nations, was legendary. Crosby Hall was the showpiece and prototype of a female academic tradition hitherto unknown in much of Continental Europe, but with roots stretching back to the nineteenth century in the United States and Britain.

With the institution of an international fellowship program in 1922, the IFUW acquired additional credibility in both academic and political circles. In no small measure, the program created the conditions to overcome the rifts of World War I. For many recipients, an IFUW international fellowship unlocked opportunities to establish academic careers in their own countries. IFUW fellows made good use of the female network's personal and professional connections during their research abroad, while also benefiting from the chance to work alongside the leading male proponents of their research fields.

The IFUW encouraged academic women to set up national organizations and join a first-of-its-kind international network. Nowhere was the challenge to its internationalist ethos greater than in post–World War I Germany. The vanquished nation was initially held at arm's length, and rapprochement with the Anglo-American organization did not begin until 1923. Thereafter, however, events moved quickly, with the DAB forming in 1926 and immediately joining the IFUW. University-trained women in Germany, much more readily than their male counterparts, seized upon scientific internationalism as an opportunity. Prewar contacts within the international women's movement facilitated this development, for the women's movement had maintained the ethos of internationalism throughout the war to a greater extent than had the scientific organizations and academies of men. In Weimar Germany, the IFUW offered women graduates (and especially women academics) a template for a new form of identification that was explicitly both female and educated. For the first time, women were encouraged to feel part of an academic elite defined by gender, a group that regarded itself not as a professional lobby but as the representative of a social and intellectual caste that laid claim to a leadership role in society. Equally important were the academic and political links with other countries that the IFUW provided. German university women used the IFUW for scholarly exchange and research, but also as a political platform to test a new understanding of women's politics. Now that women had won the vote, the pursuit of "cultural politics" in the German mold, frequently in confrontation with the IFUW, seemed an excellent way to fulfill their new role as citizens. Here, a transnational perspective reveals the full dynamism of the history of the women's movement in Weimar Germany.

The DAB's formation in the Weimar Republic also drew inspiration from Anglo-American efforts to cultivate a female academic tradition, modeled so strikingly by the IFUW. The new German organization campaigned hard for the interests of women students in the male-dominated universities, and sought to follow the example of their US and British sisters by setting up "women's halls" where students could meet undisturbed to work, rest, converse, and get to know their seniors on more collegial terms. The Berlin chapter of the DAB was in the vanguard of such endeavors. Its Helene Lange Home for women students in the City Palace, close to the city's oldest university, created a space designed to help a female academic tradition unfurl across the generations.

In May 1933, the DAB's Nazification undermined German women's attempts to redefine themselves and network on the Anglo-American model. The association's last democratically minded president tried to deploy international connections to prevent its alignment with and consequent destruction by the regime, but her strategy was vetoed by the DAB board. Younger members of the leadership, especially those who had been propounding nationalist conservative and anti-Semitic positions since the mid-1920s, hoped to continue the organization's work under National Socialist auspices, and in October 1933, they implemented the regime's directive to expel the members now defined as "Jewish." My study of the DAB's Nazification indicates how closely women's politics in this period was attuned to the international context: even the Nazi women of the DAB initially placed a high premium on maintaining membership in the IFUW and continuing to benefit from its international resources, connections, and reputation.

However, the change of power within the German association and its National Socialist reorganization generated bitter disputes within the IFUW over the network's underlying ethical principles and its policy of research promotion. German women were particularly angered by the IFUW's heightened support for the members who had been expelled; in turn, the IFUW responded to the DAB's anti-Semitic policy not only by offering help to the persecuted, but also by changing its own constitution. With an amendment proposed by the IFUW Council in 1934 and finally approved by the general meeting shortly before the outbreak of the war in 1939, the constitution now ruled that a national association must be barred from the umbrella organization if it refused women membership on the grounds of race, politics, or religion. The IFUW was the only international academic organization to react with such determination to the Nazi policy of exclusion and persecution.

In 1936, the DAB preempted its expulsion from the IFUW by resigning voluntarily, and was then speedily dissolved within Germany as well. This did not, however, mean that academic women's networking vanished completely from Nazi Germany. My study of organizational and biographical continuities indicates that in the Nazi period, women often successfully defended the ground they had previously gained as female university and college graduates. Working shoulder to shoulder with the National Socialist German Women's Agency, they recreated their networking structures within the Nazi organizations that were supposed to assume the functions of the dissolved DAB. With the help of those structures, former DAB members hoped to hold their own in their professions. Their career paths show that women physicians, lawyers, and academics were indeed able to pursue their professional ambitions, especially if they attached themselves to the Nazi women's organizations, championed racial science, and sought their niche within the racialized and gender-segregated national community. Some younger female academics among this group developed concepts of a woman's place in Nazi science, explicitly distancing themselves from the values and traditions of the international female community to which they had formerly so proudly belonged.

Many university women who were excluded from the racial community in Germany after 1933 drew invaluable benefit from their Weimar-era links to the IFUW and its British and American member associations. The agenda, character, and resources of the IFUW sustained a network that—like other, better-known academic refugee assistance organizations in Britain and the United States—immediately condemned the new German government's anti-Semitic policies and offered persecuted members help in escaping from Germany. The IFUW's sustained presence and its efforts to aid emigration, beginning as early as 1933, were critically important to women academics dismissed from teaching and research and forced to leave the country. Especially in Britain, the BFUW supported many women with professorial qualifications, enabling them to continue their research.

Although academics were initially the main beneficiaries, after 1938 IFUW refugee assistance was extended to all members. Particularly committed support from the AAUW, BFUW, and IFUW went to those members of the Austrian, German, and other now dissolved national organizations who had previously played an active role in the federation and were personally known to their colleagues abroad. This book reveals for the first time the existence of a vibrant female network that operated globally as an academic refugee assistance organization. It was a network that expanded women's resources beyond close family, relations, and immediate colleagues as they struggled to escape Germany and rebuild their lives in emigration. As their stories make clear, it is not the case that graduate women in exile generally failed in their careers because they were so much less likely than men to be awarded grants or financial aid by the well-known organizations, or indeed slipped completely under those organizations' radar. In fact, educated female refugees received a high level of support from the IFUW's assistance programs, despite its limited financial means. The existing, predominantly pessimistic reviews of the exodus of university-trained women from Germany thus require significant modification.

This study also challenges certain assumptions of exile studies concerning education, age, and gender. The tension between family ties and career orientation is in particular need of revision, as is these women's success in reestablishing their careers after their enforced emigration from Germany. Certainly, women researchers and professionals, most of whom were unmarried, were typically much more involved in family obligations than their male colleagues; such bonds kept them in Germany for longer because they often lived with their mothers or both parents. However, the timing of their decision to leave also depended on how they assessed their future career prospects in the German Reich. After being dismissed from state positions, many women moved into the new, segregated Jewish educational and health sectors, which were expanding due to the regime's anti-Semitic policies. They left their homeland only when they saw no further chance of qualified employment. Once in exile, these women clung to the goal of returning to their vocation. Older women, especially, were more likely to succeed in that objective than is commonly assumed. Historians have been wrong to conclude that women's professional identity was not pronounced enough to withstand the challenges of exile; arguably, in many cases it was even stronger than that of their male colleagues, honed by the multiple obstacles they had already faced as women battling to enter their chosen fields. Support from the IFUW, and especially from the American and British coordinators of refugee aid, proved to be crucial, whether in finding affidavits and sponsors, providing short-term financial assistance in emergencies, or helping to build a new social circle.

Clearly, there were significant differences in how the IFUW's national organizations experienced the end of Nazism and World War II. Most of the European member associations that had been closed down or brought to a standstill during the Nazi occupation were able to resume work as soon as the war ended, if not before. With the founding of the United Nations in summer 1945, a new era also began for the IFUW as a whole. The organization was granted permanent observer status in all UN bodies and an advisory role in UNESCO, a role that soon shifted the network's agenda from transatlantic to genuinely global dimensions.

In stark contrast to the rapid reconstruction of the IFUW's transnational networks elsewhere in Europe, the task of creating a new, democratic network in Germany seemed almost impossible after the country's defeat. It was only the renewed encounters between the DAB's Weimar-era protagonists and their British colleagues in Crosby Hall in fall 1946, along with individual personal contacts with American university women stationed with the occupying forces, that gradually set in motion the process of rebuilding German university women's networks. Among the most surprising lines of continuity is the support that Berlin women received from the BFUW's refugee office and its London manager. The combination of regular aid consignments and a steady stream of visits to Crosby Hall renewed and reinforced the friendships without which the new DAB could never have been founded in 1949 and admitted to the IFUW in 1951.

German academic women's networking was only revitalized in the western zones, which together became the Federal Republic of Germany in 1949. It was driven by women who had graduated from university or college before World War I and had forged extensive transnational connections during Weimar. Members of this generation assumed the role of societal facilitators, mediating between the past and the present—with a special interest in convincing younger colleagues to engage in the democratic reconstruction of West Germany. They shied away from attributions of personal blame, particularly on the question of which individuals had advocated the DAB's Nazification in 1933 or had accepted the Nazi regime's conditions in order to defend their professional gains. In the 1950s, new narratives arose to describe women's careers during Nazism and the association's Nazification in May 1933, which was now redefined as a voluntary dissolution. Not until 1968, when the IFUW conference was held on German soil for the first time and an international historical retrospect celebrated the network's first fifty years, were West German women forced to confront their association's past and interrogate the myth of self-dissolution.

The DAB women's incapacity to acknowledge the historical facts echoes mechanisms for coping with the past that have been identified for male university graduates and other sections of West German society. It also reflected a provincialization of the academic community occurring at that time on both sides of the Atlantic. If international bloc formation favored the Federal Republic's integration into the Western alliance, the shadow of the Cold War had a disintegrative effect on the transnational networking of academic women. All across Central and East-Central Europe, efforts to reestablish and fortify the old connections ground to a halt. The American association, and those of its members who had worked with such dedication for their colleagues fleeing Germany, came under attack from anticommunist denunciation campaigns; in response, many of these women stepped back from the internationalist principles that their own association had crafted after World War I. The reasons for that departure were more than political—they also reflected the passing of a generation. The expansion of the US higher education system after World War II was accompanied by the silent elimination of women from academic leadership roles; male successors were found for almost all the female college presidents who had left such a deep imprint on the IFUW's politics up to 1945. When that pioneer generation of leading women academics retired from the top ranks of the IFUW, their key concern disappeared along with them: the pursuit of international understanding, both as a political mission and as a transnational strategy for the advancement of women in higher education and research. The Cold War brought that era of internationalism to an end.

Notes

1 Introduction

1. *Report of the British Educational Mission* (Manchester: Morris & Yeaman, 1919), 84–6.
2. See Eckhardt Fuchs and Matthias Schulze, "Globalisierung und transnationale Zivilgesellschaft in der Ära des Völkerbundes," *Zeitschrift für Geschichtswissenschaft* 54, no. 10 (2006): 837–40. On the concept of civil society, see Jürgen Kocka, "Zivilgesellschaft als historisches Problem und Versprechen," in *Europäische Zivilgesellschaft in Ost und West: Begriff, Geschichte, Chancen*, ed. Manfred Hildermeier, Jürgen Kocka, and Christoph Conrad (Frankfurt/Main: Campus, 2000), 13–39. For the history of international organizations more generally, see Akira Iriye, *Global Community: The Role of International Organizations in the Making of the Contemporary World* (Berkeley: University of California Press, 2002).
3. On US historians' interest in transnational historiography, see especially Thomas Bender's groundbreaking study *Rethinking American History in a Global Age* (Berkeley: University of California Press, 2002); and Mary Nolan, *The Transatlantic Century: Europe and America, 1890–2010* (Cambridge: Cambridge University Press, 2012). The impact of growing international interest in the historical study of transnational "flows, ties, and appropriations" can be seen in Akira Iriye and Pierre-Yves Saunier, *The Palgrave Dictionary of Transnational History: From the Mid-19th Century to the Present Day* (New York: Palgrave Macmillan, 2009). An extensive literature exists on the debate over the benefits and limitations of transnational historiography in Germany. See, especially, Sebastian Conrad and Jürgen Osterhammel, eds., *Das Kaiserreich transnational: Deutschland in der Welt 1871–1914* (Göttingen: Vandenhoeck & Ruprecht, 2006); Gunilla-Friederike Budde, *Transnationale Geschichte: Themen, Tendenzen und Theorien* (Göttingen: Vandenhoeck & Ruprecht, 2006); Berthold Unfried, Jürgen Mittag, and Marcel van der Linden, eds., *Transnationale Netzwerke im 20: Jahrhundert historische Erkundungen zu Ideen und Praktiken, Individuen und Organisationen* (Leipzig: Akademische Verlags-Anstalt, 2008). See also the series of papers on transnational history in the specialist forum established within Clio online and the H-Soz-u-Kult list: http://geschichte-transnational.clio-online.net/?pn=texte&id=584 (accessed January 18, 2013). For an overview of gender-historical approaches in Germany, see the report by Nele Albrecht, "Gender History in a Transnational Perspective," http://h-net.msu.edu/cgi-bin/logbrowse.pl?trx=vx&list=h-soz-u-kult&month=0712&week=b&msg=LnsuGFwJSC4xvTgbAez37w&user=&pw= (accessed January 18, 2013).
4. See the essay collections edited by Lee Ann Banaszak, *The U.S. Women's Movement in Global Perspective* (Lanham, MD: Rowman & Littlefield, 2006) and by Kimberly Jensen and Erika Kuhlman, *Women and Transnational Activism in Historical Perspective* (Dordrecht: Republic of Letters Publishing, 2010); also Rumi Yasutake, *Transnational Women's Activism: The United States, Japan, and Japanese Immigrant Communities in California, 1859–1920* (New York: New York University Press, 2004) and Susan Zimmermann, "The Challenge of Multinational Empire for the International Women's Movement: The Habsburg Monarchy and the Development

of Feminist Inter/National Politics," *Journal of Women's History* 17, no. 2 (2005): 87–117.

5. See Marie Sandell, "'Truly International'? The International Federation of University Women's Quest for Expansion in the Interwar Period," *History of Education Researcher* 82 (2008): 74–83; Christy Jo Snider, "Creating a Transnational Identity: The International Federation of University Women Confronts Racial and Religious Membership Restrictions in the 1930s," in *Women and Transnational Activism in Historical Perspective*, ed. Kimberly Jensen and Erika Kuhlman (Dordrecht: Republic of Letters Publishing, 2010), 193–218; Joyce Goodman, "International Citizenship and the International Federation of University Women Before 1939," *History of Education* 40, no. 6 (2011): 701–21; Joyce Goodman, "Women and International Intellectual Co-operation," *Paedagogica Historica* 48, no. 3 (2012): 357–68.
6. Here, mention must be made primarily of the groundbreaking studies by Margaret W. Rossiter, *Women Scientists in America*, vol. 1: *Struggles and Strategies, 1870–1940* (Baltimore, MD: Johns Hopkins University Press, 1982) and vol. 2: *Before Affirmative Action, 1940–1972* (Baltimore, MD: Johns Hopkins University Press, 1995). For Britain, see Carol Dyhouse, *No Distinction of Sex? Women in British Universities, 1870–1939* (London: UCL Press, 1995). For the Netherlands, see Mineke Bosch, *Het Geslacht van de wetenschap* (Amsterdam: Sua, 1994); for Germany, Patricia M. Mazon, *Gender and the Modern Research University: The Admission of Women to German Higher Education, 1865–1914* (Stanford, CA: Stanford University Press, 2003); Trude Maurer, ed., *Der Weg an die Universität: Höhere Frauenstudien vom Mittelalter bis zum 20. Jahrhundert* (Göttingen: Wallstein, 2010). For an overview of research in the history of science from a gender and women's perspective, see Sally Gregory Kohlstedt and Helen Longino, "The Women, Gender, and Science Question: What Do Research on Women in Science and Research on Gender and Science Have to Do with Each Other?" in *Women, Gender, and Science: New Directions*, ed. Sally Gregory Kohlstedt and Helen E. Longino (Chicago, IL: University of Chicago Press, 1997): 3–15.
7. In the German literature, especially, points of contact are rare between the histories of gender, science, and women's movements. On the history of the German women's movement, see Barbara Greven-Aschoff, *Die bürgerliche Frauenbewegung in Deutschland 1894–1933* (Göttingen: Vandenhoeck & Ruprecht, 1981); Ute Gerhard and Ulla Wischermann, *Unerhört: Die Geschichte der deutschen Frauenbewegung* (Reinbek: Rowohlt, 1996); Iris Schröder, *Arbeiten für eine bessere Welt: Frauenbewegung und Sozialreform 1890–1914* (Frankfurt/Main: Campus, 2001); Angelika Schaser, *Frauenbewegung in Deutschland 1848–1933* (Darmstadt: WBC Wissenschaftliche Buchgesellschaft, 2006). For Britain: Linda Walker, *The Women's Movement in Britain, 1790–1945* (London: Routledge, 2006); Elizabeth Crawford, *The Women's Suffrage Movement in Britain and Ireland: A Regional Survey* (London: Routledge, 2006). For the United States: Ellen Carol DuBois, *Feminism and Suffrage: The Emergence of an Independent Women's Movement in America, 1848–1869* (Ithaca, NY: Cornell University Press, 1999); Eleanor Flexner and Ellen F. Fitzpatrick, *Century of Struggle: The Woman's Rights Movement in the United States* (Cambridge, MA: Belknap Press of Harvard University Press, 1996). For France: Christine Bard, "Feministinnen in Frankreich. Frauenstimmrecht und Frieden, 1914–1940," in *Feminismus und Demokratie: Europäische Frauenbewegungen der 1920er Jahre*, ed. Ute Gerhard (Königstein: Helmer, 2001), 84–103, and *Les femmes dans la société francaise au 20e siècle* (Paris: Colin, 2001).
8. See Karen Offen, *European Feminisms, 1700–1950: A Political History* (Stanford, CA: Stanford University Press, 2000); Leila Rupp, *Worlds of Women: The Making of an International Women's Movement* (Princeton, NJ: Princeton University Press, 1997); Margaret H. McFadden, *Golden Cables of Sympathy: The Transatlantic Sources of Nineteenth-Century Feminism* (Lexington: University of Kentucky Press, 1999).

9. An excellent exception is Molly M. Wood, "'Commanding Beauty' and 'Gentle Charm': American Women and Gender in the Early Twentieth-Century Foreign Service," *Diplomatic History* 31, no. 3 (2007): 505–30. On transnational aspects of higher education, see Anja Werner, *The Transatlantic World of Higher Education: Americans at German Universities, 1776–1914* (London: Berghahn, 2013). Christophe Charle, Jürgen Schriewer, and Peter Wagner, eds., *Transnational Intellectual Networks: Forms of Academic Knowledge and the Search for Cultural Identities* (Frankfurt/Main: Campus, 2004) focuses exclusively on male networks.
10. Other examples of a cultural history of international relations include Akira Iriye, *Cultural Internationalism and World Order* (Baltimore, MD: Johns Hopkins University Press, 2000); Jessica C. E. Gienow-Hecht, *Sound Diplomacy: Music and Emotions in Transatlantic Relations, 1850–1920* (Chicago, IL: University of Chicago Press, 2009).
11. Carroll Smith-Rosenberg, "The Female World of Love and Ritual: Relations between Women in Nineteenth-Century America," *Signs* 1 (1975): 1–29. See also Blanche Wiesen Cook, "Female Support Networks and Political Activism: Lillian Wald, Crystal Eastman, Emma Goldman," in *A Heritage of Her Own: Toward a New Social History of American Women*, ed. Nancy F. Cott and Elizabeth H. Pleck (New York: Simon and Schuster, 1979), 412–44; Mineke Bosch and Annemarie Kloosterman, eds., *Politics and Friendship: Letters from the International Woman Suffrage Alliance, 1902–1942* (Columbus: Ohio State University Press, 1990); Leila J. Rupp, "Women's History in the New Millennium: Carroll Smith-Rosenberg's 'The Female World of Love and Ritual' after 25 Years," *Journal of Women's History* 12, no. 3 (2000): 8–9.
12. Edith Saurer, "Frauenbewegung und soziale Netzwerke. Kommentar," in *Das Jahrhundert des Feminismus: Streifzüge durch nationale und internationale Bewegungen und Theorien*, ed. Anja Weckwert and Ulla Wischermann (Königstein: Helmer, 2006), 77–94. On the ways that women took on this originally masculine concept of friendship during the nineteenth century, see Kirsten Heinsohn, *Politik und Geschlecht: Zur politischen Kultur bürgerlicher Frauenvereine in Hamburg* (Hamburg: Verein für Hamburgische Geschichte, 1997).
13. From 1919 to 1936, Virginia Gildersleeve spent almost every summer in England with Caroline Spurgeon, until Spurgeon moved to Phoenix, Arizona, for health reasons. In August 1925, the two women together bought "Old Postman's Cottage" at the edge of the village of Alciston in Sussex, on the southeast coast of England. After rebuilding and adding a library, the house became a meeting point for the circle of intellectual women friends. In turn, Spurgeon followed Gildersleeve's invitation to spend several semesters as a guest professor at Barnard College in New York. See Virginia Crocheron Gildersleeve, *Many a Good Crusade: Memoirs* (New York: Macmillan, 1954), 135, 189.
14. With regard to the established international academic organizations, Elizabeth Crawford describes academic internationalism as a "casualty of the war." What survived World War I, in her view, were the prewar period's international organizations and international academic cooperation, but not its academic internationalism. Elizabeth Crawford, "The Universe of International Science, 1880–1939," in *Solomon's House Revisited: The Organization and Institutionalization of Science*, ed. Tore Frängsmyr (Canton, MA: Science History Publications, 1990), 251–69, here 261; also Eckhardt Fuchs, "Wissenschaftsinternationalismus in Kriegs- und Krisenzeiten. Zur Rolle der USA bei der Reorganisation der internationalen 'Scientific Community,' 1914–1925," in *Wissenschaft und Nation in der europäischen Geschichte*, ed. Ralph Jessen and Jakob Vogel (Frankfurt/Main: Campus, 2002), 263–84. With respect to the IFUW, I would dispute Crawford's view.
15. The founders of the IFUW were thus among what Akira Iriye has described as "new crusaders" who were willing to overcome national interests and to spread

"mutual understanding" after World War I. Iriye, *Cultural Internationalism and World Order*, 61.

16. David N. Livingstone, *Putting Science in Its Place: Geographies of Scientific Knowledge* (Chicago, IL: University of Chicago Press, 2003), 89.
17. Maria Rentetzi, *Trafficking Materials and Gendered Experimental Practices* (New York: Columbia University Press, 2007).
18. Elise Richter, *Summe des Lebens, hg. vom Verband der Akademikerinnen Österreichs* (Vienna: WUV Universitätsverlag, 1997).
19. On the suspension of access to this archival material, see also Susan Cohen, "'Now You See Them, Now You Don't': The Archives of the Refugee Committee of the British Federation of University Women," in *Refugee Archives: Theory and Practice*, ed. Andrea Hammel, Anthony Grenville, and Sharon Krummel (Amsterdam: Rodopi, 2008), 109–28.

2 Global War, Global Citizens, Global Mission: The Anglo-American Project of an International Federation of University Women

1. On the United States' entry into the war, see David M. Kennedy, *Over Here: The First World War and American Society*, 25th anniversary ed. (Oxford: Oxford University Press, 2004); Kimberly Jensen, *Mobilizing Minerva: American Women in the First World War* (Urbana: University of Illinois Press, 2008); with reference to the colleges and universities, see John R. Thelin, *A History of American Higher Education* (Baltimore, MD: Johns Hopkins University Press, 2004), 199–204; on American literary mobilization, see Keith Gandal, *The Gun and the Pen: Hemingway, Fitzgerald, Faulkner, and the Fiction of Mobilization* (Oxford: Oxford University Press, 2008).
2. See Marion Talbot and Lois Kimball Mathews Rosenberry, *The History of the American Association of University Women, 1881–1931* (Boston, MA: Houghton & Mifflin, 1931), 242; ACA War Service Committee to the members of the ACA, n.d. [fall 1917], Barnard College Archives, Dean's Office 1917–1918, Box 2.
3. V. Gildersleeve, draft of "Barnard Women's Service Pamphlet," n.d. [December 1918], p. 2, Barnard College Archives, Dean's Office, 1918–1919, Box 4.
4. On the work of the CPI, see *Complete Report of the Chairman of the Committee on Public Information: 1917–1918–1919* (Washington, DC: Public Information Committee, 1920); James R. Mock and Cedric Larson, *Words That Won the War: The Story of the Committee on Public Information, 1917–1919* (Princeton, NJ: Princeton University Press, 1939); Stephen L. Vaughn, *Holding Fast the Inner Lines: Democracy, Nationalism, and the Committee on Public Information* (Chapel Hill: University of North Carolina Press, 1980). A short overview is offered by Kennedy, *Over Here*, 60–7. See also Thomas A. Hollihan, "Propagandizing in the Interest of War. A Rhetorical Study of the Committee on Public Information," *Southern Speech Communication Journal* 49 (1984): 241–57.
5. Arthur Eugene Bestor, "To men and women engaged in patriotic education," transcript, n.d. [September 1917], AAUW Archives, Series II, Reel 5/37, Box 777, War Service Committee.
6. "War Service Committee report on the ACA's mobilization to support state propaganda for entry into World War I," n.d. [late 1917], AAUW Archives, Series II, Reel 5/37, Box 777.
7. ACA War Service Committee's circular to all members, n.d. [April 1917], Barnard College Archives, Deans Office, 1917–1918.
8. Ibid.

9. One member of the ACA, Mrs. Morgan, joined the Advisory Committee of the CPI's Speaking Division for this purpose. See Talbot and Rosenberry, *History of the AAUW*, 243.
10. Jörg Nagler's extensive study offers an informative analysis of the "state of social upheaval" in America in 1917: Jörg Nagler, *Nationale Minoritäten im Krieg: "Feindliche Ausländer" und die amerikanische Heimatfront während des Ersten Weltkriegs* (Hamburg: Hamburger Edition, 2000), 12–48. However, from the very start there was also resistance to America joining the war; see Jeanette Keith, *Rich Man's War, Poor Man's Fight: Race, Class, and Power in the Rural South during the First World War* (Chapel Hill: University of North Carolina Press, 2004); Kathleen Kennedy, *Disloyal Mothers and Scurrilous Citizens: Women and Subversion during World War I* (Bloomington: Indiana University Press, 1999).
11. See David O. Levine, *The American College and the Culture of Aspiration, 1915–1940*, 3rd ed. (Ithaca: Cornell University Press, 1986); also Christoph Strupp, "Perceptions of German Science and Research in the USA during World War I," in *Science, Technology, and Society in Russia and Germany during the First World War*, ed. Eduard Kolchinsky, Dietrich Beyrau, and Julia Lajus (St. Petersburg: Nestor Istorija, 2007), 114–35.
12. Levine, *The American College and the Culture of Aspiration*, especially the chapter "The College Goes to War." See also Thelin, *A History of American Higher Education*, 200.
13. ACA, "Special Note to Branch Presidents," n.d. [spring 1917], AAUW Archives, Series II, Reel 5/37, Box 777, War Service Committee.
14. "Report of the War Service Committee on the ACA's mobilization to support government propaganda for joining World War I," n.d. [late 1917], p. 7, AAUW Archives, Series II, Reel 5/37, Box 777, War Service Committee.
15. At the beginning of the war, the CPI itself, via its local committees, selected seventy-five thousand men who were to make four-minute addresses to promote support for the United States' war objectives. Each of these men had to present references from three local authority figures (e.g., a bank manager, an academic, or an entrepreneur). See Kennedy, *Over Here*, 61.
16. "Report of the War Service Committee on the ACA's mobilization to support government propaganda for joining World War I," n.d. [late 1917], p. 4, AAUW Archives, Series II, Reel 5/37, Box 777, War Service Committee.
17. Talbot and Rosenberry, *History of the AAUW*, 244.
18. "Report of the War Service Committee on the ACA's mobilization to support government propaganda for joining World War I," n.d. [late 1917], p. 4, AAUW Archives, Series II, Reel 5/37, Box 777, War Service Committee. Many Germans were virtually forced to buy these loans in order to prove they were patriotic Americans. See Paul Finkelman, "The War on German Culture, 1917–1925," in *Confrontation and Cooperation: Germany and the United States in the Era of World War I, 1900–1924*, ed. Hans-Jürgen Schröder (Providence, RI: Berg, 1993), 177–206.
19. The most important national body, created in 1916–17 for the central coordination of the sciences in the United States, was the National Research Council. It was set up after the sinking of the *Lusitania* by German U-boats, and in 1917 was incorporated into the National Council of Defense as the section dealing with science and research. Among its most important responsibilities was to keep an overview not only of the United States' own research resources, but also of those of the Allies, especially Britain and France. See Eckhardt Fuchs, "Wissenschaftsinternationalismus in Kriegs- und Krisenzeiten. Zur Rolle der USA bei der Reorganisation der internationalen 'Scientific Community,' 1914–1925," in *Wissenschaft und Nation in der europäischen Geschichte*, ed. Ralph Jessen and Jakob Vogel (Frankfurt/Main: Campus, 2002), 266.

20. The most important organizations were the Association of American Universities, National Organization of State Universities, Association of American Colleges, the national organizations of the Professional Schools (medicine, law, engineering), the National Association of Agricultural Colleges, and the Catholic Educational Association. The ACA became a member in 1919. See Minutes of the American Council on Education, Hotel Washington, Washington, DC, February 5, 1919, Columbia University, Gildersleeve Collection, International Federation of University Women, ACA, 1918–1928, Box 44, Folder 1919a.
21. "American University Union in Europe," *Institute of International Education Bulletin* (November 1927), 3.
22. Minutes of the Meeting of the American Council on Education, Hotel Washington, Washington, DC, February 5, 1919, Columbia University, Gildersleeve Collection, International Federation of University Women, ACA, 1918–1928, Box 44, Folder 1919a.
23. Keith Clark, "French Girls in American Colleges," press release, Publicity Committee, War Work Council, YMCA, n.d. [1918], Barnard College Archives, Dean's Office, 1918–1919, Box 3.
24. Martha Hanna points out that the proportion of foreign students in France had been high since the beginning of the twentieth century, although up to 1915 the majority of these students came from Eastern Europe. The shift in orientation across the Atlantic was, in Hanna's view, mainly driven by financial motivations (it was hoped that rich American students would save the Sorbonne from imminent ruin) as well as an element of cultural imperialism: "French scholars were also keen to extend the cultural influence of France within the United States in order to undermine the dominant role Germany had played since at least the mid-nineteenth century in American academic life." Martha Hanna, "French Women and American Men. 'Foreign' Students at the University of Paris, 1915–1925," *French Historical Studies* 22, no. 1 (1999): 87–112, here 87–9, 97.
25. This included, in particular, the very successful one-year "Cours de civilisation française"; the "Certificats d'études supérieures" that were awarded for individual courses after an examination and attracted credits at American colleges and universities; the "Junior Year Abroad," first offered in 1923 and enjoying greater numbers of participants every subsequent year under the auspices of the Institute for International Education; and finally the "licence française," a course in French literature, culture, geography, and history that could be pursued without extensive knowledge of Latin and which thus facilitated access to university not only for Americans, but also for French women students (it did not, however, lead to an authorization to teach at higher schools). Ibid., 98, 108.
26. John H. Wigmore, "A Proposal to Restore the True Status of French Science and Learning in America," *Journal of the Association of American Colleges*, n.d. [1917]: 1–8, Barnard College Archives, Dean's Office 1917–1918, Box 2. A lengthy study from a history-of-science perspective, covering numerous disciplines and providing both information about the traditions of French scholarship and practical tips on research opportunities in France, appeared in 1917 as a "national homage, offered from the Universities of America to the Universities of France." Several hundred American professors, male and female, are listed as sponsors of the book. John Henry Wigmore, *Science and Learning in France, with a Survey of Opportunities for American Students in French Universities: An Appreciation by American Students* (Chicago, IL: Society for American Fellowships in French Universities, 1917), ix, xiii–xxxviii.
27. John Henry Muirhead, "Miss Rose Sidgwick. An Address delivered to the Guild of Undergraduates on January 31, 1919," Barnard College Archives, Dean's Office 1918–1919, Box 4.

28. "Report of the Committee on International Relations," April 1919, Columbia University, Butler Library, Rare Books Collection, Gildersleeve Collection, IFUW, ACA, 1918–1928, Folder 1919c.
29. The president of Milwaukee-Downer College, Wisconsin, and Dean Lucy W. Stebbins of the University of California represented the Midwest and the West Coast; Mabel H. Skinner of Victoria College, Toronto University, represented Canada. See "Report of the Committee on International Relations," April 1919, Columbia University, Gildersleeve Collection, IFUW, ACA, 1918–1928, Folder 1919c.
30. Minutes of the Meeting of the Committee on International Relations of the ACA, November 6, 1918, AAUW Archives, Series I, Reel 6/39, Box 777, Founding Materials.
31. Minutes of the Meeting of the Committee on International Relations of the Association of Collegiate Alumnae, June 11, 1918, Gildersleeve Collection, International Federation of University Women, ACA, 1918–1928, Box 44, Folder 1919b.
32. See, for example, Carl Frederick Wittke, *The Anti-German Hysteria of World War One* (Munich: Saur, 1994); also Nagler, *Nationale Minoritäten im Krieg*; Finkelman, "The War on German Culture."
33. See Kennedy, *Over Here*, 57–9; Thelin, *A History of American Higher Education*, 199–204; for Britain, see especially Stuart Wallace, *War and the Image of Germany: British Academics, 1914–1918* (Edinburgh: Donald, 1988).
34. The existing literature has mainly focused on the continued existence of connections with Germany and the circumstance that American researchers, in their desire to participate in war research, oriented themselves on Germany and tried hard to catch up with the head start held by the laboratories in German universities and Kaiser Wilhelm Institutes. See Strupp, "Perceptions of German Science and Research in the USA during World War I."
35. After visiting the University of Madison, Spurgeon and Sidgwick returned to the East Coast while their male colleagues traveled to the universities of the southern states. The group met up again in early December in Boston, the Mission's last stop. *Report of the British Educational Mission*, 11, Barnard College Archives, Dean's Office, 1918–1919, Box 4.
36. Jürgen Herbst, *The German Historical School in American Scholarship: A Study in the Transfer of Culture* (Ithaca, NY: Cornell University Press, 1965), chapter 1.
37. American Council on Education, Conference on After-War Problems in the Higher Education of Women in Great Britain and the United States, June 21, 1918, Report, p. 8, Columbia University, Gildersleeve Collection, IFUW, ACA 1918–1928, Box 44, Folder 1918.
38. This official intervention in the universities' affairs was prompted by an interview with the popular poet Alfred Noyes, who, in June 1916, had returned from a long lecture trip around the United States. He reported: "The enormous educational machine in America, which is so much more important than anything of the same kind we have in Europe, is almost entirely run from Germany.... A very large proportion of the staffs of the colleges and Universities has received all its educational training, or at any rate its postgraduate training, in Germany.... The feeling is that we have held aloof in educational facilities, and this feeling really accounts for a good deal of anti-English bias." Alfred Noyes, "America and the War," *The Observer*, June 18, 1916. Within a few days of the interview being published, writes Renate Simpson, the Foreign Office opened the first file under the heading "German educational influence in America." Simpson notes that the file includes the beginnings of the debate on the introduction of PhD in Britain. Renate Simpson, *How the PhD Came to Britain: A Century of Struggle for Postgraduate Education* (Guildford: Society for Research into Higher Education, 1983).

39. Lord Bryce, "As to American Students at British Universities," Memorandum of October 31, 1916, for the Foreign Office, cited in Simpson, *How the PhD Came to Britain*, 116–17.
40. University Registry Oxford, Statement, September 19, 1918, Archives, Royal Holloway, University of London (RHUL Archives), P67/6/1/2. On Cambridge, see Simpson, *How the PhD Came to Britain*, 47, 59, 140.
41. American Council on Education, Conference on After-War Problems in the Higher Education of Women in Great Britain and the United States, June 21, 1918, Report, p. 8, Columbia University, Gildersleeve Collection, IFUW, ACA 1918–1928, Box 44, Folder 1918.
42. Ibid., p. 9.
43. Ibid., p. 9. At this time, the best-known discussion of the situation and methods of American girls' education available in Britain was Sara Burstall's *Impressions of American Education in 1908*. In her travel report, Burstall, the headmistress of Manchester High School, had emphasized the improved professional opportunities opening up for women in the colleges of Vassar, Bryn Mawr, Smith, or Wellesley, and the uniqueness of these institutions. She was also struck by the enviably widespread, unshakable faith in education. However, she made no mention of the magnificent facilities offered by the American colleges and universities—it seems that at this point, the differences were not yet so striking. See Sara Annie Burstall, *Impressions of American Education in 1908* (London: Longmans, 1909), 258–83, 302. It becomes clear from other reports that the standing of American universities within the British university scene was very low until well into World War I. Derogatory claims were made that the American universities, apart from isolated exceptions such as Harvard, were of the lowest standard. See Simpson, *How the PhD Came to Britain*, 118.
44. "The British Educational Mission to the United States, October–December 1918. Supplementary Report on Women's University Education," p. 6, Barnard College Archives, Dean's Office, 1918–1919, Box 4.
45. Ibid., p. 9.
46. Ibid., p. 12.
47. Ibid.
48. "The British Educational Mission to the United States, October–December 1918. Supplementary Report on Women's University Education," p. 12, Barnard College Archive, Dean's Office, 1918–1919, Box 4. Oxford University accepted women as full members from 1920. However, Cambridge women, although allowed to take all degrees from 1921, did not receive the associated privileges granted to men, namely participation in University government. Full membership was not granted to Cambridge women until 1947. On Oxford, see Annie Mary Anne Henley Rogers, *Degrees by Degrees: The Story of the Admission of Oxford Women Students to Membership of the University* (London: Oxford University Press, 1938), 113–24; on Cambridge, see Felicity Hunt and Carol Barker, *Women at Cambridge: A Brief History* (Cambridge: University of Cambridge Publications Office, 1998).
49. American Council on Education, Conference on After-War Problems in the Higher Education of Women in Great Britain and the United States, June 21, 1918, Report, p. 11, Columbia University, Gildersleeve Collection, IFUW, ACA 1918–1928, Box 44, Folder 1918.
50. "The British Educational Mission to the United States, October–December 1918. Supplementary Report on Women's University Education," p. 15, Barnard College Archives, Dean's Office, 1918–1919, Box 4.
51. American Council on Education, Conference on After-War Problems in the Higher Education of Women in Great Britain and the United States, June 21, 1918, Report,

p. 2, Columbia University, Gildersleeve Collection, IFUW, ACA 1918–1928, Box 44, Folder 1918.

52. Scholarships for Women of the Empire and the United States of America. Meeting at the invitation of the Duchess of Marlborough, July 18, 1918, RHUL Archives, P 67/6/1/1. See also Spurgeon's idea for an "Associated Board for the Promotion of Imperial-American Interchange for Woman University Students," n.d. [September 1918], RHUL Archives, P 67/6/1/2. This committee would bring together information from all English-speaking countries on the funding possibilities open to women, coordinate the exchange of information between universities, and serve as a contact point for all interested women: exactly the same services for female students and researchers, in other words, as were provided by the AUUE and the new Institute of International Education in New York, founded with assistance from the Carnegie Foundation, for students in general.
53. Talbot and Rosenberry, *History of the AAUW*, 278.
54. Virginia Gildersleeve's opening address in IFUW, *Report of the First Conference, London 1920* (London: The Federation, 1920), 18.
55. Point 12 of the report's recommendations runs: "Memorials to those who have fallen in the War should take form of Scholarships not less than of Buildings and old members of the Universities, including the Fraternities and Sororities in America, should take action in this matter." *Report of the British Educational Mission*, 16–17, Barnard College Archive, Dean's Office, 1918–1919, Box 4.
56. Since 1906, Jackson (1839–1921) had been regius professor of Greek at Cambridge University. He was one of the professors who had voted for the admission of women to Oxford and Cambridge degrees; see Lilly Library Manuscript Collection, University of Indiana, Henry Jackson MSS, Introduction, www.indiana.edu/~liblilly/lilly/mss/html/jacksnh.html (accessed January 18, 2013).
57. John Henry Muirhead, "Miss Rose Sidgwick. An Address delivered to the Guild of Undergraduates on January 31, 1919," p. 4, Barnard College Archive, Dean's Office, 1918–1919, Box 4.
58. Virginia Crocheron Gildersleeve, *Many a Good Crusade: Memoirs* (New York: Macmillan, 1954), 130.
59. John Henry Muirhead, "Miss Rose Sidgwick. An Address delivered to the Guild of Undergraduates on January 31, 1919," p. 2, Barnard College Archive, Dean's Office, 1918–1919, Box 4.
60. Mabel Choate's mother had sat on the Barnard College board from its beginning, and Mabel herself was elected to the committee in 1918. See Marian Churchill White, *A History of Barnard College* (New York: Columbia University Press, 1954), 200. In this capacity, she worked for the acquisition of a one-million-dollar endowment for building work to extend Barnard College, a goal achieved in 1920. See "An Appeal for Barnard College," June 21, 1918, Barnard College Archives, Dean's Office 1918–1919, Box 4. Mabel Choate accompanied Gildersleeve, Thomas, and their right-hand woman Helen Taft on the journey to the first IFUW meeting with the British women in summer 1919. She appears to have supported the founding of the IFUW in other ways as well; see Gildersleeve, *Many a Good Crusade*, 131: "Travelling with us was Miss Mabel Choate, daughter of that very distinguished American Ambassador to Britain, Joseph H. Choate, so greatly loved on both sides of the Atlantic. Though not a university woman herself, she was much interested in our plans, had served as a treasurer of the Rose Sidgwick Memorial Fund, and encouraged and helped us in many ways."
61. Colleges and universities undertook similar initiatives to draw the war deaths of male students into the service of education's national and international obligations. In memory of the 121 young fallen Americans who had been deployed in France

as medics even before the United States joined the war, the American Field Service in 1921 established the "American Field Service Fellowships for French Universities Incorporation" as an "enduring memorial for the...men who gave their lives in the great war." These grants, wrote the Institute of International Education in 1933, had not only promoted the cause of international understanding, but had also created "a better realization of the eminent place France holds in every field of science and learning." *Institute of International Education, Bulletin*, 14th Series, April 1933, 5. On the homage paid to fallen college students, especially by the naming of sports stadia, see Thelin, *A History of American Higher Education*, 201–4.

62. The Rose Sidgwick Memorial Fund, February 17 to May 28, 1919, Barnard College Archives, Dean's Office, 1918–1918, Box 3.
63. In memory of her late husband, Mrs. Willard Straight paid for a new Student Union building to be erected at Cornell University. Named the "Willard Straight Hall," it was to commemorate his former students. See Rebecca H. Cofer, *The Straight Story: An Informal History of Willard Straight Hall, 1925–1990* (Ithaca, NY: Cornell University Press, 1990).
64. This, at least, is how the executive interpreted the great attention attracted by the appeal.
65. Margaret E. Maltby, *History of the Fellowships Awarded by the American Association of University Women, 1888–1929, with the Vitas of the Fellows* (Washington, DC: American Association of University Women, 1929), 8.
66. AAUW Rose Sidgwick Memorial Fellowship, see www.finaid.wwu.edu/scholarships/pages/general_information/how_to/international_students.php (accessed January 18, 2013).
67. Gildersleeve, *Many a Good Crusade*, 130. On the relationship of Gildersleeve and Spurgeon, see also chapter 1 of the present volume.
68. Mount Holyoke was another of the women's colleges successfully working for the academic training of women. See Miriam R. Levin, *Defining Women's Scientific Enterprise: Mount Holyoke Faculty and the Rise of American Science* (Hanover, NH: University Press of New England, 2004).
69. Especially the president of Wellesley, Ellen Fitz Pendleton, and Ada Comstock (since 1912 the dean of Smith College for Women; from 1923 to 1943 the first woman president of Radcliffe College in Cambridge, MA). On Comstock, see Susan Margot Smith, "Ada Comstock Notestein. Educator," in *Women of Minnesota: Selected Biographical Essays*, ed. Barbara Stuhler and Gretchen Kreuter (St. Paul: Minnesota Historical Society Press, 1998), 208–25. On Pendleton, see John A. Garraty, ed. *American National Biography* (New York: Oxford University Press, 1999), vol. 14, 278–9.
70. In Emily S. Rosenberg's view, with this commitment they helped to draw the contours of later US foreign policy. Emily S. Rosenberg, *Spreading the American Dream: American Economic and Cultural Expansion, 1890–1945* (New York: Hill and Wang, 1982), 108–21, especially 21.
71. Renate Haas, "Caroline Spurgeon: English Studies, the United States, and Internationalism," *Studia Anglica Posnaniensia: International Review of English Studies* (2002), 1–15.
72. IFUW, *Report of the First Conference, London 1920*, 10.
73. "We admit to membership only national associations of federations, not one of several groups within a country—and never more than one federation from any one country, so that it will be left to the university women of each country to agree among themselves on their own form of association or federation, to get together and then apply to our International Council for membership." Ibid., 39.
74. Because the Southern Association of College Women had only 800 members, the merger did not alter the balance of power within the web of American associations.

International matters remained entirely in the hands of former ACA women. The greatest difficulties caused by the fusion arose from the difference in the standards applied to membership of colleges south and north of the Mason-Dickson Line. See Talbot and Rosenberry, *History of the AAUW*, 59–60.

75. "Four Important Questions to be Considered at the ACA Biennial," n.d. [March 1919], Columbia University, Gildersleeve Collection, IFUW, ACA, 1918–1928, Folder 1919b.
76. IFUW, *Report of the First Conference, London 1920*, 3.
77. Virginia Gildersleeve, "What the AAUW Has Meant to the IFUW," *AAUW Journal* (1941): 13–15. The governments of other countries, such as Spain, France, and Czechoslovakia, contributed to the costs of the IFUW by reimbursing their delegates' expenses for traveling to the IFUW conferences.
78. Virginia Gildersleeve to Carey Thomas, April 8, 1919, Bryn Mawr College Archives, M. Carey Thomas Official Papers, Incoming Correspondence, July-December 1919, n.p. Gildersleeve writes: "As you have already been informed, the Biennial designated you and Dean Taft as representatives of the International Committee to make investigations abroad regarding relations with foreign institutions and to report to the committee.... You could, I am sure, acquire a great deal of valuable information in the course of your tour." A similar course of action was followed by the American medic Esther Lovejoy in the mid-1920s. She traveled right across Europe for the Medical Women's International Association encouraging women to establish national groups and to collaborate as an international association. Claudia Huerkamp, *Bildungsbürgerinnen: Frauen im Studium und in akademischen Berufen 1900–1945* (Göttingen: Vandenhoeck & Ruprecht, 1996), 250.
79. "Smith College Tradition. Honorary Degrees from Smith College," Maria de Maeztu, LLD 1919, www.smith.edu/collegerelations/honorary.php (accessed January 8, 2008).
80. On the debates around the reform of girls' schooling in France, see Karen Offen, "The Second Sex and the Baccalauréat in Republican France," *French Historical Studies* 13, no. 2 (1983): 252–86; Françoise Mayeur, *L'Enseignement secondaire des jeunes filles sous la Troisième République* (Paris: Presses de la Fondation nationale des sciences politiques, 1977).
81. Alongside the intensified teaching of Latin and Ancient Greek and a greater role for the natural sciences, this involved the creation of an "école d'application" (including kindergarten, elementary school, and junior high school), next door to the École normale, where students could practice their teaching skills from an early stage in their training. *Le Cinquantenaire de l'École Normale Supérieure de Sèvres, 1881–1931*, (Paris: Printory, 1932), 192–4. On the role of French women teachers in the reform process in general, see also Jo Burr Margadant, *Madame le Professeur: Women Educators in the Third Republic* (Princeton, NJ: Princeton University Press, 1990).
82. Carey Thomas, January 23, 1920, from Monte Carlo, to Helen Taft in Bryn Mawr, Bryn Mawr College Archives, Carey Thomas Papers, Official Outgoing Correspondence, Reel 151.
83. From 1906 onward, the Club Autour du Monde maintained its own clubhouse on Albert Kahn's estate in Boulogne. It was intended to serve as a venue and meeting point for a cosmopolitan audience. Albert Kahn, *1860–1940: Réalités d'une Utopie [Exposition, Musée Départemental Albert-Kahn, Boulogne, 28 novembre 1995–18 septembre 1996]* (Boulogne: Musée Albert-Kahn, 1995). In 1913 Mespoulet, together with Madeline Mignon and at the request of Albert Kahn, assembled a remarkable social-anthropological series of color photographs about Ireland for Kahn's transnational "Archive de la Planète." These photographs are now stored in the Musée Albert Kahn in Paris.

84. Nicole Fouché, "Reid Hall, l'Association française des femmes diplômées des universités et la Fédération internationale des femmes diplômées des universités, 1919–1993," *Diplômées* 178 (1994): 190–200.
85. IFUW, *Report of the First Conference, London 1920*, 35.
86. See Ruth Crawford Mitchell, *Alice Garrigue Masaryk, 1879–1966: Her Life as Recorded in Her Own Words and by Her Friends* (Pittsburgh, PA: University Center for International Studies, 1980), xix–xxiv. In 1912, Lathrop was appointed the first director of the US government's newly established Children's Bureau. She had been one of those who—with support from the ACA—had worked doggedly for Masaryk's release when she was arrested in 1915. For many ACA members, harking back to that time, Alice Masaryk's name was synonymous with the struggle for independence in the smaller Central European states. Ibid.
87. The effective press work and correspondingly wide public impact of this tour is demonstrated by the many newspaper cuttings to be found among Spurgeon's papers, held at Royal Holloway College, London, P67/6/1/4 RHUL Archives, University of London.
88. In this, they differed from most of the German male professors of their generation. See, for example, Gabriele Metzler, *Internationale Wissenschaft und nationale Kultur: Deutsche Physiker in der internationalen Community, 1900–1960* (Göttingen: Vandenhoeck & Ruprecht, 2000).
89. This also explains why the IFUW did not expand beyond North America, Europe, and the British Commonwealth during the interwar period. See Marie Sandell, "'Truly International'? The International Federation of University Women's Quest for Expansion in the Interwar Period," *History of Education Researcher* 82 (2008).
90. Nicole Fouché, "Des Américaines protestantes à l'origine des 'University Women' françaises, 1919–1964," *Bulletin de la Société d'histoire du protestantisme français* 146 (2000): 133–52.
91. Carey Thomas, January 23, 1920, from Monte Carlo, to Helen Taft in Bryn Mawr, Bryn Mawr College Archives, Carey Thomas Papers, Official Outgoing Correspondence, Reel 151. Diverging from Leila Rupp's analysis, I would not speak of an "international identity" that grew almost of its own accord out of the encounters between women of different nations. Rather, I would argue that an already extant internationality in female academic identities made possible the rapid construction of the IFUW. See Leila J. Rupp, *Worlds of Women: The Making of an International Women's Movement* (Princeton, NJ: Princeton University Press, 1997).
92. This is another reason why the IFUW's orientation on the promotion of science was so energetically advocated within its leadership. See, for example, Lise Meitner's address in Edinburgh, "Some International Aspects in the Development of Science," in IFUW, *Report of the Sixth Conference, Edinburgh 1932* (London: The Federation, 1932), 49–51.

3 Female Networks for Science: Programs and Politics

1. See IFUW, *Report of the First Conference, London 1920* (London: The Federation, 1920), 10. On the project of a transnational civil society, see Akira Iriye, *Global Community: The Role of International Organizations in the Making of the Contemporary World* (Berkeley: University of California Press, 2002); Helmut Anheier, Marlies Glasius, and Mary Kaldor, eds., *Global Civil Society* (Oxford: Oxford University Press, 2001); Eckhardt Fuchs and Matthias Schulze, "Globalisierung und transnationale Zivilgesellschaft in der Ära des Völkerbundes," *Zeitschrift für Geschichtswissenschaft* 54, no. 10 (2006); Anna-Katharina Wöbse, "'To Cultivate the International Mind.'

Der Völkerbund und die Förderung der globalen Zivilgesellschaft," *Zeitschrift für Geschichtswissenschaft* 54, no. 10 (2006); Thomas Richard Davies, "The Rise and Fall of Transnational Civil Society: The Evolution of International Non-Governmental Organizations Since 1839," City University Working Papers on Transnational Politics 003 (2008), www.staff.city.ac.uk/tom.davies/CUWPTP003.pdf (accessed January 18, 2013).

2. Shirley M. Eoff, *Viscountess Rhondda. Equalitarian Feminist* (Columbus: Ohio State University Press, 1991).
3. Also present was the General Secretary of the League of Nations, Inazo Nitobe; see Nitobe to Theodora Bosanquet, IFUW, November 9, 1920, League of Nations Archives, Sec. 13, International Bureau, Series 7759, File 4459, IFUW Submits Report of First Conference of Federation.
4. IFUW, *Report of the First Conference, London 1920*, 15.
5. See, for example, Pat Thane and Gisela Bock, eds., *Women and the Rise of the European Welfare State, 1880s–1950s* (London: Routledge, 1991); Sara Delamont, ed., *A Women's Place in Education: Historical and Sociological Perspectives in Gender and Education* (Aldershot: Avebury, 1996); Barbara Greven-Aschoff, *Die bürgerliche Frauenbewegung in Deutschland 1894–1933* (Göttingen: Vandenhoeck & Ruprecht, 1981); Claudia Opitz and Elke Kleinau, eds., *Geschichte der Mädchen- und Frauenbildung*, vol. 2 (Frankfurt/Main: Campus, 1996).
6. See Elizabeth Crawford, "The Universe of International Science, 1880–1939," in *Solomon's House Revisited: The Organization and Institutionalization of Science*, ed. Tore Frängsmyr (Canton, MA: Science History Publications, 1990); for the German context, see Gabriele Metzler, *Internationale Wissenschaft und nationale Kultur: Deutsche Physiker in der internationalen Community, 1900–1960* (Göttingen: Vandenhoeck & Ruprecht, 2000), especially Chapter 1. Joyce Goodman emphasizes the influence of "conversations that circulated within discourses of liberal peace, cultural internationalism and developments in the discipline of international relations." Joyce Goodman, "International Citizenship and the International Federation of University Women before 1939," *History of Education* 40, no. 6 (2011): 708.
7. On scientific internationalism before and after 1914, see Eckhardt Fuchs, "Wissenschaftsinternationalismus in Kriegs- und Krisenzeiten. Zur Rolle der USA bei der Reorganisation der internationalen 'Scientific Community,' 1914–1925," in *Wissenschaft und Nation in der europäischen Geschichte*, ed. Ralph Jessen and Jakob Vogel (Frankfurt/Main: Campus, 2002), 265.
8. David N. Livingstone, *Putting Science in Its Place: Geographies of Scientific Knowledge* (Chicago, IL: University of Chicago Press, 2003), 89.
9. IFUW, *Report of the First Conference, London 1920*, 15.
10. Ibid., 20.
11. As an example of this phenomenon, Carey Thomas said she had asked a male colleague in France about the number of female university graduates, and his answer had been "one." In fact, she said, there had been 200. Ibid.
12. Ibid., 13.
13. Kristine Bonnevie, "The Development of the International Mind in the Universities," in IFUW, *Report of the Third Conference, Oslo 1924* (London: The Federation, 1924), 76. A very similar point is made by current research literature, which notes that "transnational forms of knowledge have to be understood against a background of profoundly national structures of scholarly work." Peter Wagner, "Introduction to Part I," in *Transnational Intellectual Networks: Forms of Academic Knowledge and the Search for Cultural Identities*, ed. Christophe Charle, Jürgen Schriewer, and Peter Wagner (Frankfurt/Main: Campus, 2004), 20.
14. "International Federation of University Women, Constitution and Bye-Laws," in IFUW, *Report of the First Conference, London 1920*, Appendix.

15. IFUW, *Report of the First Conference, London 1920*, 56.
16. Carey Thomas, "Next Steps for University Women," in IFUW, *Report of the First Conference, London 1920*, 56.
17. IFUW, *Report of the First Conference, London 1920*, 66.
18. In the interwar period, the Council meeting was held only once in the United States. Out of consideration to the far more modest financial means of their European colleagues, the American women held back, despite the fact that they dominated the Federation in terms of numbers. They used the working meetings and conferences as an opportunity for lengthy tours around Europe.
19. Virginia Gildersleeve, closing address, in IFUW, *Report of the Third Conference, Oslo 1924*, 95. On the importance of the international gatherings for the members of national associations, see also chapter 4.
20. Report on the activities of the DAB from June 1930 to June 1931, LAB, B Rep. 235-5, 1/1, HLA 3631. This meeting resulted in an invitation to a three-month lecture tour through the United States, which was adopted by Maria Schlüter-Hermkes, standing in for Zahn-Harnack. Schlüter-Hermkes spoke on the political, cultural, and economic situation in Germany, with special emphasis on the difficult situation of women students. As a result of her words, several American women personally offered financial assistance to selected students. Report on the activities of the DAB from June 1, 1930, to October 1, 1932, p. 12, LAB, B Rep. 235-5, 1/1, HLA 3633.
21. See Bosanquet to George Oprescu, the League of Nations' secretary for Intellectual Cooperation, October 6, 1924. In this letter, Bosanquet suggested the following as a mode of collaboration: "Our use would be to diffuse knowledge of your work." League of Nations Archives, Geneva, Sec. 13, Series 34468, File 36139x, Correspondence with IFUW, London.
22. Margarete Rothbarth, "Internationale geistige Zusammenarbeit," in *Wörterbuch des Völkerrechts und der Diplomatie*, vol. 3 (Berlin: Vereinigung Wissenschaftlicher Verleger, 1928), 1–12.
23. The other two women delegates to the League of Nations assembly were Anna Bugge-Wicksell from Sweden and Henni Forchhammer from Denmark. On the establishment and remit of the IFUW Committee, see Bonnevie, "The Development of the International Mind in the Universities," 79.
24. See IFUW, *Report of the Eleventh Council Meeting, Vienna 1927* (London: The Federation, 1927), 24; IFUW, *Report of the Seventh Conference, Cracow 1936* (London: The Federation, 1936), 91.
25. For example, there were hardly any requests for translations, and the unanswered question of how the continuity of employees' social insurance in their home country should be secured during their exchange visit abroad meant that longer-term exchanges were virtually impossible. In response to this problem, the IFUW called on its members to use their vacations to pay short visits to foreign partner organizations. IFUW, *Report of the Sixth Conference, Edinburgh 1932*, 105; IFUW, *Report of the Seventh Conference, Cracow 1936*, 92.
26. *International Glossary of Academic Terms* (Paris: International Federation of University Women, 1939). A detailed report on the background to and work on this study can be found in IFUW, *Report of the Eighth Conference, Stockholm 1939* (London: The Federation, 1939), 56–8.
27. On the legal work around married women's citizenship, see Goodman, "International Citizenship," 714–18. Cooperation with other international women's organizations was limited to IFUW representation on the Joint Standing Committee of Women's International Organizations, which had existed since 1925 and in 1931 was reorganized to form the Liaison Committee of Women's Organizations. Essentially, their shared activity extended only to the right to nominate women to the various

bodies of the League of Nations and the International Labour Organization. On the composition of these committees, see Karen Offen, *European Feminisms, 1700–1950: A Political History* (Stanford, CA: Stanford University Press, 2000), 347.

28. IFUW, *Report of the Fourth Conference, Amsterdam 1926* (London: The Federation, 1926), 55; IFUW, *Report of the Eleventh Council Meeting, Vienna 1927*, 21.
29. IFUW, *Report of the Twelfth Council Meeting, Madrid 1928* (London: The Federation, 1928), 37. The person concerned was the German teacher Dr. Vester, who spent one year in London at Clapham School. "Was will der DAB?" information leaflet, n.d. [1931], p. 2, BAK, N 1151, no. 281.
30. IFUW, *Report of the Fifteenth Council Meeting, Prague 1930* (London: The Federation, 1930), 40.
31. After attending the IFUW's Edinburgh conference, the Section delegate from the League of Nations said the key reason was that the economic crisis was making it impossible to adjust teachers' salaries to match the requirements of their stay abroad. See the report by Armi Inkeri Hallsten-Kallia on the Edinburgh conference, submitted on August 31, 1932, p. 2, League of Nations Archives, Geneva, Sec. 5 B, 1928–1932, Series 651, File 6323, Correspondence with IFUW.
32. It is true that numbers rose impressively fast. In the years 1937–39, 132 exchange partnerships were set up between American and British schools. IFUW, *Report of the Eighth Conference, Stockholm 1939*, 66.
33. Report by Armi Inkeri Hallsten-Kallia on the conference in Edinburgh, submitted on August 31, 1932, p. 3, League of Nations Archives, Geneva, Sec. 5 B, 1928–1932, Series 651, File 6323, Correspondence with IFUW.
34. Between Britain and Germany, the first arrangement was an exchange of posts in 1928: the Berlin high school teacher Dr. Vester went to London for one year, while her British colleague Miss Dallas taught in the Schöneberg district of Berlin for the same period. Report on the activities of the DAB from June 1928 to June 1929, LAB, B Rep. 235-5, 1/1, HLA 3634. The change in the law was achieved with the help of the Anglo-German Bureau in London and the Deutsche Pädagogische Auslandsstelle in Berlin. Report on the activities of the DAB from June 1929 to June 1930, LAB, B Rep. 235-5, 1/1, HLA 3630.
35. IFUW, *Report of the Fourth Conference, Amsterdam 1926*, 54.
36. IFUW, *Report of the Nineteenth Council Meeting, Budapest 1934* (London: The Federation, 1934), 32–3. On the reception of Arató's study, see, for example, Vernon Mallinson, "Some Sources for the History of Education in Belgium," *British Journal of Educational Studies* 4, no. 1 (1955): 62–70, here 64.
37. Mallinson, "Some Sources for the History of Education." During a journey to investigate Italian schooling, Charles E. Little witnessed a confrontation of this kind with the principal of the Liceo Scientifico C. Cavour in Rome, Professor Oliviero: "An Italian Lady, Dottoressa Isabella Grassi, one of the high officials of the association of Italian university women and with her, visiting the liceo, a lady from Hungary, Amelie Arato, Docteur des Lettres (seemingly from a French University) who lives at Budapest, were holding a spirited discussion with Signore Oliviero, asking why there were no women allowed to occupy important positions like principals and inspectors of schools, and when such a reform would be likely to come in Italy." Charles E. Little, "The Italians and Their Schools," *Peabody Journal of Education* 10, no. 2 (1932): 77–86, here 81–2.
38. IFUW, *Report of the First Conference, London 1920*, 86.
39. Ibid., 11–12.
40. Ibid., 13.
41. Laura P. Morgan, President of the Washington Branch, to Mrs. Martin, Secretary of ACA, August 2, 1919, Columbia University, Gildersleeve Collection, IFUW, ACA, 1918–1928, Folder 1919b.

42. "Four Important Questions to be Considered at the ACA Biennial," n.d. [spring 1919], Columbia University, Gildersleeve Collection, IFUW, ACA, 1918–1928, Folder 19b.
43. American Council on Education, Officers 1919–1920, Barnard College Archives, Dean's Office, 1919–1920, Box 3.
44. On Suffrage House, see Jacqueline van Voris, *Carrie Chapman Catt: A Public Life* (New York: Feminist Press, 1996), 142. Suffrage House, near the White House, was taken on in December 1916 by the North American Woman Suffrage Alliance (NAWSA). It was a handsome complex with guestrooms and communal rooms, the existence of which was intended to convey the message that the women's suffrage movement had sufficient financial reserves to remain unmoved from the President's doorstep until women's suffrage was introduced. Ibid.
45. Most members of the AAUW lived in the Midwest; on the East Coast, the graduates of the "Seven Sisters," in particular, were less willing to join the AAUW in view of the well-organized alumnae services they already enjoyed. See Laura P. Morgan, President of the Washington Branch, to Mrs. Martin, Secretary of the ACA, August 2, 1919, Columbia University, Gildersleeve Collection, IFUW, ACA, 1918–1928, Folder 1919b.
46. The new premises had previously housed the Men's City Club of Washington. After extensive renovation, including the installation of an elevator, they opened their doors in December 1923. Marion Talbot and Lois Kimball Mathews Rosenberry, *The History of the American Association of University Women, 1881–1931* (Boston, MA: Houghton & Mifflin, 1931), 260.
47. IFUW, *Report of the Third Conference, Oslo 1924*, 37; Talbot and Rosenberry, *History of the AAUW*, 261.
48. On this, see the excellent study by Helen Lefkowitz Horowitz, *Alma Mater: Design and Experience in the Women's Colleges from Their Nineteenth-Century Beginnings to the 1930s* (New York: Knopf, 1984).
49. Thus the journalist Richard Kluger says in his carefully researched and well-written history of the *New York Herald Tribune*: Richard Kluger, *The Paper: The Life and Death of the New York Herald Tribune* (New York: Knopf, 1986), 171–206.
50. Marian Churchill White, *A History of Barnard College* (New York: Columbia University Press, 1954), 200.
51. Kluger, *The Paper*, 141.
52. Mariea Caudill Dennison, "The American Girls' Club in Paris. The Propriety and Imprudence of Art Students, 1890–1914," *Woman's Art Journal* 26, no. 1 (2005): 32–7.
53. Ibid., 35.
54. IFUW, *Report of the Second Conference, Paris 1922* (London: The Federation, 1922), 25. After the death of Helen Rogers and Virginia Gildersleeve, Reid Hall was left to Columbia University, New York. Even today, the building houses "Columbia University Paris," where students both male and female are offered courses and conferences, and researchers from all over the world can find overnight accommodation. See the Reid Hall homepage, http://columbiaprograms.reidhall.net/ (accessed January 18, 2013).
55. The basis of the contract for the building's use was altered in 1928, from a lease with particular AAUW members to a right of use without a time limit. This dramatically changed the Center's financial situation: "Since the upkeep of the Club building is very expensive in proportion to the number of residents accommodated and the rates charged, every possible economy has to be practised in order to make ends meet and an attempt is being made to raise a fund of 30,000 dollars as a guarantee against future deficits and to provide resident scholarships at the clubhouse and assist in international activities." IFUW, *Report of the Twelfth Council*

Meeting, Madrid 1928, 43. Also in 1928, Elizabeth Reid donated $60,000 for an extensive renovation of Reid Hall's entire roof structure, saving the enterprise from bankruptcy. See IFUW, *Report of the Fifth Conference, Geneva 1929* (London: The Federation, 1929), 63. See also Daniele Pitavy-Souques, "Reid Hall. Petite Histoire d'une maison d'excellence," *Diplômées* 223 (December 2007).

56. Annabel Robinson, *The Life and Work of Jane Ellen Harrison* (Oxford: Oxford University Press, 2002), 291–2. (This is also the source of the quotations.)
57. In the past five years alone, the women reported, more than $20 million had been donated for higher education there: mostly by men and by women who had not enjoyed a higher education themselves. IFUW, *Report of the Second Conference, Paris 1922*, 65.
58. "The quest for funds would not be as easy in Great Britain as in America; there was not the same material to draw upon, and unfortunately rich people in Great Britain were not so much interested in women's educational enterprises.... The university women were much hampered by the fact that they were almost all of them very busy, working early and late, and with hardly any time over to do the necessary social work. They had, however, determined to do their best." Ibid., 66–7.
59. See Willy Cohen, "Crosby Hall," *New York Times*, February 3, 1986; Henry James, "Refugees in Chelsea," in *Within the Rim, and Other Essays 1914–1915* (London: W. Collins, 1918).
60. Some board members considered it a hopeless undertaking to try to raise the necessary funds in view of the grim economic context, but their concerns were overridden in the trust that "many interested in international... matters would be prepared to give substantial help." Report of the debate on Crosby Hall at the Annual General Meeting, July 1, 1922, RHUL Archives, PP7/6/5/4.
61. In 1922, the University and City Association, a private association promoting the provision of residences for students, agreed to sell Crosby Hall to the BFUW for only £10,000, inclusive of a 500-year lease from the London County Council. See "Future of Crosby Hall," *The Observer*, December 3, 1922.
62. Miss Nettlefold to Caroline Spurgeon, July 28, 1922, RHUL Archives, PP7/7/1/1.
63. IFUW, *Report of the Second Conference, Paris 1922*, 67.
64. Marguerite Bowie-Menzler, *Founders of Crosby Hall* (London: Privately published, 1981).
65. IFUW, *Report of the Second Conference, Paris 1922*, 66.
66. Margaret G. Blaine, "Crosby Hall, The Opening." Report, n.d. [August 1927], AAUW Archives, Series IX, Reel 149/9, Box 832.
67. "International Clubhouses," in IFUW, *Report of the Third Conference, Oslo 1924*, 39.
68. Bertrand Russell and Alys Russell, *German Social Democracy: Six Lectures; With an App. on Social Democracy and the Woman Question in Germany* (London: Longmans, Green, 1896). A German translation by Achim von Borries appeared with the title *Die deutsche Sozialdemokratie. Mit einem Anhang von Alys Russell: Die Sozialdemokratie und die Frauenfrage in Deutschland* (Berlin: Dietz, 1978).
69. On Roper, see John Guy, *A Daughter's Love: Thomas More and his Dearest Meg* (Boston: Houghton Mifflin Harcourt, 2009).
70. Bowie-Menzler, *Founders of Crosby Hall*, 10–13. On the Russell family, see Paul Scherer, *Lord John Russell: A Biography* (Selinsgrove/London: Susquehanna University Press/Associated University Presses, 1999).
71. "International Clubhouses," in IFUW, *Report of the Third Conference, Oslo 1924*, 39.
72. Report by the warden of Crosby Hall, in IFUW, *Report of the Twelfth Council Meeting, Madrid 1928*, 51–2.
73. Paris, in contrast, remained more clearly American in its orientation, prompting the club management to hope in 1928 that in future it would become "even more international than in the past." Ibid., 43.

74. Berta Karlik, "Bericht über mein Stipendienjahr in England 1930/31," lecture manuscript, March 17, 1932, private archives Anna Zemann, Vienna. Translator's note: Here and throughout, all translations from the German are my own unless otherwise attributed.
75. Ibid.
76. Both of these cases were not so much national as Anglo-American initiatives: "American University Women were going to build a hostel in Athens, where there would be international privileges, and in Rome Mrs. Arthur Strong, Assistant Director of the British School in Rome, had formed a committee interested in opening an international Clubhouse, an idea which had met with great sympathy from the Italian Federation. The Italian Government had promised to give some land for the clubhouse as soon as the IFUW could give an assurance of being able to go forward with the building." "International Clubhouses," in IFUW, *Report of the Third Conference, Oslo 1924*, 39.
77. IFUW, *Report of the Eleventh Council Meeting, Vienna 1927*, 7.
78. See "List of International Clubhouses and Hotels," in IFUW, *Report of the Sixth Conference, Edinburgh 1932*, 15.
79. IFUW, *Report of the First Conference, London 1920*, 71.
80. On the beginnings of the AAUW's research promotion policy, see Rossiter, *Women Scientists in America*. Vol. 2, *Before Affirmative Action, 1940–1972* (Baltimore, MD: Johns Hopkins University Press, 1995), 268–9.
81. Most private elite universities were closed to women at this time; the first woman succeeded in taking her doctorate at Yale in the early 1890s. Ibid., 37.
82. These were the physicist Margaret Maltby, the mathematician Mary Winston (later Nelson), and the physiologist Ida Hyde. Ibid., 38–42.
83. Virginia Gildersleeve "was constantly alert to find among the undergraduates those rare minds which are capable of truly great scholarship. She saw to it that the whole resources of College and University were made available to them. She watched, encouraged, and stimulated them, and for them she would take up the cudgels with anyone at all. The long line of women scholars whom her faculty graduated is an eloquent testimonial to her, and their passionate interest in real scholarship." White, *A History of Barnard College*, 113.
84. "Report of the Committee on Fellowships, March 1922," *AAUW Journal* 25 (1922): 121.
85. Hanna Rydh submitted her doctoral thesis, on Gotlandic women's clothing of the Viking era, in 1919, and subsequently worked in Sweden as the first woman archaeologist to hold a PhD. She saw herself as a "public historian," and most of her work was written for a broad-based audience. She was known for her excellent travel and children's books on pre- and early historical themes. In all her studies, the status of women in society played a key role. See Ewa Ryberg, "Hanna Rydh—Förmedlare av Förhistorien," *Fornvännen* 85 (1990): 303–9; Elisabeth Arwill-Nordbladh, "Hanna Rydh. Eine Forscherin mit vielen Dimensionen," in *Eine Dame zwischen 500 Herren: Johanna Mestorf, Werk und Wirkung*, ed. Julia K. Koch and Eva-Maria Mertens (Münster: Waxmann, 2002), 281–94.
86. IFUW, *Report of the Second Conference, Paris 1922*, 23.
87. There were similarly high numbers of applicants for the Swedish "Prize Fellowship." It was awarded to the British biochemist Wheldale Onslow of Cambridge, for her work on the membranes of the plant cell; IFUW, *Report of the Fifth Council Meeting, London 1923* (London: The Federation, 1923), 60.
88. Ibid., 56.
89. Ibid., 63.
90. This was the view of Agnes L. Rogers, the chair of the American Fellowship Committee; see IFUW, *Report of the Fourth Conference, Amsterdam 1926*, 116.

91. See, especially, Brecher's work on the pupal coloration of the Cabbage White butterfly: *Pieris brassicae, Erster, zweiter und dritter Teil* (Leipzig: W. Engelmann, 1917).
92. As a later letter from Brecher indicates, in the end she spent the period in Germany at the University of Rostock. See Brecher to Erna Hollitscher, April 6, 1938, from Vienna, WL@LSE, 5BFW (old Box 31), Refugees.
93. Cecilia Dentice di Accadia, *Tommaso Campanella* (Florence: Vallecchi, 1921).
94. Christine Touaillon, *Der deutsche Frauenroman des 18: Jahrhunderts* (Vienna: Braumüller, 1919). For Touaillon's biography, see especially Hanna Bubenicek, "Wissenschaftlerin auf Umwegen. Christine Touaillon, geb. Auspitz (1878–1928), Versuch einer Annäherung," in *Über Frauenleben, Männerwelt und Wissenschaft: Österreichische Texte zur Frauenforschung*, ed. Beate Frakele (Vienna: Verlag für Gesellschaftskritik, 1987), 5–17. The other prizewinners were Geneviève Bianquis of the University of Paris with her study of Caroline von Günderode; Anna Söderhjelm, the first woman professor at the University of Helsingfors, with her historical studies of the French Revolution; and Agnes Langenskjöld, again from the University of Helsingfors, with her work on the Swedish writer Victor Rydberg. See IFUW, *Report of the Fifth Council Meeting, London 1923*, 59.
95. According to Stuart Wallace, academics' communication with German (and in many cases also Austrian) colleagues was limited to personal contact; until the late 1920s, the international organizations of science were very cautious in their official contacts with the former Central Powers. The Union Académique Internationale and the International Council of Scientific Unions, the organizations replacing the prewar International Association of Academies, remained closed to citizens of the Central Powers right up to 1935. See Stuart Wallace, *War and the Image of Germany: British Academics, 1914–1918* (Edinburgh: Donald, 1988), 198.
96. IFUW, *Report of the Fifth Council Meeting, London 1923*, 58. It is well known that the award of the Nobel Prize to Max Planck in 1919 gave rise to a similar effect; see Metzler, *Internationale Wissenschaft und nationale Kultur*, 76.
97. IFUW, *Report of the Third Conference, Oslo 1924*, 63.
98. IFUW, *Report of the Fifth Council Meeting, London 1923*, 64.
99. Ibid., 65.
100. IFUW, *Report of the Third Conference, Oslo 1924*, 112.
101. Ibid., 36.
102. Further members of the committee were Ellen Gleditsch (Norway), Mrs. Howard Vernon (United States), and Professoressa Margareta Loschi (Italy). Virginia Gildersleeve (United States) was attached to the committee *ex officio*. Ibid., 64.
103. *AAUW Journal* 26 (1932): 161. In 1929, the AAUW was spending $14,000 a year on fellowships; the branches raised an additional $25,000 per year for scholarships awarded to undergraduates. See Susan Levine, *Degrees of Equality: The American Association of University Women and the Challenge of Twentieth-Century Feminism* (Philadelphia, PA: Temple University Press, 1995), 19.
104. Margaret Rossiter makes a very similar point for the American context. Until 1920, she says, colleges did little to ensure that their most promising women scientists had sufficient time for research. Rossiter, *Women Scientists in America. Before Affirmative Action*, 23.
105. IFUW, *Report of the Fourth Conference, Amsterdam 1926*, 114.
106. IFUW, *Report of the Third Conference, Oslo 1924*, 110.
107. IFUW, *Report of the First Conference, London 1920*, 14.
108. IFUW, *Report of the Third Conference, Oslo 1924*, 114. Gleditsch had spent several years working in the Paris laboratory of Marie Curie. See Annette Lykknes, Lise Kvittengen, and Anne Kristine Børresen, "Ellen Gleditsch: Duty and Responsibility in a Research and Teaching Career, 1916–1946," *Historical Studies in the Physical Sciences* 36, no. 1 (2005): 131–88.

109. IFUW, *Report of the Third Conference, Oslo 1924*, 112.
110. IFUW, *Report of the Fourth Conference, Amsterdam 1926*, 114.
111. IFUW, *Report of the Third Conference, Oslo 1924*, 110. On the benefits and necessity of international research, see also Lise Meitner's lecture in 1932 in Edinburgh, IFUW, *Report of the Sixth Conference, Edinburgh 1932*, 51. Agnes L. Rogers, who taught at Bryn Mawr and chaired the AAUW's selection committee, expressed the obligation to promote research as follows: "While doing many things that formed part of the programme of other organisations, they had, as university women, one distinct duty of their own. They had to do all they could to aid scientific study." IFUW, *Report of the Fourth Conference, Amsterdam 1926*, 117.
112. This donation had been acquired through the daughter of one of the elderly men. See IFUW, *Report of the Third Conference, Oslo 1924*, 62.
113. Ibid., 95.
114. IFUW, *Report of the Fifteenth Council Meeting, Prague 1930*, 23.
115. Ibid., 74.
116. Ibid., 99. Mecenseffy published the findings of this research in a 1929 essay: "Philipp IV. von Spanien und seine Heirat mit Maria Anna von Österreich," in O. Brunner et al. eds, *Historische Studien: A.F. Pribram zum 70. Geburtstag dargebracht* (Vienna: Steyrermühl, 1929), 49–70.
117. IFUW, *Report of the Fifteenth Council Meeting, Prague 1930*, 99. In her book *The Background of the Revolution for Mexican Independence*, which appeared in Boston in 1934, Fisher thanked the Spanish university women for the fellowship and "for their kindness to me while in Spain." She did not, however, fulfill their political hopes: Fisher's book traced the failure of Spanish imperialism in the New World.
118. Report on the activities of the DAB from June 1928 to June 1929, LAB, B Rep. 235-5, 1/1, HLA 3634.
119. IFUW, *Report of the Fifteenth Council Meeting, Prague 1930*, 78. In 1930, the Yugoslav association offered a prize that would be awarded with the aim of stimulating research on the legal situation of women in Yugoslavia, women writers and artists, or women in public social work. Those members who were still students were to be attracted by a prize for the best thesis on women writers in Yugoslav literature or on the portrayal of women in modern Yugoslav novels and drama. Ibid., 85.
120. After World War I, Britain, unlike Germany and France, still maintained a relatively well-functioning network of private patrons of the arts and sciences, even if this by no means matched the much more favorable situation in the United States. See Peter Alter, *Wissenschaft, Staat, Mäzene: Anfänge moderner Wissenschaftspolitik in Großbritannien 1850–1920* (Stuttgart: Klett-Cotta, 1982).
121. IFUW, *Report of the Eleventh Council Meeting, Vienna 1927*, 19.
122. IFUW, *Report of the Fifteenth Council Meeting, Prague 1930*, 63–4.
123. *AAUW Journal* 26 (1932): 161.
124. IFUW, *Report of the Third Conference, Oslo 1924*, 40. Between 1920 and 1930, the number of female college graduates rose from 283 thousand to 480 thousand. See Barbara Miller Solomon, *In the Company of Educated Women: A History of Women and Higher Education in America* (New Haven, CT: Yale University Press, 1985), 142.
125. In 1929, the AAUW took on a well-remunerated general secretary, Kathryn McHale: it had become unrealistic to expect a college president, working for the association on a volunteer basis, to cope with all the practical tasks of the association's presidency. See Levine, *Degrees of Equality*, 21.
126. Talbot and Rosenberry, *History of the AAUW*, 139; see also Levine, *Degrees of Equality*, 19.
127. Atkinson was married to a businessman, but she herself was the daughter of the president of Hamline University; Levine, *Degrees of Equality*, 20. In the United

States in this period, there was a tradition of defining philanthropy to include both monetary gifts and volunteer work. As Kathleen McCarthy argues, to adopt the lens of this nineteenth-century definition of philanthropy is to extend our view beyond the exclusive world of the rich donors and onto the "public stage, on which men and women, rich and poor, black and white, publicly contested for authority and power during the nation's youth." Kathleen McCarthy, *American Creed: Philanthropy and the Rise of Civil Society 1700–1865* (Chicago, IL: University of Chicago Press, 2003), 3.

128. *AAUW Journal* 26 (1932): 161.
129. Thus, in 1930, the Austrian association expressed its regret at not having participated in the work for the IFUW Fellowship Fund, the reason having been that its members were barely in a position even to pay their own annual dues: "We regret very much that we have not yet been able to make even a small addition to our quota to the International Fellowship Fund, a branch of work that which is felt to be of highest interest." IFUW, *Report of the Fifteenth Council Meeting, Prague 1930*, 58.
130. On the close intertwining of philanthropy and civil society in the United States, see McCarthy, *American Creed*. McCarthy describes the history of American voluntarism from the eighteenth century until the Civil War as "America's associational revolution." In the course of this "quiet revolution," she argues, giving, voluntary work, and collecting donations emerged as a basic model of social participation and reform: "White women, free blacks, and white male elites collectively built institutions and participated in public policymaking, providing a yardstick for measuring levels of public influence over time." Ibid., 3. The collecting of tiny amounts of money, especially for the education of the poor, must, she says, be considered a key factor in social reform, as is shown by the "truly amazing" result of the efforts of former slaves to create their own infrastructure after the end of the Civil War: in the mere four years between 1866 and 1870, collections totaled what would be around $10 million in today's money. The donations created an enduring foundation for a system of schooling for black Americans in the southern states. See ibid., 198.
131. IFUW, *A List of International Fellowships for Research* (London: The Federation, 1934), 10–12.
132. IFUW, *A List of International Fellowships* (London 1928, 1930, 1934). The booklet arose from an idea proposed by Winifred Cullis. In 1922, at the Paris conference, Cullis had appealed to all members to help create an international list of this kind in order to clarify what sources of research funding were actually open to women and encourage women to apply for them. See IFUW, *Report of the Second Conference, Paris 1922*, 60. Under the heading "Germany" there are 18 foreign scholarships, among the smallest number for any country listed. Ibid., 52–6.
133. IFUW, *Report of the Seventh Conference, Cracow 1936*, 92.
134. On the following points, see Bettina Vincenz, *Biederfrauen oder Vorkämpferinnen? Der Schweizerische Verband der Akademikerinnen (SVA) in der Zwischenkriegszeit, 1924–1939* (Baden: hier + jetzt, 2011), 104–6.
135. Ibid. Vincenz refers to a dossier drawn up by Du Bois in 1948 for the IFUW on the scientific work she undertook after completing her fellowship; Vincenz, *Biederfrauen oder Vorkämpferinnen*, 105, n. 529.
136. On Karlik's biography, see Brigitte Bischof, *"...Junge Wienerinnen zertrümmern Atome...." Physikerinnen am Wiener Institut für Radiumforschung* (Mössingen: Talheimer-Verlag, 2004), 133–57.
137. In her explanation of the high proportion of women at the Institute for Radium Research, working on a scientifically equal basis with men, Maria Rentetzi focuses on the personnel policy of the director, Stefan Meyer, who specifically aimed to achieve

a gender-political impact. See Maria Rentetzi, "Gender and Radioactivity Research in Interwar Vienna. The Case of the Institute for Radium Research," in *Women Scholars and Institutions: Proceedings of the International Conference (Prague, June 8–11, 2003)*, ed. Soňa Štrbáňová, Ida H. Stamhuis, and Katarina Mojsejová (Prague: Archiv Akad. Věd České Republiky v Nakl. Arenga, 2004), 611–38, here 627.

138. Letter from Karlik to her thesis adviser, Stefan Meyer, on the occasion of a job offer from the institute in Lainz, April 12, 1931, Archiv Radiumsforschung ÖAW, cited in Bischof, *"...Junge Wienerinnen,"* 137, n. 219.
139. It may have been significant that one of the members of the IFUW's Fellowship Committee was physicist Lise Meitner, someone fully able to judge the groundbreaking importance of Karlik's work. The Norwegian radiochemist Ellen Gleditsch, who had been in close contact with Meyer and the Viennese Institute since World War I and frequently visited the Institute in the course of her own research, probably knew Karlik already and may also have argued in her favor. In later years, Gleditsch and Karlik shared a cordial friendship. See Lykknes, Kvittengen, and Børresen, "Ellen Gleditsch."
140. See Maureen M. Julian, "Women in Crystallography," in *Women of Science: Righting the Record*, ed. Gabriele Kass-Simon, Deborah Nash, and Patricia Farnes (Bloomington: Indiana University Press, 1990), 335–83. The later Nobel laureate Dorothy Hodgkin, as well, started out in the circle around Bragg. Ibid.
141. See Isabel Ellie Knaggs and Berta Karlik, *Tables of Cubic Crystal Structures of Elements and Compounds* (London: Hilger, 1932). On the reception of this 90-page work, see Karl Lintner, "Nachruf auf Berta Karlik," *Almanach der Österreichischen Akademie der Wissenschaften* 140 (1989–90): 306.
142. After World War I, Rutherford supplied the Institute for Radium Research with international scientific literature that the Institute could not afford from its own resources. Stefan Meyer also set up Karlik's contact with Bragg. See Rentetzi, "Gender and Radioactivity Research," 618, 625.
143. Because of her high regard for Meyer, Curie looked after his protégée "to an extraordinary extent," as Karlik herself put it. See Bischof, *"...Junge Wienerinnen,"* 136.
144. Cited in Lintner, "Nachruf auf Berta Karlik," 309. On the international community of physicists and their German-speaking representatives before and after World War I, see Metzler, *Internationale Wissenschaft und nationale Kultur*. According to Metzler, study visits abroad for German-speaking physicists were limited to a small elite both before and after the war. This was, of course, partly due to the fact that physics in Germany enjoyed a worldwide reputation and itself attracted numerous students from abroad. Ibid., 50.
145. Friedrich Hernegger and Berta Karlik, "Uranium in Sea-Water," *Meddelanden från Göteborgs Högskolas Oceanografiska Institution* 12 (1935).
146. On genetics, see Helga Satzinger, "Women's Places in the New Laboratories of Genetic Research in Early 20th Century: Gender, Work, and the Dynamics of Science," in *Women Scholars and Institutions: Proceedings of the International Conference (Prague, June 8–11, 2003)*, ed. Soňa Štrbáňová, Ida H. Stamhuis, and Katarina Mojsejová (Prague: Archiv Akad. Věd České Republiky v Nakl. Arenga, 2004), 265–94; Ida Stamhuis and Arne Monsen, "Kristine Bonnevie, Tine Tammes and Elisabeth Schiemann in Early Genetics: Emerging Chances for a University Career for Women," *Journal of the History of Biology* 40, no. 3 (2007): 427–66. On nuclear physics, see Maria Rentetzi, *Trafficking Materials and Gendered Experimental Practices* (New York: Columbia University Press, 2007).
147. The Rockefeller Foundation presents a similar picture. Here, too, a close relationship existed between the post-1918 program of promoting science, the transatlantic connections arising from that program, and the commitment to working

for academic refugees after 1933. See G. Falk, "The Reaction of Americans to the Persecution of the European Scholars during the Nazi Era," *International Review of History and Political Science* 9 (1972): 78–98; Helke Rausch, "US-Amerikanische 'Scientific Philanthropy' in Frankreich, Deutschland und Großbritannien zwischen den Weltkriegen," *Geschichte und Gesellschaft* 33 (2007): 73–98, here 89.

4 Reactions in Central Europe: The German Case

1. The term "atrocities" alluded to the German war crimes against the civilian populations of Western European countries, above all Belgium and France. For a historical analysis of these, see John Horne and Alan Kramer, *German Atrocities, 1914: A History of Denial* (New Haven, CT: Yale University Press, 2001). The London resolution is reproduced in full in "International Scientific Organization," *Science* 48, no. 1247 (November 22, 1918): 509–10, here 510. On the politics of the IRC, see Gabriele Metzler, *Internationale Wissenschaft und nationale Kultur: Deutsche Physiker in der internationalen Community, 1900–1960* (Göttingen: Vandenhoeck & Ruprecht, 2000), 121–2. Eckardt Fuchs argues that a relatively moderate position prevailed at this meeting thanks to Anglo-American efforts and the stance of the neutral countries. In his view, Germany's renewed integration into the international scientific community was made conditional on changes in the policies of the Central Powers, not on personal recantations by individual academics. See Eckhardt Fuchs, "Wissenschaftsinternationalismus in Kriegs- und Krisenzeiten. Zur Rolle der USA bei der Reorganisation der internationalen 'Scientific Community,' 1914–1925," in *Wissenschaft und Nation in der europäischen Geschichte*, ed. Ralph Jessen and Jakob Vogel (Frankfurt/Main: Campus, 2002)," 276.
2. Margarete Rothbarth, "Die deutschen Gelehrten und die internationalen Wissenschaftsorganisationen," in *Volkstum und Kulturpolitik: Eine Sammlung von Aufsätzen, gewidmet Georg Schreiber zum 50. Geburtstage*, ed. Heinrich Konen and Johann Peter Steffes (Cologne: Gilde-Verlag, 1932), 143–57, here 143.
3. From the perspective of international academic organizations, see Fuchs, "Wissenschaftsinternationalismus"; Metzler, *Internationale Wissenschaft und nationale Kultur*, especially 138–44. According to Metzler, Einstein was "the only physicist working in Germany who was willing and able to break through international isolation"; ibid., 141. On the exclusion and self-isolation of German physicists, see also Michael Desser, *Zwischen Skylla und Charybdis: Die "scientific community" der Physiker, 1919–1939* (Vienna: Böhlau, 1991), 11–25.
4. "Too proud to repudiate the political errors of the wartime period and too confident of the importance and status of their research work," as Metzler puts it, the Germans "curtly rejected the offers of reconciliation." Gabriele Metzler, "Nationalismus und Internationalismus in der Physik des 20. Jahrhunderts," in *Wissenschaft und Nation in der europäischen Geschichte*, ed. Ralph Jessen and Jakob Vogel (Frankfurt/Main: Campus, 2002), 287. German physicists remained excluded from the most important international physics conferences until 1927; Metzler, *Internationale Wissenschaft und nationale Kultur*, 139. It is possible that the dynamics of other disciplines differed from this particular development.
5. Metzler, *Internationale Wissenschaft und nationale Kultur*, 164.
6. The conflicts between the DAB and the IFUW after 1933 are analyzed in detail in chapter 5.
7. Hoping to emphasize its close identification with the German sciences, the Austrian National Council resolved in 1921 to rename the "Austrian Academy

of Sciences" the "Academy of Sciences in Vienna." See Gerhard Oberkofler and Eduard Rabofsky, *Studien zur Geschichte der österreichischen Wissenschaft zwischen Krieg und Frieden* (Vienna: Edition Fortschrittliche Wissenschaft, 1987), 19.

8. The IFUW's contacts with Richter came about through the Vienna-based British art historian Alice Levetus. See Report on the Activities of the Verband der akademischen Frauen Österreichs, Columbia University, Butler Library, Rare Books Collection, Gildersleeve Collection, IFUW, ACA, 1918–1928, Box 44, Folder 22–23. Alice Levetus was a member of the BFUW and ran an English-language course at Vienna's municipal night school that had become an institution. She probably met Elise Richter through Elise's older sister Helene, who also taught night-school classes. On Levetus, see Wilhelm Filla, "Miss A. Levetus—Kunsthistorikerin und Volksbildnerin. Portrait einer grenzüberschreitenden Pionierin," *Spurensuche: Zeitschrift für Geschichte der Erwachsenenbildung und Wissenschaftspopularisierung* 12, no. 1 (2001): 24–39.
9. As late as 1918, Elise Richter joined the new Bourgeois Freedom Party (later to be named the Bourgeois-Democratic Workers' Party), and for a time was part of the party's national committee. However, she ceased her party work a short time later, concerned about her academic future at the university. See Elise Richter, *Summe des Lebens, hg. vom Verband der Akademikerinnen Österreichs* (Vienna: WUV Universitätsverlag, 1997), 205–9.
10. Ibid., 117.
11. Ibid.
12. Ibid. Until 1930, Elise Richter succeeded in maintaining a difficult equilibrium between the Austrian association's internal political animosities; she then gave up and resigned. A close relationship with the international federation was forged by her student Hedwig Kuranda, who was responsible for corresponding with the IFUW. This can be seen in the detailed and very frank reports Kuranda sent to the IFUW about the situation in Austria and the progress of the VAÖ. See, in particular, IFUW, *Report of the Fifth Council Meeting, London 1923*, 17–21; IFUW, *Report of the Third Conference, Oslo 1924*, 4–5; IFUW, *Report of the Ninth Council Meeting, Brussels 1925* (London: The Federation, 1925), 1–2.
13. See Alice Salomon, "Elisabeth Altmann-Gottheiner zum Gedächtnis," *Die Österreicherin* 9 (1930): 4.
14. Quoted in Caroline Spurgeon to Mme Cestre, June 15, 1922, Archives nationales de France, Fontainebleau, AFFDU, Relations internationales FIFDU, questions de l'admission des déléguées allemandes.
15. Caroline Spurgeon to Mme Cestre, June 15, 1922, ibid.
16. Altmann-Gottheiner to Spurgeon, June 18, 1922, ibid.
17. Proceedings of the Third Council Meeting in Paris, July 15, 1922, AAUW Archives, Series IX, Reel 150/28, Box 832.
18. Helene Lange, "Internationale Tagungen von Akademikerinnen," *Die Frau* 31 (1923/24).
19. "Ein internationaler Verband der Akademikerinnen," *Die Frau* 30 (1922): 30.
20. Ibid.
21. This view was widely held throughout German academia. See Metzler, *Internationale Wissenschaft und nationale Kultur*, especially 121–44.
22. In these articles, Bäumer explains why the board of the BDF federation, for many decades a member of the International Council of Women, resolved in 1919 not to attend the first international convention after the war, held in Christiania. Because Bäumer, like other leading BDF members, also played a crucial part in founding the DAB, we may also read her commentary as casting light on the attitude of the academic women who would later organize within the DAB. Gertrud Bäumer, "Prinzipienfragen des Frauenweltbundes," *Die Frau* 28 (1920): 1–4.

23. Gertrud Bäumer, "Rechtsfrieden? Erklärung des Bundes Deutscher Frauenvereine zur Gründung des Völkerbundes," *Die Frau* 26 (1918): 37–8. This declaration by the BDF articulated the new view of politics held by "women who have become aware of their citizenly responsibility in a new way." Helene Lange, "Nachtrag zur Erklärung," *Die Frau* 26 (1918): 38. Lange closed her commentary with the words: "The idea of the community of nations is not alien to the German spirit. Those from whom we still draw our strength today, the leading lights of German idealism, believed in it. They combined this belief with another: that cultural exchange between human beings is the sphere in which the German idea must fulfill its special mission—as other nations must fulfill theirs. It has been held against us that we share this belief, indeed that we have put our national goals respecting the life of humanity at the service of this idea. We are aware that a certain, all-too-German brusqueness has contributed to this mission being misunderstood in the world at large. Nevertheless, we cleave to it today, in the form that our intellectual leaders desired."
24. Bäumer, "Prinzipienfragen," 2.
25. On the outrage at the "shameful diktat of Versailles," see Vanessa Conze, "'Unverheilte Brandwunden in der Außenhaut des deutschen Volkskörpers.' Der deutsche Grenzdiskurs in der Zwischenkriegszeit," in *Ordnungen in der Krise: Zur Politischen Kulturgeschichte Deutschlands, 1900–1933*, ed. Wolfgang Hardtwig (Munich: Oldenbourg, 2007), 21–48; on isolationist sentiment in Weimar nationalism, see Stefan Breuer, "Der Neue Nationalismus in Weimar und seine Wurzeln," in *Mythos und Nation*, ed. Helmut Berding (Frankfurt/Main: Suhrkamp, 1996), 257–74; Andreas Hillgruber, "Unter dem Schatten von Versailles—Die außenpolitische Belastung der Weimarer Republik. Realität und Perzeption bei den Deutschen," in *Weimar: Selbstpreisgabe einer Republik. Eine Bilanz heute*, ed. Karl-Dietrich Erdmann and Hagen Schulze (Düsseldorf: Droste, 1980), 51–67.
26. The International Woman Suffrage Alliance (IWSA) was founded in Berlin in 1904 to fight for the vote for women. It was reorganized in the 1920s and, in 1926, took the new name "International Alliance of Women for Suffrage and Equal Citizenship," abbreviated as IAWSEC or simply as IAW. Karen Offen, *European Feminisms, 1700–1950: A Political History* (Stanford, CA: Stanford University Press, 2000), 346. See also Mineke Bosch and Annemarie Kloosterman, eds., *Politics and Friendship: Letters from the International Woman Suffrage Alliance, 1902–1942* (Columbus: Ohio State University Press, 1990), especially the introduction to the second section, "Reconstruction in the Twenties," 175–83. For the sake of simplicity, I use the abbreviation IAW also, below, for the period before the organization was restructured.
27. A vivid description of her first impressions in the international arena can be found in the memoirs of Dorothee von Velsen, who attended her first international women's conference in 1920, aged 27. See Dorothee von Velsen, *Im Alter die Fülle: Erinnerungen* (Tübingen: Wunderlich, 1956), 247.
28. "If I were to describe the many small and great moments when this feeling suddenly arose during debates or personal conversations," Bäumer recapitulated in *Die Frau*, "it would seem rather sentimental. It is better to remain silent in the knowledge that everyone who was there experienced something of this kind many times." Gertrud Bäumer, "Der Frauenweltbund im Haag," *Die Frau* 29 (1921/22): 264–70.
29. On the Ruhr crisis and its impact on domestic and foreign policy, see Conan Fischer, *The Ruhr Crisis, 1923–1924* (Oxford: Oxford University Press, 2003); Gerd Krumeich and Joachim Schröder, *Der Schatten des Weltkriegs: Die Ruhrbesetzung 1923* (Essen: Klartext, 2004).
30. In the negotiations for readmission into all the international associations, the "language question" played an outstanding role. See Irmgard Rathgen,

"Meinungsaustausch. Zur Frage des Internationalen Bundes Akademischer Frauen," *Die Studentin* 1, no. 12 (1924).
31. Velsen, *Im Alter die Fülle*, 259.
32. Offen, *European Feminisms*, 344; Bäumer, "Der Frauenweltbund im Haag," 264.
33. IFUW, *Report of the Fifth Council Meeting, London 1923*, 39.
34. On Bäumer's activity in the German women's movement and politics, see the excellent double biography by Angelika Schaser, *Helene Lange und Gertrud Bäumer: Eine politische Lebensgemeinschaft* (Cologne: Böhlau, 2000).
35. See Gertrud Bäumer, "Die Not der geistigen Arbeiterin," *Die Frau* 30 (1923): 204–10, here 210.
36. In this text, Irmgard Rathgen addresses the generation conflict within the women's movement. Irmgard Rathgen, "Vom gemeinsamen Aufbau," in *Dritte Generation: Für Gertrud Bäumer*, ed. Hilde Lion, Irmgard Rathgen, and Else Ulig-Beil (Berlin: F. A. Herbig, 1923), 86–8. In the wider milieu of this group of younger women (i.e., women born after 1890) were the historians Margarete Rothbarth (born 1887) and Dorothee von Velsen (born 1883). Rothbarth had been working in Paris at the Institute of Intellectual Cooperation since 1926; von Velsen was elected the new president of the ADF in 1923. Both contributed to the Festschrift for Bäumer.
37. Rathgen discussed this, her first experience abroad, in Irmgard Rathgen, "Die Kopenhagener Tagung des Internationalen Frauenbundes," *Die Frau* 31 (1923/24): 311–13.
38. See Emma Alp, "Völkerbund und Pazifismus," *Die Studentin* 2, no. 1/2 (1924): 3–6.
39. Rathgen, "Meinungsaustausch. Zur Frage des Internationalen Bundes Akademischer Frauen," 45.
40. IFUW, *Report of the Third Conference, Oslo 1924*, 61.
41. In 1904, at only 22, Corbett Ashby had accompanied her mother to the founding congress of the IAW in Berlin, and, in 1919, she was Britain's only female delegate to the Paris peace conferences that prepared the Versailles Treaty. Corbett Ashby retained her office as the president of the IAW from 1923 to 1946. On Corbett Ashby, see especially Offen, *European Feminisms*, 367–76; also Bosch and Kloosterman, eds., *Politics and Friendship*, 179–85.
42. IFUW, *Report of the Third Conference, Oslo 1924*, 61.
43. Emma Alp, "Meinungsaustausch. Ein internationaler Studentinnenkongress in Kristiania," *Die Studentin* 1, no. 1 (1924): 14–15.
44. Alp, "Völkerbund und Pazifismus," 4. Alp's pacifism becomes evident here: "Being a pacifist means, instead, being internationalist on one's own account; it means striving for a cosmopolitanism of the highest and most mature consequence, means seeing a brother in everyone."
45. Emma Alp, "Meinungsaustausch. Noch etwas über den Internationalen Verband Akademischer Frauen," *Die Studentin* 1, no. 12 (1924): 65. The following points also come from this source.
46. Theodora Bosanquet, "Der Internationale Akademikerinnen-Verband," *Die Studentin* 2, no. 1/2 (1925): 7.
47. Rathgen, "Meinungsaustausch. Zur Frage des Internationalen Bundes Akademischer Frauen." The following points also come from this source.
48. Gabriele Humbert, "Der DAB," *Die Studentin* 2, no. 11/12 (1925): 82.
49. Gertrud Kuchel, "Die deutsche Akademikerin und der Internationale Verband Akademischer Frauen," *Die Studentin* 1, no. 2/3 (1924): 29.
50. Lange, "Internationale Tagungen von Akademikerinnen," 366.
51. IFUW, *Report of the Fourth Conference, Amsterdam 1926*, 9. The complete text of Simons's German welcome is reproduced in a commemorative edition of the conference report that was presented to, among others, the members of the AAUW's International Relations Committee. AAUW Archives, Series IX, Reel 150/28,

Box 832. In the official report on the conference this passage is missing, and is replaced with the summary: "The speech closed with some words of welcome spoken in French and in German, addressed to the representatives of the Latin countries and to Germany"; IFUW, *Report of the Fourth Conference, Amsterdam 1926*, 5–7. According to a report by Marie-Elisabeth Lüders in the newspaper *Vossische Zeitung*, the mayor of Amsterdam also welcomed the delegation in German, and at the evening festivities a one-act play in German was performed. See Marie-Elisabeth Lüders, "Die geistige Internationale der Frauen," *Vossische Zeitung*, August 12, 1926.

52. IFUW, *Report of the Fourth Conference, Amsterdam 1926*, 34.
53. In 1926, the DAB's paying members numbered 3,787, but by 1929 this figure had dropped to 3,117. See the DAB's business report for 1930, LAB, B Rep. 235–5, 1/1, HLA 3634.
54. After the conference in Amsterdam, Agnes von Zahn-Harnack reported rather scathingly that it had quickly been possible "to establish through a few conversations or observations that, for example, the standard of the more than 100 American women who had come over to the conference was in many cases far below what one generally understands by an academic educational background." Agnes von Zahn-Harnack, "Die vierte Tagung des Internationalen Akademikerinnenbundes," *Die Frau* 33 (1925/26): 732.
55. On Rosa Kempf, see Elke Reining, "Rosa Kempf (1874–1948). Ihr Kampf für die Rechte der Frauen," *Zeitschrift für bayerische Landesgeschichte* 64 (2001): 149–68; on Margarete von Wrangell, see Agnes von Zahn-Harnck, "Margarete Fürstin Andronikow-Wrangell," *Die Frau* 39 (1931/32): 571–3, especially on the matter of Wrangell's work for the DAB.
56. On Linden, see Patricia M. Mazon, *Gender and the Modern Research University: The Admission of Women to German Higher Education, 1865–1914* (Stanford, CA: Stanford University Press, 2003), 180–90; on Altmann-Gottheiner, see Iris Schröder, *Arbeiten für eine bessere Welt: Frauenbewegung und Sozialreform 1890–1914* (Frankfurt/Main: Campus, 2001), 76–7.
57. IFUW, *Report of the Eleventh Council Meeting, Vienna 1927*, 61–2; IFUW, *Report of the Fifth Conference, Geneva 1929*, 5–7; IFUW, *Report of the Sixth Conference, Edinburgh 1932*, 5.
58. In her memoirs, Elise Richter mentions that in Vienna she worked with Agnes von Zahn-Harnack, whom she liked very much, for the "adoption of German as the third working language in sessions and reports. With success." Richter, *Summe des Lebens*, 118.
59. IFUW, *Report of the Eleventh Council Meeting, Vienna 1927*, 13.
60. IFUW, *Report of the Fifth Conference, Geneva 1929*, 57–61.
61. Ibid., 112.
62. Ibid., 114.
63. Marie-Elisabeth Lüders, address on the occasion of Winifred Cullis's visit to the DAB in Berlin, October 31, 1931, LAB B Rep, 235, 1/1, HLA 3631.
64. In 1929, the Deutscher Philologinnenverband's 2,150 members made up more than two-thirds of the DAB's then 3,117 members. See List of membership numbers in the DAB's member associations, 1926–29, n.d. [1929], LAB, B Rep. 235–5, 1/1, HLA 3634.
65. Minutes of the meeting of the DAB board on June 10, 1930, at the Lingner Schloss near Dresden, n.d. [July 1930], p. 3, LAB, B Rep. 235–5, 1/1, HLA 3630.
66. Deutscher Philologinnenverband to the executive of the DAB (Lüders), July 2, 1931, LAB, B Rep. 235–5, 1/1, HLA 3631. See also a letter from the Deutscher Philologinnenverband to Lüders, November 6, 1930, which noted that the association could "not with good conscience accept its members remaining in the

international federation of university women for the long term if the language question is not resolved speedily to the satisfaction of its members and negotiations are not accelerated." LAB, B Rep. 235–5, 1/1, HLA 3630.

67. This was Lüders's comment in her address on the occasion of Winifred Cullis's visit to the DAB in Berlin on October 31, 1931. LAB B Rep, 235, 1/1, HLA 3631.
68. Record of the session of the DAB board to discuss the language question with Prof. Cullis at Harnack House, Berlin-Dahlem, on October 31, 1931, at 3 p.m. (preparatory discussion), p. 1, LAB, B Rep. 235–1/1, HLA 3631. Translation was provided by Johanna Willich, a senior high school teacher who would later take on the presidency of the DAB when it was aligned with the Nazi regime in 1933. See chapter 5.
69. Record of the session of the DAB board to discuss the language question with Prof. Cullis at Harnack House, Berlin-Dahlem, on October 31, 1931, LAB, B Rep. 235–1/1, HLA 3631.
70. Ibid.
71. Ibid.
72. The text resolved ran: "English, French, and German may be spoken at the meetings for delegates and at the members' meetings.... In view of the need for making an effective and economical use of the time at the disposal of the Conference, members are asked to simplify the proceedings as much as possible by reducing the necessity for translations to a minimum." IFUW, *Report of the Sixth Conference, Edinburgh 1932*, 175.
73. For this reason, the German delegation consisted of only three members. The three attendees were Anna Schönborn, physician Ilse Szagunn, and physicist Lise Meitner, who held one of the opening papers on the topic of "The Promotion of Science as a Problem of Our Era." See Report on the activities of the DAB from June 14, 1930, to October 1, 1932, p. 7, LAB, B Rep. 235–5, 1/1, HLA 3633.
74. Anna Schönborn (Berlin) to Frau Grave, August 15, 1932, LAB, B Rep. 235–1/1, HLA 3632.
75. Ibid.
76. The DAB could not fulfill its promise to host the conference, having been taken over by the Nazi regime, and neither did the agreement on the language question reached in Edinburgh ever come to fruition. See Marie-Elisabeth Lüders's report to the Foreign Office, "Bericht zu den Tagungen internationaler Frauenorganisationen zwischen dem 21. und 31.9.1933 in Genf an das Auswärtige Amt," BArch Koblenz, N 1151, no. 8.
77. For the case of physicists in the Weimar Republic, Gabriele Metzler has observed a similar tendency for politics and science to move closer together. See Metzler, *Internationale Wissenschaft und nationale Kultur*, 140–4.
78. Eckard Michels, "Deutsch als Weltsprache? Franz Thierfelder, the Deutsche Akademie in Munich and the Promotion of Language Abroad, 1923–1945," *German History* 22, no. 2 (2004): 206–28.
79. Irene Stoehr finds evidence for this in the international work of the ADF, which became the German branch of the IAW in 1923 but had previously worked primarily on a local level. See Irene Stoehr, "Bedingt beglückend: Die internationale Kooperation der deutschen Frauenbewegung vor 1933," in *Über Grenzen hinweg: Zur Geschichte der Frauenstimmrechtsbewegung und zur Problematik der transnationalen Beziehungen in der deutschen Frauenbewegung*, ed. Irina Hundt (Berlin: Deutscher Staatsbürgerinnenverband e.V., 2007), 128–41, here 131–2.
80. On the debates around the "work abroad" of the conservative, nationalist women's organizations, see Dorothee von Velsen, "Voraussetzungen und Möglichkeiten internationaler Arbeit," *Die Frau* 29 (1922): 353–9, here 355; and "Deutsche Interessen auf internationalen Frauentagungen," *Die Frau* 30 (1922/23): 43–6.

81. On Rosa Kempf's energetic efforts to counter the repercussions of the Versailles Treaty, see Reining, "Kempf," 160–3.
82. More will be said in chapter 5 about Szagunn's biographical and social background, and on her role in the later Nazification of the DAB.
83. This is true not only for the American women who helped found the IFUW, but also, and in particular, for Caroline Spurgeon, who played such an exceptional role in mobilizing British women academics against Germany. See Renate Haas, "Caroline Spurgeon: English Studies, the United States, and Internationalism," *Studia Anglica Posnaniensia: International Review of English Studies* (2002): 1–15.
84. See the proposal drafted by Caroline Spurgeon for an "Associated Board for the Promotion of Imperial-American Interchange for Women University Students," n.d. [September 1918], PP67/6/1/2, RHUL Archives.
85. One last report on this theme appears in IFUW, *Report of the Third Conference, Oslo 1924*, 72–6.
86. The use of English as the negotiating language of the Paris peace conferences was implemented by the US president Woodrow Wilson. It is regarded as an innovation and as the beginning of the English language's dominance in international negotiations. See Haas, "Caroline Spurgeon," 5.
87. As Eckart Fuchs puts it, "the American initiatives on the international stage fulfilled…a double role: the promotion of international academic cooperation and the strengthening of America's own national scholarship." Fuchs, "Wissenschaftsinternationalismus," 267.
88. Offen, *European Feminisms*, 344.
89. See Brian Harrison, *Prudent Revolutionaries: Portraits of British Feminists between the Wars* (Oxford: Clarendon Press, 1987); Richard J. Evans, *Comrades and Sisters: Feminism, Socialism, and Pacifism in Europe, 1870–1945* (Brighton, Sussex: Wheatsheaf Books, 1987). A critique of this approach can be found in Carol Miller, "'Geneva—the Key to Equality': Inter-War Feminists and the League of Nations," *Women's History Review* 3, no. 2 (1994): 219–45.
90. See, for example, Barbara Greven-Aschoff, *Die bürgerliche Frauenbewegung in Deutschland 1894–1933* (Göttingen: Vandenhoeck & Ruprecht, 1981); Hiltraud Schmidt-Waldherr, *Emanzipation durch Professionalisierung? Politische Strategien und Konflikte innerhalb der bürgerlichen Frauenbewegung während der Weimarer Republik und des Nationalsozialismus* (Frankfurt/Main: Materialis, 1987); Elizabeth Harvey, "The Failure of Feminism? Young Women and the Bourgeois Feminist Movement in Weimar Germany 1918–1933," *Central European History* 28, no. 1 (1995): 1–28. Angelika Schaser has pointed out that various organizations within the women's movement, including the DAB, did, in fact, develop extensive activities that have only begun to be investigated, their existence and significance having been largely forgotten after 1933. Schaser, *Frauenbewegung in Deutschland 1848–1933* (Darmstadt: WBC Wissenschaftliche Buchgesellschaft, 2006), 97–8.
91. Bäumer, "Die Not." The following points also come from this source.
92. Ibid., 207.
93. Ibid., 209–10.
94. On the achievements of women in universities and research during the 1920s, see Annette Vogt, *Vom Hintereingang zum Hauptportal? Lise Meitner und ihre Kolleginnen an der Berliner Universität und in der Kaiser-Wilhelm-Gesellschaft* (Stuttgart: Steiner, 2007); Claudia Huerkamp, *Bildungsbürgerinnen: Frauen im Studium und in akademischen Berufen 1900–1945* (Göttingen: Vandenhoeck & Ruprecht, 1996).
95. Marianne Weber, "Vom Typenwandel der studierten Frau," 1917, in *Frauenfragen und Frauengedanken: Gesammelte Aufsätze* (Tübingen: Mohr, 1919), 179–200. An excellent overview of the situation of women students in the Weimar Republic is provided by Angelika Schaser, "Die 'undankbaren' Studentinnen. Studierende

Frauen in der Weimarer Republik," in *Frauen auf dem Weg zur Elite*, ed. Günther Schulz (Munich: Boldt, 2000), 97–116.

96. Elisabeth Altmann-Gottheiner, *Die Berufsaussichten der deutschen Akademikerinnen* (Halle: Niemeyer, 1921), 3.
97. Ibid.
98. This section, which described itself as the association of female academically educated and student teachers, was re-formed in 1925 as the Deutscher Philologinnenverband, the German association of women philologists. The new association had almost 1,650 members, and was the strongest of the various professionally based associations for women. See Huerkamp, *Bildungsbürgerinnen*, 221.
99. Britta Lohschelder, *"Die Knäbin mit dem Doktortitel": Akademikerinnen in der Weimarer Republik* (Paffenweiler: Centaurus, 1994), 119.
100. See Marie-Elisabeth Lüders, Retrospect and assessment of the association of women economists on the occasion of its dissolution in December, 1934, n.d. [1934], BArch Koblenz, N 1151, no. 255. The organization only offered full membership to women with doctorates, but it did extend a special invitation to three scholars without the title, "on the basis of their scholarly contributions": Elisabeth Gnauck-Kühne, Gertrud Dyhrenfurth, and Helene Simon. In the early 1920s, members rejected a proposal to open up the association generally for "women economists without doctorates but whose scholarly work is very eminent or who hold outstanding public positions." On this matter, see an article by Helene Simon, "Geistige Arbeiter," *Die Frau* 28 (1920/21): 45.
101. The federation of German women doctors, BDÄ, was founded in 1924 on the initiative of the American Esther Pohl Lovejoy. See Christine Eckelmann, *Ärztinnen in der Weimarer Zeit und im Nationalsozialismus: Eine Untersuchung über den Bund Deutscher Ärztinnen* (Wermelskirchen: WFT Verlag für Wissenschaft, Forschung und Technik, 1992).
102. The number of female students rose from 1,132 in the winter semester of 1908 to 4,053 in 1914, accounting for 6 percent of the total student numbers of around sixty thousand. During World War I, the proportion of female students officially rose to 9 percent, but, in practice, in many institutions they made up well over 80 percent of the students present, male students being away at the front. In 1919, the number of students overall initially rose to ninety thousand, dropping back in the subsequent years to below sixty thousand in 1925/26. The number of women students was 8,269 in the summer semester of 1921, and by the winter semester of 1925/26 had fallen to 6,760. See Huerkamp, *Bildungsbürgerinnen*, 76–7.
103. Altmann-Gottheiner, *Berufsaussichten*, 2.
104. Ibid., 9. A vivid portrait of the employment situation of women high school teachers in Berlin and Prussia is provided by Katharina Bieler, *Im preußischen Schuldienst: Arbeitsverhältnisse und Berufsbiographien von Lehrerinnen und Lehrern in Berlin-Schöneberg 1871–1933* (Cologne: Böhlau, 2007).
105. Altmann-Gottheiner, *Berufsaussichten*, 12. Instead, Masling argued, women should study pharmacy. Judging by the rapid and substantial rise in the number of female pharmacy students, it seems that statements of this kind were taken seriously. Ibid., 13.
106. Altmann-Gottheiner, *Berufsaussichten*, 16.
107. Ibid., 15.
108. However, in 1924, the journal *Die Studentin* published a warning against studying economics. The association of German women economists issued the admonition, arguing that "this whole field, which just a few years ago was very much open to new recruits," was now almost completely out of the reach of "any woman

without a strong sense of self-sacrifice or an independent income." "Warnung vor dem Studium der Nationalökonomie," *Die Studentin* 12, no. 3 (1924): 46.

109. Altmann-Gottheiner, *Berufsaussichten*, 21.
110. Gertrud Bäumer, "Die Akademikerin und die Volkskultur. Zur Gründung des Verbandes Deutscher Akademikerinnen," *Die Frau* 33 (1926): 513–7, here 517.
111. On the founding of the AAUW, see Marion Talbot and Lois Kimball Mathews Rosenberry, *The History of the American Association of University Women, 1881–1931* (Boston, MA: Houghton & Mifflin, 1931).
112. Barbara Miller Solomon, *In the Company of Educated Women: A History of Higher Education in America* (New Haven, CT: Yale University Press, 1985), 43–7.
113. The campaign was pursued primarily by the ADF, founded in 1865 in Leipzig by Louise Otto-Peters. James Albisetti, *Schooling German Girls and Women: Secondary and Higher Education in the Nineteenth Century* (Princeton, NJ: Princeton University Press, 1988), 96–7; Carol Diethe, *The Life and Work of Germany's Founding Feminist, Louise Otto-Peters (1819–1895)* (Lewiston, NY: Edwin Mellen Press, 2002). See also Ilse Brehmer, ed., *Frauen in der Geschichte IV. "Wissen heisst Leben...": Beiträge zur Bildungsgeschichte von Frauen im 18. und 19. Jahrhundert* (Düsseldorf: Schwann, 1983).
114. Ulrike Bussemer, *Frauenemanzipation und Bildungsbürgertum: Sozialgeschichte der Frauenbewegung in der Reichsgründungszeit* (Weinheim: Beltz, 1985). Albisetti concludes that "the first concerted drive to improve girls' schools and to open up new educational and employment opportunities lost momentum by the mid-70s and nearly died out by 1880. Most organizations and institutions continued to exist, but by 1879 women had explicitly been barred from entering all German universities and from being certified as physicians." Albisetti, *Schooling German Girls*, 93.
115. Helene Lange, *Lebenserinnerungen* (Berlin: Herbig, 1921/30), 161–9, here 62.
116. Lily Braun, *Die Frauenfrage, ihre geschichtliche Entwicklung und wirtschaftliche Seite* (Leipzig: Europäischer Hochschulverlag, 1901), 76. Lily Braun (1865–1916) was initially a member of the bourgeois association Frauenwohl, or "women's welfare." In 1895, she joined the "revisionist" wing of the Social Democratic Party and agitated within the party—in vain—for cooperation with the bourgeois women's movement. On Lily Braun, see Alfred G. Meyer, *The Feminism and Socialism of Lily Braun, 1865–1916* (Urbana: Indiana University Press, 1986); Dorothee Wierling, *Eine Familie im Krieg. Leben, Sterben und Schreiben 1914–1918* (Göttingen: Wallstein, 2013).
117. Up to that point, women had only been able to gain admission to a university by personal invitation, as was the case for almost all of the pioneer generation. Trude Maurer, ed., *Der Weg an die Universität: Höhere Frauenstudien vom Mittelalter bis zum 20. Jahrhundert* (Göttingen: Wallstein, 2010); Mazon, *Gender and the Modern Research University*.
118. Mazon, *Gender and the Modern Research University*, 184.
119. Ibid., 199.
120. Ibid., 190.
121. Annette Vogt, *Elsa Neumann: Berlins erstes Fräulein Doktor* (Berlin: Verlag für Wissenschafts- und Regionalgeschichte, 1999), 20–1.
122. As the daughter of the wealthy Berlin banker Max Neumann, this physicist was not dependent on earning her own living, and after her doctorate she was able to continue her scientific work in the private Berlin laboratory of the chemists Carl Friedheim and Arthur Rosenheim. Although the professors maintained the laboratory mainly for the purpose of training their doctoral students, they also rented out working space to private scholars. Ibid., 31.
123. Ibid., 28.
124. On Wegscheider, see Mazon, *Gender and the Modern Research University*, 199–206.

125. The complete list of the association's members can be found in Vogt, *Elsa Neumann*, 26–30. Researchers joining the association at a later date were the archaeologist Margarete Bieber, the geneticist Elisabeth Schiemann, and the physician Paula Hertwig. Ibid.
126. See Katharina Freifrau von Kuenssberg, *Vom langsamen Aufstieg der Frau. Dr. Katharina Freifrau von Kuenssberg erzählt aus ihrem Leben*, ed. Katharina Holger (Heidelberg: DAB, 1973), 18–19. Kuenssberg here reports that she attended lectures at the Lyceum Club on "all spheres of knowledge, just as we still do today in the association of university women."
127. In the postwar period, the DAB cooperated closely with Elsa Neumann's association. In 1930, Anna Schönborn, a member of the DAB executive, succeeded Lydia Rabinowitsch-Kemper as president of the association, which had, however, by this point lost its assets due to war and inflation and now saw its chief duty as chasing up the repayment of loans from former grant recipients so that further students could be funded. Report on the activities of the DAB from June 1929 to June 1930, LAB, B Rep. 235–5, 1/1, HLA 3630.
128. See Sophie Forgan, "Bricks and Bones: Architecture and Science in Victorian Britain," in *The Architecture of Science*, ed. Peter Galison and Emily Thompson (Cambridge, MA: MIT Press, 1999), 181–212.
129. Vickery illustrates this point with the case of Girton, one of Cambridge's oldest women's colleges. It was only in the 1890s, when Girton College had become an accepted institution, that the architectural allusions to domesticity became noticeably weaker and the buildings began to take on the character of classical "masculine" educational architecture. Margaret Birney Vickery, *Buildings for Bluestockings: The Architecture and Social History of Women's Colleges in Late Victorian England* (Newark: University of Delaware Press, 1999), 34–5.
130. Forgan, "Bricks and Bones"; see also Carol Dyhouse, *No Distinction of Sex? Women in British Universities, 1870–1939* (London: UCL Press, 1995); Pauline Adams, *Somerville for Women: An Oxford College, 1879–1993* (Oxford: Oxford University Press, 1996).
131. In addition, most coeducational universities had a women's dean to look after the female students. One of these deans was the founder of the ACA (later AAUW), Marion Talbot, who was appointed women's dean of the University of Chicago in 1893. On Talbot, see Andrea Glauser, *More Than a Watchdog: Marion Talbot und die Chicago Sociology* (Bern: Institut für Soziologie, 2003).
132. Women's "special characteristics" were brought into play by women academics most often when they were trying to lay claim to particular professional domains, such as medicine or social welfare. For the case of women doctors, an example is Frieda Duensing, a social reformer and the holder of a doctorate in law. See Iris Schröder, "Soziale Frauenarbeit als bürgerliches Projekt. Differenz, Gleichheit und weiblicher Bürgersinn in der Frauenbewegung um 1900," in *Wege zur Geschichte des Bürgertums: Vierzehn Beiträge*, ed. Klaus Tenfelde and Ulrich Wehler (Göttingen: Vandenhoeck & Ruprecht, 1994), 209–30. In the United States, reference to specifically feminine capacities began around 1900, when resistance to the growing numbers of female students started to emerge in many universities. The introduction of degree courses in "home economics" was an expression of this tendency. See Sarah Stage and Virginia Bramble Vincenti, eds., *Rethinking Home Economics: Women and the History of a Profession* (Ithaca, NY: Cornell University Press, 1997).
133. Lange, *Lebenserinnerungen*, 165.
134. This argument can be traced back to the classic text by Hedwig Dohm, *Die wissenschaftliche Emancipation der Frau* (Berlin: Wedekind und Schwieger, 1874).
135. In the United States, there was a whole spectrum of women's colleges as well as coeducational, mostly public, universities that were based only partially on the

principle of a "value-free" science in this sense. The boundaries between strictly "scientific, masculine" and "feminine" education were thus blurred. See Solomon, *Company*, 44–63.

136. This did not, in my view, apply to the courses offered by Helene Lange. They prepared women to sit the university entrance exams as external candidates at boys' high schools. Luise Berthold, who attended the courses from 1907 to 1909 before studying German language and literature in Berlin and Marburg and, in 1923, becoming the University of Marburg's first female university lecturer, described these courses as still deeply marked by the unity of advanced education and the women's movement ethos: "As a result, something of the wonderfully fresh breeze of the 'classic' women's movement still drifted through the courses." Luise Berthold, *Erlebtes und Erkämpftes: Ein Rückblick* (Marburg: Privately published, 1969), 14.
137. See, in particular, James C. Albisetti, "American Women's Colleges through European Eyes, 1865–1914," *History of Education Quarterly* 32, no. 4 (1992): 439–58.
138. Dorothee Wierling points out that even after women students were admitted, the universities themselves ensured that "the whole of student life" was geared to the needs of male students. Dorothee Wierling, "Studentinnen in der Weimarer Republik," in *Bürgerliche Gesellschaft in Deutschland: Historische Einblicke, Fragen, Perspektiven*, ed. Lutz Niethammer (Frankfurt/Main: Fischer, 1990), 364–82, here 370.
139. Katharina Belser, *Ebenso neu als kühn: 120 Jahre Frauenstudium an der Universität Zürich* (Zurich: eFeF-Verlag, 1988).
140. Anja Burchardt, *Blaustrumpf, Modestudentin, Anarchistin? Deutsche und russische Medizinstudentinnen in Berlin, 1896–1918* (Stuttgart: Metzler, 1997), 146.
141. Lilly Zarncke and Sophie Apolant, "Verband der Studentinnenvereine Deutschlands (V.St.D.)," in *Das akademische Deutschland*, ed. Michael Doeberl et al. (Berlin: C. A. Weller, 1931), 589. Cited in Burchardt, *Blaustrumpf*, 146.
142. Burchardt, *Blaustrumpf*, 148.
143. Ibid., 147. See also Tatjana Behr, "Der Deutsche Verband Akademischer Frauenvereine," *Die Studentin* 2, no. 9/10 (1925).
144. In 1914, the DVAF had 180 members, while the federation of women students' organizations Verband der Studentinnenvereine Deutschlands had 500; in contrast, in 1928, around 500 women students belonged to the DVAF, while the number of students in the Verband had dropped to 300. Huerkamp, *Bildungsbürgerinnen*, 144, 145.
145. The two organizations had been negotiating this merger since 1916. The liberal Verband der Studentinnenvereine Deutschlands wanted "to move closer only in respect to purely academic matters,...these being separate from the introduction of political and religious tendencies, which the federation deplores." Cited in Burchardt, *Blaustrumpf*, 147.
146. Behr, "Der Deutsche Verband Akademischer Frauenvereine," 69. Behr writes that the merger was originally planned as a cartel of all student organizations, joined by nonstudent associations of women. The members of the merged organization were to have been the Catholic senior teachers gathered in the Verband katholischer Oberlehrerinnen Deutschlands; the associations of Catholic women students federated in the Verband katholischer Studentinnenvereine Deutschlands; the conservative federation DVAF; the Christian association of women students Deutsche Christliche Vereinigung studierender Frauen; the economists of the Vereinigung der Nationalökonominnen; the federation of women students' associations Verband der Studentinnenvereine Deutschlands; the jurists' association Juristinnenverein; and the chemists' organization Verein deutscher Chemikerinnen.

147. Gabriele Humbert, "Der DAB," *Die Studentin* 2, nos. 11/12 (1925), 82.
148. It was Elisabeth Altmann-Gottheiner, the treasurer of the women economists' association, who pointed out this problem, although she argued that the women's professional organizations were in a comparatively strong position in that the wages and salaries of white-collar workers and public servants had kept pace with the inflationary price rises relatively well. In Altmann-Gottheiner's view, it was the "purely feminist associations" that were suffering worst—groups which most employed women regarded as considerably less important than the professional associations. Elisabeth Altmann-Gottheiner, "Die Krisis im Vereinsleben. Gedanken einer Schatzmeisterin," *Die Frau* 29 (1921/22): 231–5.
149. Cited in Eckelmann, *Ärztinnen*, 21.
150. Willy Hellpach, *Die Wesensgestalt der deutschen Schule* (Leipzig: Quelle & Meyer, 1925), cited in Schaser, "Die 'undankbaren' Studentinnen," 107.
151. Charlotte Leubuscher, "Die Berufslage der deutschen Hochschuldozentinnen," *Die Frau* 33 (1925/26): 669–73.
152. List of membership numbers of the DAB's member organizations, n.d. [1929], LAB, B Rep. 235-5, 1/1, HLA 3634.
153. On Erdmann, see Sabine Koch, "Leben und Werk der Zellforscherin Rhoda Erdmann (1870–1935)," PhD diss., University of Marburg, 1985; Brigitte Hoppe, "Die Institutionalisierung der Zellforschung durch Rhoda Erdmann (1870–1935)," *Biologie heute: Beilagen zur Zeitschrift Naturwissenschaftliche Rundschau* 366, no. 7 (1989): 2–4, 9. In the late 1920s, the oceanographer Lotte Möller was made chair of the association.
154. Leubuscher, "Berufslage," 673.
155. Ibid.
156. Ibid., 672.
157. In particular, the monthly *Die Frau* published articles attacking the ban on women's habilitation and giving detailed portrayals of women researchers' activities. See Trude Maurer, "Von der Gleichzeitigkeit des Ungleichzeitigen. Das deutsche Frauenstudium im internationalen Kontext," in *Der Weg an die Universität: Höhere Frauenstudien vom Mittelalter bis zum 20. Jahrhundert*, ed. Trude Maurer (Göttingen: Wallstein, 2010), 18, n. 46.
158. Leubuscher, "Berufslage," 672.
159. Rhoda Erdmann, "Typ eines Ausbildungsgangs weiblicher Forscher," in *Führende Frauen Europas: In sechzehn Selbstschilderungen*, ed. Elga Kern (Munich: E. Reinhardt, 1928), 35–54.
160. Ibid., 53.
161. Ibid., 52.
162. Ibid., 54.
163. See, especially, Gertrud Bäumer, "Berufsorganisationen und Frauenbewegung," *Die Frau* 32 (1925): 97–100.
164. Harvey, "The Failure of Feminism?"; Renate Bridenthal, "Professional Housewives: Stepsisters of the Women's Movement," in *When Biology Became Destiny: Women in Weimar and Nazi Germany*, ed. Renate Bridenthal, Atina Grossmann, and Marion A. Kaplan (New York: Monthly Review Press, 1984), 153–73. The same has been observed for the case of the women physicians' federation BDÄ. This body covered an enormous spectrum of views, from members of the association of socialist doctors like Laura Turnau or Flora Chajes, to the group around Agnes Bluhm, with its interest in nationalist racial hygiene, and the decidedly anti-Semitic Edith Lölhöffel, an early National Socialist. In order to function politically as an organization, the BDÄ had to keep strictly to its officially neutral stance and its carefully delineated program of social hygiene. Johanna Bleker and Christine Eckelmann, "'Der Erfolg der Gleichschaltungsaktion kann als durchschlagend bezeichnet

werden.' Der 'Bund Deutscher Ärztinnen' 1933–1936," in *Medizin im Dritten Reich*, ed. Johanna Bleker and Norbert Jachertz (Cologne: Deutscher Ärzte-Verlag, 1989), 87–96, here 87–8.

165. On Ilse Szagunn, see Louisa Sach, "'Gedenke, dass du eine deutsche Frau bist.' Die Ärztin und Bevölkerungspolitikerin Ilse Szagunn (1887–1971)," PhD diss., University of Berlin (Charité Universitätsmedizin), 2006.
166. Address list of the Deutscher Akademikerinnenbund, c. 1930, LAB, Rep. 235–5, 1/1, HLA 3629.
167. Ibid.
168. Ilse Szagunn to Frau Dr. Matz, September 30, 1929, LAB, B Rep. 235–5, 1/2, HLA 3635, asking for contacts for a woman researcher based in Dessau who "feels very lonely."
169. Information sheet, "Lecture fees for academic women," n.d. [1926], LAB, B Rep. 235–5. 1/2, HLA 3635.
170. This was often done in cooperation with other associations. See, for example, a letter to the Reich ministries protesting the dismissal of married women civil servants, April 26, 1931, LAB, B Rep. 235–5, 1/1, HLA 3631.
171. Agnes von Zahn-Harnack, Bibliography of the woman question and the women's movement, MS, 5 pp., n.d. [1929], LAB, B Rep. 235–5, 1/1, HLA 3629, p. 1. As so often, Zahn-Harnack quoted from Goethe when justifying this project: "Bleibt im Dunkel, unerfahren, muß von Tag zu Tage leben" ("Let him who fails to learn and mark... still stay/Void of experience, in the dark,/And live from day to day"). Ibid., p. 5.
172. The difficulties arose partly from the fact that several of the DAB's member associations showed less support for this project than Zahn-Harnack had hoped. See Zahn-Harnack's confidential memorandum to the DAB associations, March 10, 1931. In this note, she emphasized that "the amounts raised by the women's organizations are so small that we have decided not to publish them, since we must always stress to the Emergency Association that the work is substantially funded by the women themselves." LAB, B Rep. 235–5, 1/2, HLA 3636.
173. Hans Sveistrup and Agnes von Zahn-Harnack, eds., *Die Frauenfrage in Deutschland: Strömungen u. Gegenströmungen 1790–1930. Sachl. geordn. u. erl. Quellenkunde* (Burg: Hopfer, 1934). Up to 1984, the work was reissued many times, along with a total of 12 follow-up volumes. See Deutscher Akademikerinnenbund, ed., *Die Frauenfrage in Deutschland 1951–1960* (Cologne: Deutscher Akademikerinnenbund, 1961); *Die Frauenfrage in Deutschland, Bd. IV, 1961–65* (Cologne: Deutscher Akademikerinnenbund, 1966); *Die Frauenfrage in Deutschland, Bd. V, 1966–70* (Cologne: Deutscher Akademikerinnenbund, 1972). From 1981 to 1991, the bibliography was continued by the Frau und Gesellschaft institute in Hanover. The last volume appeared in 1991.
174. Report on the activities of the DAB from June 1930 to June 1931, p. 9, LAB, B Rep. 235–5, HLA 3630.
175. Ibid., p. 10. This sentence is struck through in the manuscript, probably because it was so undiplomatically frank.
176. Ibid.
177. Irmgard Rathgen, "Generationen," in *Dritte Generation*, ed. Hilde Lion, Irmgard Rathgen, and Else Ulig-Beil (Berlin: F. A. Herbig, 1923), 86. See also Irmgard Rathgen, "Wissen und Praxis," *Die Frau* "Sondernummer April 1932 zu Alice Salomons 60. Geburtstag" (1932).
178. Irmgard Rathgen, "Praktische Aufgaben des deutschen Akademikerinnenbundes," *Die Studentin* 2, no. 7/8 (1925): 57–8, here 57. Rathgen's second suggestion, and the main request directed at the leadership, was succinct: "fellowships, studies abroad." It was, she said, well known "that the state has taken the place of the charitable

foundations ruined by inflation." She called on the DAB to ensure that women were considered when the new state funding was being distributed, and to disseminate information among women students, urging them to make use of such financial opportunities. A systematic policy of information was required in this matter, she said. Chances to study abroad, as well, were currently ruled by "pure chance," so that the DAB, with its wealth of foreign connections, had an important task to fulfill in advertising opportunities to as many women students as possible. Ibid.

179. Rosa Kempf, "Aussprache. Lokale oder fakultätsweise Zusammenfassung der Akademikerinnen?" *Die Frau* 32 (1924/25): 120.
180. Charlotte Lorenz, "Wieviel Frauen studieren in Deutschland?" *Die Frau* 34 (1926/27): 240.
181. Ibid.
182. Margot Melchior, "Deutschlands Studentinnenheime," *Die Frau* 36 (1928/29): 568–92. Since 1922, Leipzig had had a small day center founded as a "special facility offering economic assistance for women students." However, this only had space for 11 students, and admitted only women who came "particularly highly recommended in terms of their character and academic ability." Catholic hostels for women students were more common, offering a total of 550 places and thus covering around 16 percent of all Catholic women students. All the approximately ten residences were managed by nuns, and tried to maintain the character of "cultivated family life."
183. Marie-Elisabeth Lüders had been involved in preparations for this facility during her own student years. See Studentenwerk Berlin, ed., *Ein Jahr Studentinnentagesheim (Helene-Lange-Heim), 1928–1929* (Berlin: Studentenwerk, 1929), 7.
184. Specifically, the DAB's plans for the women students' residence alluded to the IFUW's residences and guesthouses in Madrid, Paris, and London. See Report on the activities of the DAB from June 1928 to June 1929, LAB, B, Rep 235–5, 1/1, HLA 3634.
185. Ilse Szagunn, "Crosby Hall, Ein internationales Clubhaus für Akademikerinnen," *Die Frau* 34 (1926/27): 721.
186. Ibid.
187. Ibid.
188. Establishing a women students' day home in Berlin. Appeal for donations, n.d. [summer 1927], BArch Koblenz, N 1151, no. 284.
189. List of donors I (cash donations) and II (donations in kind), in "Zu den Spendern gehörten: AEG, Bewag, die Berliner Blindensanstalt, wie auch die Kaufhäuser Israel, Herzog, Tietz, Wertheim und andere," Studentenwerk Berlin, *Ein Jahr Studentinnentagesheim*, 25–7.
190. Ibid., 11.
191. Ibid.
192. On Gertrud Hamer-von Sanden and her relationship with Gertrud Bäumer, see Schaser, *Helene Lange und Gertrud Bäumer*, 193–5.
193. Studentenwerk Berlin, *Ein Jahr Studentinnentagesheim*, 9.
194. Ibid., 19.
195. Helene-Lange-Heim. Manager's report on the winter semester 1931/32. BArch Koblenz, N 1151, no. 284.
196. Report on the activities of the DAB from June 1928 to June 1929, LAB, B Rep. 235–5, 1/1, HLA 3634; Report on the activities of the DAB from June 1929 to June 1930, LAB, B Rep. 235–5, 1/1, HLA 3630.
197. Studentenwerk Berlin to Marie-Elisabeth Lüders, April 7, 1933, BArch Koblenz, N 1151, no. 284.
198. Both women are included in the list of donors and as members of the supporting association; Studentenwerk Berlin, *Ein Jahr Studentinnentagesheim*, 17.

199. See Huerkamp, *Bildungsbürgerinnen*, 145.
200. Gertrud Jung, "Dies academicus der Frau," *Die Frau* 37 (1929/30): 614.
201. Hildegard Gallmeister, "Die Studentin im akademischen Leben," *Die Frau* 37 (1929/30), 623–30, here 625.
202. Ibid., 629.

5 World Community under Threat

1. The debate on the Nazification of the German women's movement has long been influenced by Richard J. Evans's study *The Feminist Movement in Germany 1894–1933* (London: Sage, 1976). Evans notes a strong nationalistic attitude within the BDF even in the Weimar era, and deduces from this an affinity with ideas of *Volksgemeinschaft* that he believes facilitated the bourgeois women's rapid cooperation with the Nazi state after the dissolution of the BDF. See also Jill Stephenson, *Women in Nazi Germany* (Harlow: Longman, 2001). For more recent summaries of the debates around the BDF's dissolution and the stance of its leaders, see Angelika Schaser, *Helene Lange und Gertrud Bäumer: Eine politische Lebensgemeinschaft* (Cologne: Böhlau, 2000), 283–99; Gisa Bauer, *Kulturprotestantismus und frühe bürgerliche Frauenbewegung in Deutschland: Agnes von Zahn-Harnack (1884–1950)* (Leipzig: Evangelische Verlagsanstalt, 2006), 270–8. On the Nazification of the BDÄ, see Johanna Bleker, "Anerkennung durch Unterordnung? Ärztinnen und Nationalsozialismus," in *Weibliche Ärzte: Zur Durchsetzung eines Berufsbildes*, ed. Eva Brinkschulte (Berlin: Edition Hentrich, 1995), 126–35; Christine Eckelmann, *Ärztinnen in der Weimarer Zeit und im Nationalsozialismus: Eine Untersuchung über den Bund Deutscher Ärztinnen* (Wermelskirchen: WFT Verlag für Wissenschaft, Forschung und Technik, 1992).
2. Circular from Lüders to the local chapters, May 10, 1933, DAB Karlsruhe, cited in Barbara Guttmann, *Den weiblichen Einfluss geltend machen...Karlsruher Frauen in der Nachkriegszeit, 1945–1955* (Karlsruhe: Badenia Verlag, 2000), 56.
3. Marie-Elisabeth Lüders to Dorothee von Velsen, April 19, 1933, BArch Koblenz, N 1151, no. 256.
4. Ibid.
5. The German version of the present study provides a more detailed description of this strategy: Christine von Oertzen, *Strategie Verständigung: Zur transnationalen Vernetzung von Akademikerinnen, 1917–1955* (Göttingen: Wallstein, 2012), 184–7.
6. Anja Burchardt, *Blaustrumpf, Modestudentin, Anarchistin? Deutsche und russische Medizinstudentinnen in Berlin, 1896–1918* (Stuttgart: Metzler, 1997), 147–8; see also Louisa Sach, "Gedenke, dass du eine deutsche Frau bist": Die Ärztin und Bevölkerungspolitikerin Ilse Szagunn (1887–1971)," PhD diss., University of Berlin (Charité Universitätsmedizin), 2006, 40–3.
7. Circular from Szagunn to the local chapters, May 11, 1933, cited in Guttmann, *Weiblicher Einfluss*, 57.
8. According to an article in the journal of the women physicians' association, the vote was then carried out "under the aspect of *Gleichschaltung*." "Die Gleichschaltung des Deutschen Akademikerinnenbundes," *Die Ärztin* 9 (1933): 129–30, here 129.
9. Ibid. For information on Matthias's own political struggle within the Nazi organizations, see Claudia Koonz, *Mothers in the Fatherland: Women, the Family, and Nazi Politics* (London: Methuen, 1987), 202–4. On Thimm, see Johanna Bleker and Sabine Schleiermacher, *Ärztinnen aus dem Kaiserreich: Lebensläufe einer Generation* (Weinheim: Deutscher Studien-Verlag, 2000), 165–6, 297.

10. They were Dr. Lotte [Charlotte] Möller, Berlin; Dr. Lotte Liebenam, Halle; Dr. Editha von Moers, Wuppertal; Dr. med. dent. Karin Fritzsche, Berlin; Cand. Phil. Ilse Balg, Berlin. See "Die Gleichschaltung des Deutschen Akademikerinnenbundes," *Die Ärztin* 9 (1933): 129–30.
11. See Johanna Willich, "Das internationale Akademikerinnen-Heim: Crosby Hall," *Die Studentin* 4, no. 4 (1927/28): 54.
12. BArch Berlin (formerly Berlin Document Center), NSDAP personnel files, Möller, Lotte, b. June 17, 1893, joined NSDAP April 30, 1933.
13. Liebenam made use of her contacts to benefit from a partnership between university women in Bristol and Halle that was initiated in the early 1930s by the Halle chapter's president, Indologist Betty Heimann. See J. H. Sondheimer, *History of the British Federation of University Women 1907–1957* (London: British Federation of University Women, 1957), 32; AAUW Archives, Series IX, Reel 149/4, Box 832.
14. "Die Gleichschaltung des Deutschen Akademikerinnenbundes," *Die Ärztin* 9 (1933): 129–30, here 130.
15. Ibid.
16. DAB circular to the member associations, June 12, 1933, cited in "Rundschreiben an die Vereinigung der Nationalökonominnen Deutschlands," n.d. [December 1933–January 1934], BArch Koblenz, N 1151, no. 255; DAB circular to the local chapters, June 12, 1933, cited in Guttmann, *Weiblicher Einfluss*, 57–8.
17. DAB circular to member associations and local chapters, October 14, 1933, DAB Karlsruhe, cited in Guttmann, *Weiblicher Einfluss*, 60. Most of the "non-Aryan" male civil servants whose dismissal was deferred were frontline veterans of World War I.
18. Johanna Willich, "Bericht über die neuen Ziele und die Umgestaltung des Deutschen Akademikerinnenbundes v. 13.12.1933," DAB Karlsruhe, cited in Guttmann, *Weiblicher Einfluss*, 58.
19. "Die Arbeit des Deutschen Akademikerinnenbundes im Neuen Deutschland: Denkschrift v. Dr. Lore Liebenam, 13.10.1933," DAB Karlsruhe, cited in ibid., 59.
20. Lüders to Frohwein, Foreign Office, February 26, 1933, BArch Koblenz, N 1151, no. 8.
21. Marie-Elisabeth Lüders's report to the Foreign Office on the Geneva conferences of international women's organizations, "Bericht zu den Tagungen internationaler Frauenorganisationen zwischen dem 21. und 31.9.1933 in Genf an das Auswärtige Amt," BArch Koblenz, N 1151, no. 8, n.p.
22. Lüders found Lise Meitner's decision to resign from the Fellowship Committee in summer 1933 particularly regrettable: she feared that the anti-German climate "will be even more unfavorably influenced if German women are completely absent." Ibid.
23. Marie-Elisabeth Lüders, "Bericht v. 24./25.5.1933 an das Auswärtige Amt," BArch Koblenz, N 1151, no. 2, n.p.
24. Lüders to Frau von Trützschler in Geneva, May 6, 1933, BArch Koblenz, N 1151, no. 8. Trützschler lived in Geneva, and stood in for Lüders at IFUW committees when Lüders was unable to attend.
25. Lüders to von Trützschler, ibid.
26. Lüders, "Bericht an das Auswärtige Amt," n.d. [early summer 1933], BArch Koblenz, N 1151, no. 2, n.p.
27. Marie-Elisabeth Lüders, "Bericht v. 24./25.5.1933 an das Auswärtige Amt," BArch Koblenz, N 1151, no. 2, n.p.
28. Marie-Elisabeth Lüders, "Zweiter Bericht an das AA über die Genfer Abrüstungstagung v. 26.-28.5.1933," BArch Koblenz, N 1151, no. 2, n.p.
29. Marie-Elisabeth Lüders, "Bericht v. 24./25.5.1933 an das Auswärtige Amt," BArch Koblenz, N 1151, no. 2, n.p. Also the following quotations.

30. Marie-Elisabeth Lüders, "Bericht zu den Tagungen internationaler Frauenorganisationen zwischen dem 21. und 31.9.1933 in Genf an das Auswärtige Amt," BArch Koblenz, N 1151, no. 8, n.p.
31. Ibid.
32. Marie-Elisabeth Lüders, "Bericht v. 24./25.5.1933 an das Auswärtige Amt," BArch Koblenz, N 1151, no. 2, n.p.
33. Marie-Elisabeth Lüders, "Bericht zu den Tagungen internationaler Frauenorganisationen zwischen dem 21. und 31.9.1933 in Genf an das Auswärtige Amt," BArch Koblenz, N 1151, no. 8, n.p. Also the following quotations.
34. Ernst Trendelenburg, whom Lüders knew from the League of Nations and world economic conferences in Geneva, left the Foreign Office when Germany withdrew from the League of Nations in October 1933. He worked for the organization of German industry, and in 1935 became head of the Reich industry group. See Auswärtiges Amt, ed., *Biographisches Handbuch des Auswärtigen Dienstes 1871–1945* (Paderborn: Schöningh, 2000), vol. 1, 622–3.
35. Agnes von Zahn-Harnack, "Geschichte des Deutschen Akademikerinnenbundes, 1926–1933," in *Agnes von Zahn-Harnack: Schriften und Reden, 1914–1950*, ed. Marga Anders and Ilse Reicke (Tübingen: Hopfer, 1964), 1–8, here 7.
36. IFUW, *Report of the Nineteenth Council Meeting, Budapest 1934* (London: The Federation, 1934), 22.
37. Between 1932/33 and 1939, the number of female students dropped by almost two-thirds, from 17,192 to 5,777. In the same period, the number of male students fell from 75,409 to 34,939. See the statistics on university students by gender in Michael Grüttner, *Studenten im Dritten Reich* (Paderborn: Schöningh, 1995), 488. A detailed analysis of the steep decline in female students can be found in Claudia Huerkamp, *Bildungsbürgerinnen: Frauen im Studium und in akademischen Berufen 1900–1945* (Göttingen: Vandenhoeck & Ruprecht, 1996), 80–91. Statistics can also be found in Jacques Pauwels, *Women, Nazis, and Universities: Female University Students in the Third Reich, 1933–1945* (Westport, CT: Greenwood Press, 1984).
38. Maria Monheim, "Bericht über die Stimmungslage im Deutschen Ärztinnenbund nach der Gleichschaltung," *Die Ärztin* 9, no. 6 (1933): 118.
39. Ibid.
40. Circular to the DAB local chapters, June 12, 1933, cited in Guttmann, *Weiblicher Einfluss*, 58.
41. Monheim, "Bericht über die Stimmungslage," 118.
42. See IFUW, *Report of the Nineteenth Council Meeting, Budapest 1934*, 27.
43. Cited in Guttmann, *Weiblicher Einfluss*, 58.
44. President of the DAB Karlsruhe chapter to Jewish members, November 29, 1933; facsimile in Guttmann, *Weiblicher Einfluss*, 59.
45. *Die Ärztin* 9 (1933): 225.
46. For a more detailed discussion of these processes, see Oertzen, *Strategie Verständigung*, 203–4.
47. "Rundschreiben des Deutschen Frauenwerks Nr. 45 v. 9.10.1934 an die Ortsgruppen des DAB," Magda Staudinger papers, file "Zwischenfall Bowie" ("Bowie incident").
48. In late 1933, the AAUW appointed the college presidents Meta Glass, Virginia Gildersleeve, and Mary Woolley to a committee whose task would be to protest the suppression of intellectual freedom "that is rapidly passing over the European countries." The committee was to pressurize the DAB to rethink its policy, review the status of women in German universities, and work out a plan to help women who were being forced to emigrate. See AAUW, General Director's Letter, December 1933, 14, quoted in Susan Levine, *Degrees of Equality: The American Association of*

University Women and the Challenge of Twentieth-Century Feminism (Philadelphia, PA: Temple University Press, 1995), 55. Unfortunately, the AAUW archives do not contain any records of this committee's activities.

49. For more detail, see chapter 6.
50. Marie-Elisabeth Lüders, "Bericht v. 24./25.5.1933 an das Auswärtige Amt," BArch Koblenz, N 1151, no. 2, n.p.
51. "National Council for Equal Citizenship. Conference on the Status of German Women, House of Commons," May 31, 1933, Resolution to be sent to the German Ambassador, WL@LSE, 5BFW (old Box 27), Folder "Conference 1933: Jews in Germany."
52. British Federation of University Women to German Ambassador, August 2 and 5, 1933, WL@LSE, 5BFW (old Box 27), Folder "Conference 1933: Jews in Germany." The support given to Jewish German academic women by the British federation and the IFUW as a whole after 1933 is detailed in chapter 6.
53. IFUW, *Report of the Nineteenth Council Meeting, Budapest 1934*, 27.
54. Extract from the minutes of the Council meeting at Budapest, September 3, 1934, Magda Staudinger papers, file "Zwischenfall Bowie."
55. Lüders had already warned the Foreign Office that changing the constitution would entail the DAB's immediate expulsion from the international organizations. Marie-Elisabeth Lüders, "Bericht v. 24./25.5.1933 an das Auswärtige Amt," BArch Koblenz, N 1151, no. 2, n.p.
56. DAB Karlsruhe, privately held, "Rundschreiben des Deutschen Frauenwerks Nr. 45 v. 9.10.1934 an die Ortsgruppen des DAB," quoted in Guttmann, *Weiblicher Einfluss*, 59.
57. See, especially, the report by Meta Glass, "Some Highlights of the Council Meeting," *AAUW Journal* 28, no. 2 (Winter 1934/35): 121–2.
58. The president of the Palestine association was Sophie Berger Mohl, a US-born Zionist who had studied at the University of Chicago and worked as a social worker before emigrating to Palestine in 1918. See "The Palestine Association," *AAUW Journal* 39, no. 1 (Fall 1945): 48.
59. There is no evidence of this in contemporary documents, but it was mentioned by Marguerite Bowie, then the IFUW treasurer and very active in British assistance for refugees, at a 1968 meeting in Karlsruhe, the first IFUW conference to be held in Germany—to the great dismay of German members. For more detail, see chapter 8.
60. IFUW, *Report of the Nineteenth Council Meeting, Budapest 1934*, 27.
61. Virginia Crocheron Gildersleeve, *Many a Good Crusade: Memoirs* (New York: Macmillan, 1954), 149–52.
62. "No Federation shall be admitted or retained as a member of the IFUW which debars qualified university women from membership by reason of their race, religion, or political opinions." Revision of the Constitution (art. II, Membership of National Associations) as resolved by the Council meeting in Budapest. IFUW, *Report of the Nineteenth Council Meeting, Budapest 1934*, 37–8.
63. Virginia Gildersleeve was chosen to carry out this task along with Erna Patzelt of the Austrian association. In her memoirs she called it "one of the most delicate jobs I ever tackled"; Gildersleeve, *Many a Good Crusade*, 150.
64. IFUW, *Report of the Nineteenth Council Meeting, Budapest 1934*, 34.
65. "Rundschreiben des Deutschen Frauenwerks Nr. 45 v. 9.10.1934 an die Ortsgruppen des DAB," Magda Staudinger papers, n.p. Also the following quotations.
66. "Rundschreiben des Deutschen Frauenwerks Nr. 45 v. 9.10.1934 an die Ortsgruppen des DAB," DAB Karlsruhe, cited in Guttmann, *Weiblicher Einfluss*, 61.
67. See "Der Reichsbund Deutscher Akademikerinnen, 23.2.1935," *Nachrichtendienst der NS-Frauenschaft* 4 (1935): 377.

68. "Vereinbarung zur Durchführung der Verfügung der Reichsfrauenführerin v. 9.10.1934 zwischen der NS-Frauenschaft, dem Deutschen Frauenwerk und dem Reichsbund Deutscher Akademikerinnen andererseits v. 28.2.1935," BArch Berlin NS 12, no. 1340.
69. Ibid.
70. The BFUW noted: "The incorporation of this resolution in the Constitution presented many technical difficulties, and it was finally decided to adopt the principles expressed in the Budapest resolution, but to leave to the Council the onerous task of finding a suitable form in which they might be incorporated in the Constitution." *BFUW News Sheet* 20 (October 1936): 12. The revision was resolved as a supplement to Art. 1 of the IFUW's constitution in Stockholm in 1939; see IFUW, *Report of the Eighth Conference, Stockholm 1939*, 41–4. For details of the discussion, see chapter 6.
71. IFUW, *Report of the Seventh Conference, Cracow 1936*, especially "Report of the Twentieth Council Meeting, Withdrawal of Italian and German Federation," 72.
72. "Reichsbund Deutscher Akademikerinnen," *Nachrichtendienst der Reichsfrauenführung* 5 (1936), 126; "Rundschreiben Nr. FW 61/36 v. 20.5.1936 der Reichsfrauenführung," reproduced ibid., 201–2.
73. Ibid., 126.
74. Ibid., 201–2.
75. Refugee assistance is discussed in chapter 6.
76. "Reichsbund Deutscher Akademikerinnen," *Nachrichtendienst der Reichsfrauenführung* 5 (1936), 126. Christine Eckelmann describes a similar conflict between the BDÄ and its international umbrella organization, the Medical Women's International Association. In this case, too, the national association's dissolution was partly prompted by indirect criticism from the international body: in 1936, the MWIA changed its statutes to enable individual membership for German Jewish women. See Eckelmann, *Ärztinnen*, 59. Interestingly, however, the dissolution of the BDÄ, unlike that of the RDA, was not accompanied by any official explanation at all in Germany. See ibid., 60.
77. Against this background, it seems untenable to claim that the political disempowerment of the "combative fascist feminists" around 1936 ushered in a complete incorporation and subordination of Nazi women's politics to the objectives of its male protagonists. On the debate around "old" and "new fighters," see Annemarie Tröger, "Die Frau im wesensgemäßen Einsatz," in *Mutterkreuz und Arbeitsbuch: Zur Geschichte der Frauen in der Weimarer Republik und im Nationalsozialismus*, ed. Frauengruppe Faschismusforschung (Frankfurt/Main: Fischer, 1981), 246–72, here 251–8; Michael Kater, *The Nazi Party: A Social Profile of Members and Leaders, 1919–1945* (Cambridge: Harvard University Press, 1983); Huerkamp, *Bildungsbürgerinnen*, 206–311; on the tasks of the women lawyers' group, also Marion Röwekamp, *Die ersten deutschen Juristinnen: Eine Geschichte ihrer Professionalisierung und Emanzipation (1900–1945)* (Cologne: Böhlau, 2011), 748–9.
78. For example, the lawyer Maria Plum from Freiburg, the English literature specialist Lore Liebenam from Halle, and the classical philologist Gertrud Carl from Karlsruhe; see Oertzen, *Strategie Verständigung*, 215–16.
79. "Liste der dem Deutschen Frauenwerk angeschlossenen Spitzenverbände, Stand v. 1.05.1937," BArch Berlin, NS 44, no. 57; "Reichsfrauenführung, Reichsreferentin und Reichbeauftragte der Rechtswahrerinnen, Rundschreiben Nr. FW 76/37 betr. Mitgliedsbeiträge v. 12.8.1937," BArch Berlin, NS 44, no. 35. Measured against the data from the occupational census of 1939, which counted just 217 female lawyers (169 freelance and 48 employed) in the whole of the German Reich, this reflects an extraordinarily high degree of organization among women

lawyers within the Frauenwerk. On the occupational census, see Huerkamp, *Bildungsbürgerinnen*, 292.

80. Eckelmann, *Ärztinnen*, 61; Michael Kater, "Medizin und Mediziner im Dritten Reich: Eine Bestandsaufnahme," *Historische Zeitschrift* 244 (1987): 299–352, here 324.
81. Elisabeth Baecker-Vowinkel, "8. Schulungslehrgang für Ärztinnen in Alt-Rehse v. 5.–15.7.1939," *Die Ärztin* 15 (1939): 260–3, here 263.
82. Ursula Kuhlo, "Das Referat Ärztinnen," *Die Ärztin* 16, no. 5 (1940): 114.
83. Ibid. The only *Gaue* lacking a women's section by 1940 were the Warthegau and those in the "Ostmark" (Austria).
84. "Vereinbarung der Reichsfrauenführerin über die Zusammenarbeit zwischen NS-Frauenschaft–Deutsches Frauenwerk und den weiblichen Mitgliedern des NS-Lehrerbundes v. 29.6.1936," *Nachrichtendienst der Reichsfrauenführerin* 5 (1936): 279–80. The agreement laid down that the women leaders in the National Socialist Teachers' League would also head the education section of the Women's League, so as to promote "the unified philosophical, cultural-policy, and economic-policy orientation of the German woman." Ibid.
85. Gertrud Scholtz-Klink, *Einsatz der Frau in der Nation: Frauenkundgebung, Reichsparteitag der Arbeit 1937*, ed. Deutsches Frauenwerk, Nachrichtendienst der Reichsfrauenfühererin (Berlin: NSDAP, 1937), 9.
86. Ilse Eben-Servaes, "Wissen ist uns Verpflichtung," *Frauenkultur im Deutschen Frauenwerk* 4, no. 2 (1938): 13.
87. Ibid.
88. Anna Kottenhoff, "Die Deutsche Frau in Lehre und Forschung: Dozentinnentagung in der Reichsfrauenführung," *Frauenkultur im Deutschen Frauenwerk* 4, no. 2 (1938): 2–11, here 9.
89. Anna Kottenhoff, "Arbeit und Ziel der Hochschulgemeinschaft Deutscher Frauen," *Die Ärztin* 17, no. 4 (1941): 157–61.
90. *Nachrichtendienst der Reichsfrauenführerin* 6 (1937), 209.
91. Eben-Servaes, "Wissen." Also the following points.
92. At the women lecturers' conference in January 1938, Scholtz-Klink stressed that she would "never permit" the organization of women to work "*alongside* instead of *with* the organization of men. But there are matters in which women hold a shared attitude and must show that attitude clearly, not in conflict but as a harmonious complement. The aim is to shape the inner unity of all German women." Kottenhoff, "Die Deutsche Frau," 4.
93. Maria Schorn, "Frau und Hochschule," *Frauenkultur im Deutschen Frauenwerk* 1, no. 11 (1935): 1–2.
94. Gertrud Bäumer to Emmy Beckmann, March 7, 1936, BArch Koblenz, KLE 267/1, fol. 57.
95. See Irene Stoehr, "Frauenerwerbsarbeit als Kriegsfall: Marie Elisabeth Lüders: Variationen eines Lebensthemas," in *Frauen arbeiten: Weibliche Erwerbstätigkeit in Ost- und Westdeutschland nach 1945*, ed. Gunilla-Friederike Budde (Göttingen: Vandenhoeck & Ruprecht, 1997), 62–77, here 67.
96. "1. Arbeitssitzung des Sachgebietes 'Wissenschaftliche Arbeit' im Deutschen Frauenwerk," *Die Ärztin* 13 (1937): 275.
97. "Erstes Studentinnenlager des deutschen Frauenwerks," *Die Ärztin* 14 (1938): 46.
98. Kottenhoff, "Die Deutsche Frau," 4.
99. Ibid.
100. Ibid., 9.
101. Ibid., 3.
102. Letter of thanks from Dr. Doris Korn, ibid., 11.
103. Gertrud Savelsberg, "Rückblick auf die Tagung der Deutschen Dozentinnen in Berlin, 3.-6.1.1938," *Frauenkultur im Deutschen Frauenwerk* 4 (1938): 12.

104. CV Prof. Dr. Doris Schachner, née Korn, c. 1948, Hochschularchiv der RWTH Aachen, personnel file PA 5718.
105. CV Gertrud Savelsberg, card catalogue, Institut für Weltwirtschaft (Kiel Institute for the World Economy).
106. Stefanie Marggraf, "Sonderkonditionen: Habilitationen von Frauen in der Weimarer Republik und im Nationalsozialismus an den Universitäten Berlin und Jena," *Feministische Studien: Zeitschrift für interdisziplinäre Frauen- und Geschlechterforschung* 20, no. 2 (2002): 40–56, here 51–2; Hannelore Erhart, "Hanna Marie Margarete Jursch (1902–1972)," in *Biographisch-Bibliographisches Kirchenlexikon*, ed. Traugott Bautz (Herzberg: Verlag Traugott Bautz, 2000), 733–93.
107. In 1940, when Hertwig's unpaid teaching appointment was confirmed only "exceptionally, subject to revocation at any time," she decided to offer a course in "anthropology and the theory of heredity and race" together with her male colleague Schultz, an SS *Obersturmbannführer* who became head of the SS "race office" in 1943. Ute Deichmann, "Frauen in der Genetik: Forschung und Karrieren bis 1950," in *"Aller Männerkultur zum Trotz": Frauen in Mathematik und Naturwissenschaften*, ed. Renate Tobies (Frankfurt/Main: Campus, 1996): 221–51, here 239.
108. Hans-Jürgen Brosin, "Charlotte Möller (1883–1973) und die gewässerkundlichen Arbeiten am Institut für Meereskunde Berlin," *Historisch-meereskundliches Jahrbuch* 6 (1999): 19–34.
109. "For reasons of principle" put forward by the Reich Ministry of Science, the permanent professorship attached to the department went not to Möller but to a male rival. Möller had to make do with the role of head of group and a paid teaching assignment. Archiv der Humboldt-Universität Berlin, Berufungsakten Charlotte Möller, PhilFak, no. 1441. See also Brosin, "Charlotte Möller," 28. On Möller's work on the Curonian Lagoon, see Waltraut Zilius-Falkenberg, "Die Frau in der wissenschaftlichen Forschung," in *Das weite Wirkungsfeld: Frauenschaffen in Deutschland*, ed. Reichsfrauenführung, Hauptabteilung Presse/Propaganda (Berlin: Reichsfrauenführung, 1941), 119. For the wider context of Möller's work and science as part of the Nazi "Generalplan Ost," see Susanne Heim, *Plant Breeding and Agrarian Research in Kaiser-Wilhelm-Institutes, 1933–1945: Calories, Caoutchouc, Careers* (Dordrecht: Springer, 2004), especially 46–53.
110. BArch Berlin (formerly BDC), National Socialist Teachers' League (NSLB) card catalogue, Charlotte Lorenz, NSDAP membership number 2,644,763.
111. See Zilius-Falkenberg, "Die Frau in der wissenschaftlichen Forschung," 119.
112. The date when Esdorn joined the Party is recorded in BArch Berlin, BDC, PK, Esdorn, Else, b. January 8, 1897.
113. As Ute Deichmann explains, medicinal botany was an important component of the Nazi "new medicine," represented in every university from 1937 onward. Among other things, it was supposed to help increase the use of German-produced drugs. Ute Deichmann, *Biologen unter Hitler: Porträt einer Wissenschaft im NS-Staat* (Frankfurt/Main: Fischer, 1995), 91–2; see also Jörg Baten and Andrea Wagner, *Autarchy, Market Disintegration, and Health: The Mortality and Nutritional Crisis in Nazi Germany, 1933–1937* (Munich: CES, 2002); Christoph Buchheim, ed., *German Industry in the Nazi Period* (Stuttgart: Steiner, 2008).
114. Zilius-Falkenberg, "Die Frau in der wissenschaftlichen Forschung," 118.
115. Kottenhoff, "Die Deutsche Frau," 2–3.
116. Almost all the lecturers attending the conference were staged and photographed by the German Women's Agency photographer Liselotte Purper. On Purper's career in the Third Reich, see Elizabeth Harvey, "Seeing the World: Photography, Photojournalism, and Visual Pleasure in the Third Reich," in *Pleasure and Power in Nazi Germany*, ed. Pamela E. Swett, Corey Ross, and Fabrice d'Almeida (Basingstoke: Palgrave Macmillan, 2011), 177–204.

117. See especially the portraits in Zilius-Falkenberg, "Die Frau in der wissenschaftlichen Forschung," 112–26, some of which have already been cited.
118. It has not yet been possible to assess the extent of dismissals, but some indications on individual professions (women high school teachers, physicians, and lawyers) are given in Huerkamp, *Bildungsbürgerinnen*, 184–95.
119. Bauer, *Kulturprotestantismus und frühe bürgerliche Frauenbewegung in Deutschland*, 279–320.
120. Velsen to Bäumer, March 6, 1941, BArch Koblenz, KLE 269–1.
121. Christine von Oertzen, "Luise Berthold: Hochschulleben und Hochschulpolitik zwischen 1909 und 1957," *Feministische Studien: Zeitschrift für interdisziplinäre Frauen- und Geschlechterforschung* 20, no. 1 (2002); Luise Berthold, *Erlebtes und Erkämpftes: Ein Rückblick* (Marburg: Privately published, 1969).
122. Heike Drummer and Jutta Zwilling, "Elisabeth Schwarzhaupt: Eine Biographie," in *Elisabeth Schwarzhaupt (1901–1986): Portrait einer streitbaren Politikerin und Christin*, ed. Hessische Landesregierung (Freiburg: Herder, 2001), 14–115, here 37–43. On the following, see also Schwarzhaupt's autobiographical sketch: Elisabeth Schwarzhaupt, "Mein Leben," n.d. [c. 1970], BArch Koblenz, N 1177, no. 37.
123. In the last two years of the war, Hampe worked as a black-market purchaser for the German war economy in the Netherlands. Interview with Asta Hampe, September 21, 2000.
124. Elvira Scheich, "Elisabeth Schliemann (1881–1972): Patriotin im Zwiespalt," in *Autarkie und Ostexpansion: Pflanzenzucht und Agrarforschung im Nationalsozialismus*, ed. Susanne Heim (Göttingen: Wallstein, 2002), 250–79. Like Berthold, Schiemann kept her distance from the Women's Agency's attempts to "assemble" women graduates under Nazi direction. As she wrote her friend Lise Meitner, the Nazi women students invited her to the former Helene Lange Home in January 1939: "I agreed with great caution. And then it was nothing more than another attempt... to set up a new federation of academic women and draw us into it—I won't be going there again." Schiemann to Meitner, January 21, 1939, quoted in Scheich, "Elisabeth Schliemann," 265.
125. Marie-Elisabeth Lüders, *Das unbekannte Heer: Frauen kämpfen für Deutschland, 1914–1918* (Berlin: Mittler, 1936). The book was even reprinted in 1937.
126. Marie-Elisabeth Lüders, *Fürchte Dich nicht: Persönliches und Politisches aus mehr als 80 Jahren, 1878–1962* (Cologne: Westdeutscher Verlag, 1963), 138–9. In her memoirs, Lüders makes no mention of her personal connection with Möller. For Lüders's political comeback after 1945, see Keith R. Allen, *Überprüfung, Befragung, Kontrolle: Die Aufnahme von DDR-Flüchtlingen in West-Berlin bis 1961* (Berlin: Christoph Links, 2013), especially 25–51.
127. Huerkamp, *Bildungsbürgerinnen*, 310.
128. Christine Eckelmann interviewed Auguste Hoffmann in 1986 as part of her study of the BDÄ. According to Eckelmann, Hoffmann remembered the period of Nazification only "very vaguely," apart from her pronounced dislike of the association's new president. Eckelmann, *Ärztinnen*, 46.
129. For a more detailed description of Hoffmann's career under the Nazis, see Oertzen, *Strategie Verständigung*, 232–9.
130. CV Auguste Hoffmann, October 24, 1937, Archiv der Humboldt-Universität Berlin, Universitätskurator, Personalakten, Auguste Hoffmann, Prof. mit vollem Lehrauftrag und nebenamtliche Oberassistentin am Anat. Institut, no. 403, fol. 71.
131. "Bericht über die Arbeitstagung der Ärztinnen des Obergaus Berlin am 6.11.1935," *Die Ärztin* 11 (1935): 198–9; "Deutscher Sportärzte-Bund, Tagungen," *Die Ärztin* 12 (1936): 112; Elisabeth Baecker-Vowinckel, "8. Schulungslehrgang für Ärztinnen in Alt-Rehse v. 5.-15.7.1939," *Die Ärztin* 15 (1939): 260–1.

132. Hoffmann, "Die Aufgaben der Ärztin im BDM," *Die Ärztin* 10, no. 2 (1934), 32. Also cited in Eckelmann, *Ärztinnen*, 58.
133. "Ansprache von Gustel Hoffmann, Bericht über die Arbeitstagung der Ärztinnen des Obergaus Berlin v. 6.11.1935," *Die Ärztin* 11 (1935): 198.
134. Lore Hiedepriem-Friedl, "Bericht über die Tagung der Obergauärztinnen des BDM," *Die Ärztin* 10, no. 11 (1934): 196–201.
135. Eckelmann, *Ärztinnen*, 59.
136. Thus Ursula Kuhlo, "8. Schulungslehrgang für Ärztinnen in Alt-Rehse v. 5.-15.7.1939," *Die Ärztin* 15 (1939), 262–4, here 263.
137. Hoffmann to Dean of the Medical Faculty, Humboldt University, March 31, 1947, Archiv der Humboldt-Universität Berlin, Universitätskurator, Personalakten, Auguste Hoffmann, no. 403 B, fol. 17.
138. CV Auguste Hoffmann, October 24, 1937, Archiv der Humboldt-Universität Berlin, Universitätskurator, Personalakten, Auguste Hoffmann, Prof. mit vollem Lehrauftrag und nebenamtliche Oberassistentin am Anat. Institut, no. 403, fol. 71. On her work during the Berlin Olympics, see also Auguste Hoffmann, "Olympiastreiflichter," *Die Ärztin* 12 (1936): 188–92.
139. Archiv der Humboldt-Universität Berlin, Charlotte Lorenz, Berufungsakten, UK-L, no. 216. On Lorenz's research, see Zilius-Falkenberg, "Die Frau in der wissenschaftlichen Forschung," 119; Olaf Boustedt, "Prof. Charlotte Lorenz zum 65. Geburtstag," in *Allgemeines Statistisches Archiv*, ed. Karl Wagner (Heidelberg: Physica, 1960), 319–21.
140. Lorenz to Dean of the Faculty of Legal and Political Science, Berlin University, May 15, 1939, Archiv der Humboldt-Universität Berlin, Charlotte Lorenz, Berufungsakten, UK-L, no. 216.
141. See Astrid Dageförde, "Frauen an der Hamburger Universität 1933–1945: Emanzipation oder Repression?" in *Hochschulalltag im "Dritten Reich": Die Hamburger Universität 1933–1945*, ed. Eckart Krause (Berlin: Reimer, 1991), 255–70, here 262; Christian Hünemörder, "Biologie und Rassenbiologie 1933 bis 1945," in ibid., 1155–1187, here 1167.
142. CV Gertrud Savelsberg, n.d., card catalogue, Institut für Weltwirtschaft (Kiel Institute for the World Economy).
143. In German *Rechtswahrerinnen*. The Nazi regime referred to all legal professionals as *Rechtswahrer* to demarcate them from the liberal associations of the term *Jurist*. These "preservers of justice" were charged with maintaining law along Nazi principles, "in the unity of *Volk* and race." For an introduction to Nazi law and justice, see Michael Stolleis, *The Law under the Swastika: Studies on Legal History in Nazi Germany* (Chicago, IL: University of Chicago Press, 1998). For further detail on women lawyers, see Oertzen, *Strategie Verständigung*, 241–2. On their exclusion by the Nazis, see Dr. Meier-Scherling, "Die Benachteiligung der Juristin zwischen 1933 und 1945," *Deutsche Richterzeitung*, January 1975: 10–13; generally on women lawyers after 1933, see Röwekamp, *Die ersten deutschen Juristinnen*, 636–765.
144. Wiltraud von Brünneck, "Die Aufgaben der Frau im Recht," *Frauenkultur im deutschen Frauenwerk* 3, no. 11 (1937): 9–10. Also the following quotations.
145. Ibid. The young lawyer's programmatic comments are especially interesting given that she was appointed as a judge of the Federal Constitutional Court in 1963. Hans-Peter Schneider's appreciation does not mention Brünneck's early affinities with a Nazi theory of law as expressed in this article. He simply notes that she passed the two stages of the state legal examinations "summa cum laude in 1936 and 1941, against all the odds." Hans-Peter Schneider, "Im Namen des Menschen: Über Leben und Wirken einer großen Richterin," in *Verfassung und Verantwortung: Gesammelte Schriften und Sondervoten. Wiltraut Rupp-von Brünneck*, ed. Hans-Peter Schneider (Baden-Baden: Nomos, 1983), 1–46, here 17. See also chapter 8.

146. Angelika Wetterer has demonstrated this strikingly for the case of women doctors. Angelika Wetterer, "Ausschließende Einschließung—marginalisierende Integration: Geschlechterkonstruktionen in Professionalisierungsprozessen," in *Vielfältige Verschiedenheiten: Geschlechterverhältnisse in Studium, Hochschule und Beruf*, ed. Ayla Neusel and Angelika Wetterer (Frankfurt/Main: Campus, 1999), 223–54.
147. Martha Unger, "Die Frau im Nationalsozialistischen Deutschland," *NS-Frauenkorrespondenz* 17 (April 1933): n.p.
148. Schorn, "Frau und Hochschule," 1.

6 Networks in Action: Assistance to Refugees

1. On academic organizations' assistance for emigration and rescue from Nazi Europe, see Kurt Düwell, "Hilfsorganisationen für deutsche Wissenschaftler im Ausland," in *Emigration: Deutsche Wissenschaftler nach 1933: Entlassung und Vertreibung*, ed. Herbert Arthur Strauss, Tilmann Buddensieg, and Kurt Düwell (Berlin: Technische Universität, 1987), XVII–XII; Stephen Duggan and Betty Drury, *The Rescue of Science and Learning: The Story of the Emergency Committee in Aid of Displaced Foreign Scholars* (New York: Macmillan, 1948); Regine Erichsen, "Vom Nationalsozialismus vertriebene Wissenschaftler auf dem Markt: Die Arbeitsvermittlung des englischen Academic Assistance Council (SPSL) am Beispiel von Türkeiemigranten," *Berichte zur Wissenschaftsgeschichte* 19, no. 4 (1997): 219–45; Gerhard Hirschfeld, "'The Defense of Learning and Science...' Der Academic Assistance Council in Großbritannien und die wissenschaftliche Emigration aus Nazi-Deutschland," *Exilforschung* 6 (1988): 28–45. None of these studies mentions the IFUW or its national member organizations. Even in the few investigations of emigration and persecution that specifically address women scholars, there is no reference at all to the network. See, for example, Harriet Pass Freidenreich, *Female, Jewish, Educated: The Lives of Central European University Women* (Bloomington: Indiana University Press, 2002); Hiltrud Häntzschel, "Kritische Bemerkungen zur Erforschung der Wissenschaftsemigration unter geschlechterdiffenzierendem Blickwinkel," *Exilforschung* 14 (1996) 150–65; Sibylle Quack, ed. *Between Sorrow and Strength: Women Refugees of the Nazi Period* (Cambridge, MA: Cambridge University Press, 1995). One exception is the short paper by Susan Cohen on the refugee activities of the BFUW between 1938 and 1945: Susan Cohen, "How the British Federation of University Women Helped Rebuild the Lives of Academic Refugee Women in the 1930s and 40s," presented on February 26, 2008, to the Centre for the History of Women's Education, University of Winchester, available online at http://sybilcampbellcollection.org.uk/scc2/images/stories/downloads/BFUW-and-AcademicRefugeeWomen.pdf (accessed January 18, 2013). See also Susan Cohen, "'Now You See Them, Now You Don't': The Archives of the Refugee Committee of the British Federation of University Women," in *Refugee Archives: Theory and Practice*, edited by Andrea Hammel, Anthony Grenville, and Sharon Krummel (Amsterdam: Rodopi, 2008).
2. By 1938 the number of applications had risen to 89, although this was not due primarily to the rise in applications from Germany. The number of grants to be awarded rose from 7 to 14. "BFUW Annual Report 1937–38," Report of Academic Sub-Committee, p. 38, WL@LSE, printed collections.
3. "BFUW Annual Report 1937–38," p. 30, WL@LSE, printed collections.
4. Johanna Westerdyk, "Help the German University Women!" *BFUW News Sheet* no. 11 (October 1933): 4–5.
5. The report on activities in Britain was the only one at the IFUW's Budapest meeting in 1934 to discuss the plight of German academic women. No other IFUW

national organization mentioned plans to help the German women. See IFUW, *Report of the Nineteenth Council Meeting, Budapest 1934*, 42–60. In France, organized assistance focused on Spanish and later Polish colleagues and on French academic women in the occupied section of France. See Nicole Dabernat, "Marie-Louise Puech-Milhau, une femme engagée, 1876–1966," *Diplômées* 204, March (2003): 17–30. On the generally reserved attitude in France toward the German refugees, see Rita Thalheim, "Jewish Women Exiled in France after 1933," in *Beween Sorrow and Strength: Women Refugees of the Nazi Period*, ed. Sibylle Quack (Cambridge, MA: Cambridge University Press, 1995), 51–62.

6. Margaret Omerod, "An International Problem," *BFUW News Sheet* no. 11 (October 1933): 5–6. The biographical sources support this impression (see chapter 7).
7. When Marie-Elisabeth Lüders, the last pre-Nazi president of the DAB, who had stepped down on May 18, 1933, gave a lecture at Crosby Hall in July 1933 titled "The Nazis," the event was attended by 250 people, around 100 of them from abroad; see "BFUW Annual Report 1933–34," p. 33, WL@LSE, printed collections.
8. The appeal that launched the AAC on May 22, 1933, bore the signatures of 41 prominent public figures. As well as numerous Nobel laureates, the signatories included many eminent scientists, some politicians, and several university vice-chancellors and presidents. The plan, drawn up by Lord Beveridge, also attracted the support of the Royal Society. See Hirschfeld, "The Defense," 31. On the politics of the AAC, later SPSL, see also Marion Berghahn, *German-Jewish Refugees in England: The Ambiguities of Assimilation* (New York: St. Martin's Press, 1984), 78.
9. "BFUW Annual Report 1932–33," p. 1, WL@LSE, printed collections. See also Margaret Omerod, "An International Problem (Continued)," *BFUW News Sheet* no. 12 (January 1934): 6–7.
10. Hirschfeld, "The Defense," 29.
11. "The Christmas Sale," *BFUW News Sheet* no. 14 (October 1934): 7.
12. In the first two years, a total of 12 academic women from Germany were shortlisted. In 1933–34, the Spanish federation's International Fellowship was awarded to the Berlin biologist Käte Pariser, who during her studies had already spent several years attending the biological practicals of Antonio de Zulueta, one of the founders of classical genetics in Spain. Zulueta himself had studied in Berlin with Max Hartmann, the director of the Institute for Infectious Diseases; see Susana Pinar, "La introducción de la genética en España durante el primer tercio del siglo XX," *Llull* 22 (1999): 453–73. When the Spanish Civil War began, in 1936, Pariser fled to Tel Aviv, and in 1939 moved from there to Sydney. She took Australian citizenship in 1944. According to unsubstantiated reports, Pariser died in West Germany in 1953; Helga Satzinger, *Differenz und Vererbung: Geschlechterordnungen in der Genetik und Hormonforschung* (Vienna: Böhlau, 2009), 228–30, here 230.
13. Emmy Klieneberger-Nobel, *Memoirs* (London: Academic Press, 1980). Klieneberger writes that she knew "not a single person" in England but had some distant professional contacts; we can certainly assume she was carrying a letter of recommendation from established former DAB members, which will have helped her to find accommodation at Crosby Hall. Ibid., 78.
14. Extra Fellowships, "BFUW Annual Report 1933–34," p. 30, WL@LSE, printed collections.
15. "Betty Heimann," in *Biographisches Handbuch der deutschsprachigen Emigration nach 1933 International Biographical Dictionary of Central European Emigrés 1933–1945*, ed. Werner Röder and Herbert A. Strauss (Munich: Saur, 1983), vol. 2, 477; see also "Betty Heimann," www.catalogus-professorum-halensis.de/heimannbetty.html (accessed January 18, 2013).

16. BFUW, Report of the Crosby Hall Association, Ltd., to the Annual General Meeting of the British Federation of University Women on July 14, 1934, WL@LSE, printed collections; Report of Academic Sub-Committee, "BFUW Annual Report 1934–35," WL@LSE, printed collections.
17. "Rosenau, Helen," in *Biographisches Handbuch deutschsprachiger Kunsthistoriker im Exil: Leben und Werk der unter dem Nationalsozialismus verfolgten und vertriebenen Wissenschaftler*, ed. Ulrike Wendland (Munich: Saur, 1999), 563–6.
18. Report of the Committee on International Relations for the year 1934–5, "BFUW Annual Report 1934–35," p. 30, WL@LSE, printed collections; History of the Israel (Palestine) Association of University Women, Jerusalem, MS 1950, p. 6–7. AAUW Archives, Folder IFUW—National Member Histories.
19. There was a feeling of "moral obligation" to provide further support for women who had received research funding if they encountered financial difficulties when their fellowship came to an end. Thus, in 1939, the Refugee Committee planned in advance to offer funding to Adelheid Heimann if she turned out to have no means of support at the end of her Aurelia Reinhardt grant. Minutes of BFUW Emergency Sub-Committee on Refugees of September 23, 1939, WL@LSE, 5BFW (old Box 27), Sub-Committee on Refugees.
20. International AAUW Fellowship—Crusade Fellowship, 1934–5, "BFUW Annual Report 1933–34," Report of the Academic Sub-Committee, p. 31, WL@LSE, printed collections.
21. Report of the Academic Sub-Committee, Allocation of Grants, "BFUW Annual Report 1938," p. 42, WL@LSE, printed collections.
22. The original habilitation thesis appeared in London in 1934, under the title *Design and Medieval Architecture*; the doctoral thesis was titled "The Architectural Development of the Synagogue," University of London, 1940. See also chapter 7.
23. "BFUW Annual Report 1938–39," p. 40, WL@LSE, printed collections. On Adelheid Heimann, see also chapter 7.
24. A second invitation was issued to Emmy Noether, but she preferred to take up the offer of a permanent position at Bryn Mawr. See Pauline Adams, *Somerville for Women: An Oxford College, 1879–1993* (Oxford: Oxford University Press, 1996), 192.
25. Freidenreich, *Female, Jewish, and Educated*, 131. BFUW efforts to ensure that Bieber received the "Senior International Fellowship of the Arts" for 1935 were unsuccessful. Although her application was accompanied by "a letter explaining the exceptional circumstances of Dr. Bieber's career which led the British Committee to submit the name of a candidate of professional rank," a younger, less well-established applicant from the Netherlands was ultimately given preference. See "BFUW Annual Report 1934–35," p. 33, WL@LSE, printed collections.
26. Helena von Reybekiel, born in Russia in 1879, had studied psychology in Zurich and attained her doctorate in 1904. See www.matrikel.unizh.ch/pages/162.htm (accessed October 25, 2007). No information is currently available as to whether she had been employed at the Hamburg Institute of Slavic Studies since its establishment in 1914; Astrid Dageförde only mentions her dismissal in 1933. Dageförde, "Frauen an der Hamburger Universität 1933–1945. Emanzipation oder Repression?" in *Hochschulalltag im "Dritten Reich": Die Hamburger Universität 1933–1945*, ed. Eckart Krause (Berlin: Reimer, 1991), 257. Of the 22 women lecturing at Hamburg university-level institutions, five were dismissed on the basis of the "Law for the Restoration of the Professional Civil Service" that barred Jews and regime opponents from working in civil service professions—including university employment.
27. Summary of Reports of Local Associations, 1934–35, "BFUW Annual Report 1933–34," p. 16, WL@LSE, printed collections. The Birmingham and Midland Institute was formed after an Act of Parliament of 1854 required "the Diffusion

and Advancement of Science, Literature and Art amongst all Classes of Persons resident in Birmingham and the Midland Counties." See http://bmi.org.uk/about .html (accessed July 13, 2013).

28. Grant recipients were the biologists Käte Pariser and Dora Ilse, the Indologist Betty Heimann, the art historians Helen Rosenau and Adelheid Heimann, the microbiologist Emmy Klieneberger, the archaeologists Elisabeth Jastrow (whose report on her research in Italy can be found in "The Work of AAUW Fellows, 1934/5," *AAUW Journal* 29 [1936], 93) and Elise Jenny Baumgärtl (an Egyptologist; see *Who Was Who in Egyptology* [London: Egypt Exploration Society, 1995], 35–6), the physicist and photochemist Gertrud Kornfeld, and the Slavicist Helena von Reybekiel. The recipients of smaller grants included the sociologists Charlotte Leubuscher and Marie Jahoda, the jurist Lore Mann, and Ilse Kaufmann, I. Lichtenstädter, T. Hamburger, H. Mank, H. C. Graef, A. Haas, and Frieda Gossmann. Younger academic women who had graduated from British universities were eligible for the British Junior Fellowship. This applied to the historian Ursula Wassermann, originally from Hamburg, who later emigrated to the United States and became a politically active Democrat. Dismayed by the excesses of McCarthyism, she left the United States in the early 1950s. See Ursula Wassermann, *I Was an American* (London: Bodley Head, 1955).
29. On this, see Herbert A. Strauss, "The Migration of Academic Intellectuals," in *Biographisches Handbuch der deutschsprachigen Emigration nach 1933: International Biographical Dictionary of Central European Emigrés 1933–1945*, ed. Werner Röder and Herbert A. Strauss (Munich: Saur, 1980), lxxv.
30. "Crosby Hall Sale," *BFUW News Sheet*, no. 14 (October 1934): 8.
31. See "The BFUW and the Younger University Woman," *BFUW News Sheet*, no. 20 (October 1936): 8–12; "Correspondence: The BFUW and the Younger University Woman," *BFUW News Sheet*, no. 21 (February 1937): 9.
32. Between 1933 and 1936, the IFUW received 131 applications, of which 21 were from members of the former German affiliate. See "Funds for the Assistance of University Women," in IFUW, *Report of the Seventh Conference, Cracow 1936*, 100.
33. Ibid.
34. Not only Germany but also Italy, Spain, and Portugal withdrew from the IFUW in 1936.
35. "Resolution No. 13," in IFUW, *Report of the Seventh Conference, Cracow 1936*, 123. At the same time, all member associations agreed to report back regularly on their refugee aid work.
36. IFUW, *Report of the Twenty-Third Council Meeting, London 1938* (London: The Federation, 1938), 8.
37. Dora (Dorothea) Ilse had worked with Richard Goldschmidt in Berlin, and in 1933 moved to Munich to continue her work on the sensory physiology of butterflies as an assistant to Professor Karl von Frisch. She had to leave this position in 1936. Satzinger, *Differenz und Vererbung*, 215.
38. In 1934, Kornfeld received a "special monetary grant" from the BFUW, very probably because her teaching post was badly paid or even unpaid. See Extra Grants, "BFUW Annual Report 1934–35," pp. 32, 39; "BFUW Annual Report 1935–36," p. 32, WL@LSE, printed collections. See also "Assistance to Displaced German Women," *AAUW Journal* 28 (1935), 50.
39. Among the five émigré women who received small-scale funding in 1937–38 were the sociologist Marie Jahoda from Vienna and again the biologist Dora Ilse. "BFUW Annual Report 1937–38," p. 42, WL@LSE, printed collections.
40. The funding for 1937 was lowest, at only £100. The following sums were awarded in this period: 1933–34: £368; 1934–35: £263; 1935–36: £150: 1936–37: £100; 1937–38: £240. BFUW Annual Reports, 1933–38, WL@LSE, printed collections.

41. The British customs authorities imposed an import charge; Klieneberger-Nobel, *Memoirs*, 82. Klieneberger saw her mother and sister for the last time in spring 1938, after which she was no longer able to cross the German border. Ibid., 90–1. Her mother, sister, and brother all took their own lives when faced with deportation. Kleineberger facilitated the rescue of several of her nephews and nieces and later their education in England. Ibid., 91–7.
42. See Louise London, "British Immigration Control Procedures and Jewish Refugees, 1933–1939," in *Second Chance: Two Centuries of German-Speaking Jews in the United Kingdom*, ed. Werner E. Mosse (Tübingen: J. C. B. Mohr/Paul Siebeck, 1991), 485–518, here 506; also Louise London, *Whitehall and the Jews, 1933–1948: British Immigration Policy, Jewish Refugees, and the Holocaust* (New York: Cambridge University Press, 2000), especially Chapter 5.
43. "BFUW Annual Report 1938–39," p. 20, WL@LSE, printed collections. See also WL@LSE, 5BFW (old Box 27), Ad-hoc Committee on Refugees, Minutes, 1938–1950.
44. CV Hollitscher n.d. [summer 1938], WL@LSE, 5BFW (old Box 27), Ad-hoc Committee on Refugees. The minutes of the Ad-hoc Committee for Refugees of September 19, 1938, note: "Hollitscher, Dr. Erna. Permit obtained by Mrs. Omerod. Very good secretarial and translation work. Been receiving hospitality from Miss Campbell. Applying for a post in Birmingham as German correspondent, if that does not materialize suggest for temporary secretary to this Committee." WL@LSE, 5BFW (old Box 27), Committee on Refugees.
45. Report of Emergency Sub-Committee for Refugees for Year 1938/9, "BFUW Annual Report 1938–39," p. 20, WL@LSE, printed collections.
46. Up to 1950, the BFUW paid her a salary of £200 per annum, though from 1940 onward half of this sum was contributed by the British government's Central Committee for Refugees. Official funding of refugee assistance ceased in March 1950, but the federation decided to continue maintaining Hollitscher's post, among other reasons because so many inquiries from refugees were still arriving and the Canadian clothing collections needed to be distributed. Memorandum from Winifred Cullis to the members of the Refugee Sub-Committee, March 24, 1949, WL@LSE, 5BFW (old Box 27), Sub-Committee on Refugees.
47. Of particular interest here is Hollitscher's correspondence with the 39-year-old Emma Reich, a Viennese physician who, accompanied by her husband Leo, reached England in June 1939 with Erna Hollitscher's help. The Reichs confided in Hollitscher, telling her of their many difficulties in settling in; after her emigration to the United States, Emma Reich continued her correspondence with Hollitscher. Among other topics, Hollitscher and Reich exchanged notes on their experiences of continued alienation and loneliness. See, for example, Hollitscher to Reich, March 15, 1950, and Reich to Hollitscher, December 12, 1950, from Phoenix, AZ, WL@LSE, 5BFW (old Box 27), War Refugees, Folder Reich, Emma.
48. In 1939, the BFUW received additional financial support for its refugee work from the Lord Baldwin Fund for Refugees (£200) and the Council for German Jewry (£200). Summary Report of the Annual General Meeting of Council 1939, "BFUW Annual Report 1939–40," p. 29, WL@LSE, printed collections.
49. The BFUW annual report for 1938–39 states that "the time taken for formalities to be completed both in this country and abroad has been appalling." Report of Emergency Sub-Committee for Refugees, "BFUW Annual Report 1938–39," p. 20, WL@LSE, printed collections.
50. Mrs. Omerod, a member of the BFUW's Sub-Committee on Refugees and a member of staff at the government-run Bloomsbury House refugee committee, had received an assurance from the government "not to make any objection to a change of permits from domestic work to more congenial work in the case of University women." See Hollitscher to Goldschmidt, October 26, 1939, WL@LSE,

5BFW (old Box 30), Refugees, Folder Schlesinger. On the issue of domestic permits, see also Tony Kushner, "An Alien Occupation—Jewish Refugees and Domestic Service in Britain," in *Second Chance: Two Centuries of German-Speaking Jews in the United Kingdom*, ed. Werner E. Mosse (Tübingen: J. C. B. Mohr/Paul Siebeck, 1991), 553–78.

51. Report of Emergency Sub-Committee for Refugees for Year 1938–9, "BFUW Annual Report 1938–39," pp. 20–22, here p. 21, WL@LSE, printed collections.
52. See, for example, the conflict with a Mrs. Prustman: she wanted a woman aged between 20 and 30, whereas the BFUW was interested in placing with her either a 50-year-old academic from Vienna or a 15-year-old girl urgently in need of a home. See Dickson to Scruby, August 5, 1938, WL@LSE, 5BFW (old Box 27).
53. Report of the Emergency Committee for Refugees to the BFUW Executive Committee, December 2, 1938, Offers of Hospitality, WL@LSE, 5BFW (old Box 27), Sub-Committee on Refugees.
54. See, for example, the cases of Gertrud Gerstenberger, Else Ernst, and Dora Kulka from Vienna. All three women had applied to BFUW in early summer 1938 for help in finding work in Britain. During her interview, Gerstenberger did not conceal her lack of enthusiasm at the prospect of domestic work, and was thus rejected for assistance; in contrast, Else Ernst, twenty-six years old with a newly earned doctorate in biology, made a credible case for being willing to accept "any work" in a research or laboratory setting, or even in a regular, unspectacular business establishment. Because she was also interested in photography and was a licensed movie projectionist, the interviewers considered her flexible enough to find work on the British labor market. After an interview in Vienna, the BFUW tried to help Dora Kulka continue her scientific research. A biochemist, Kulka had worked at the Bacteriological Museum in Vienna and had been classified as an outstanding scientist by the SPSL. See Dickson to Horton, Society of Friends, Vienna, July 22, 1938, and Dickson to Horton, August 4, 1938, WL@LSE, 5BFW (old Box 27), Refugee Sub-Committee 1938–1950. On the support that Kulka received from the BFUW after she arrived in London, see Cohen, "'Now You See them, Now You Don't,'" 116.
55. Minutes of a Meeting of the Committee on Refugees, November 22, 1938, WL@LSE, 5BFW (old Box 27), Sub-Committee on Refugees.
56. Report of Emergency Sub-Committee for Refugees, "BFUW Annual Report 1938–39," pp. 20–22, here p. 21, WL@LSE, printed collections.
57. The correspondence in the AAUW archives can be classified by discipline as follows. Letters were written by sixteen physicians, eight physicists, seven chemists, five historians, four jurists, and four economists; the disciplinary spectrum also covered biology, mathematics, art history, archaeology, psychoanalysis, sociology, philology, and social work. In terms of professions, physicians were the largest group, followed by schoolteachers, while six women were librarians. Four university lecturers and eight students were among the women seeking assistance.
58. List of Applicants before the Committee on October 24, 1938. Report of the Sub-Committee for Refugees to the BFUW Executive Committee, October 29, 1938, WL@LSE, 5BFW (old Box 27), Sub-Committee on Refugees. A precise breakdown by age and profession has been carried out for the holdings of the AAUW archives, but the much more extensive correspondence in London has not been analyzed in full, though a complete list of names could be drawn up using the minutes of the BFUW's Sub-Committees on Refugees, which met approximately every four months; as a result, details of profession and age are available for around 60 percent of all applicants. From the available evidence it seems that the British and American correspondences contain similar proportions of the various professions.

59. Röder and Strauss's dictionary of Central European émigrés lists only 8 of the 40 women scholars who received funding from the BFUW up to 1938. Werner Röder and Herbert A. Strauss, eds., *Biographisches Handbuch der deutschsprachigen Emigration nach 1933: International Biographical Dictionary of Central European Emigrés 1933–1945* (Munich: Saur, 1983).
60. Spiegel's journey is described in more detail in chapter 7. Her correspondence with Esther Brunauer between February 17, 1940, and April 25, 1941, can be consulted in the AAUW Archives, RG Programs, International Relations, War Relief Cases, Folder 38.
61. Memorandum from Winifred Cullis to the members of the Refugee Sub-Committee, March 24, 1949, WL@LSE, 5BFW (old Box 27), Sub-Committee on Refugees.
62. Minutes of the Meeting of the Committee held on Monday, September 23, 1939, WL@LSE, 5BFW (old Box 27), Sub-Committee on Refugees. Between 1940 and 1949, 50 percent of the expenses for refugee assistance were paid by the British government's Central Committee for Refugees. In April 1950, this official support ceased. Winifred Cullis to the members of the Sub-Committee for Refugees, March 24, 1949, WL@LSE, 5BFW (old Box 27), Sub-Committee on Refugees.
63. AAUW Archives, RG Programs, International Relations, War Relief Case Files, Folder 48.
64. Larissa Bonfante, "Margarete Bieber (1879–1978): An Archaeologist in Two Worlds," in *Women as Interpreters of the Visual Arts, 1820–1979*, ed. Claire Richter Sherman and Adele M. Holcomb (Westport, CT: Greenwood Press, 1981), 239–74.
65. Helene Adolf, "In Memoriam Elise Richter," *Romance Philology* 1, no. 4 (1948): 339.
66. Leonie Spitzer was a cousin of Helene Adolf's. She had studied Ancient Greek, Gothic, and Anglo-Saxon in Vienna and written her doctoral thesis on Rainer Maria Rilke. In 1939, she settled in Oxford, earning her living as a teacher at various schools and colleges. She died of cancer in 1940. See Siglinde Bolbecher and Konstantin Kaiser, *Lexikon der österreichischen Exilliteratur* (Vienna: Deuticke, 2000), 607.
67. Renate Heuer and Michael Dallapiazza, "Vorwort," in *Adolf, Helene: Gesammelte Schriften*, ed. Renate Heuer and Michael Dallapiazza (Triest: Edizioni Parnaso, 2004), ii. In Philadelphia Adolf herself initially taught modern and classical languages in schools until 1943, when she was appointed to the German department of the State University of Pennsylvania, receiving a full professorship in 1950.
68. Minutes of the Meeting of the Committee held on Monday, September 23, 1939, WL@LSE, 5BFW (old Box 27), Sub-Committee on Refugees.
69. Hospitality at Crosby Hall, "BFUW Annual Report 1938–39," p. 22, WL@LSE, printed collections.
70. Report of the Hospitality Sub-Committee, 1932–33, "BFUW Annual Report 1932–33," pp. 37–8, WL@LSE, printed collections.
71. Ibid.
72. Klieneberger-Nobel, *Memoirs*, 79.
73. Ibid. Klieneberger moved from Crosby Hall to a Chelsea apartment very close to her friend Jean Darling, a classical philologist and social worker. The two women met frequently, often daily, until the evacuation of the Lister Institute. In 1940, Klieneberger followed the Institute to a suburb of London. Darling, who stayed in Chelsea, was killed in a German bomb attack. Ibid., 97.
74. Hollitscher to Vera Jentzsch, January 7, 1941, WL@LSE, 5BFW (old Box 28), Case Folder Jentzsch. See also the words of gratitude from Gertrud Schlesinger to Hollitscher, May 14, 1939 [in German]: "I would like to thank you and everyone at Crosby Hall once more very heartily for your kind reception and for taking such nice care of me." WL@LSE, 5BFW (old Box 30), Refugees, Folder Schlesinger.

75. In summer 1933, Marie-Elisabeth Lüders gave her talk on "The Nazis" and Mrs. Omerod from Bloomsbury House spoke on "German Refugees." "BFUW Annual Report 1933–34," pp. 32–3, WL@LSE, printed collections.
76. Guest lists are available for the years 1934–35 and 1941 to 1946; Crosby Hall Attendance List, Wartime Visitors, Hospitality At Homes, 1934–1935 and 1941–1946, WL@LSE, 5BFW (old Box 10), Folder 12 and 18.
77. J. H. Sondheimer, *History of the British Federation of University Women, 1907–1957* (London: British Federation of University Women, 1957), 32, AAUW Archives, Series IX, Reel 149/4, Box 832. See also Alys Russell, "Colloquial English for the Foreigner," *University Women's Review*, no. 40 (July 1950): 5–6.
78. An example is Alice Apt, a Germanist from Dresden who had studied in Freiburg, Munich, and Königsberg and, after earning her doctorate in 1936, taught at Dresden's Jewish Adult Education Insitute. In early March 1939, she came to London on a "domestic permit," there to take care of the young invalid daughter of a wealthy lawyer. When the girl was taken to a sanatorium and Apt lost her job, an English BFUW member put her in touch with Hollitscher. The first interview took place in the BFUW offices, but all the subsequent meetings were held during BFUW "at homes." See Apt to Hollitscher, December 7, 1941, WL@LSE, 5BFW (old Box 28), Folder Apt.
79. Apt to Hollitscher, November 11, 1943, ibid.
80. "BFUW Annual Report 1936–37," p. 35, WL@LSE, printed collections.
81. Summary of Reports of Local Associations 1934–35, Birmingham, "BFUW Annual Report 1934–35," p. 16; East Surrey, ibid., p. 21; Manchester, ibid., p. 26; Reading, ibid., p. 27, WL@LSE, printed collections. In this connection, the local associations' "Bring and Buy Sales," held every November, are of particular relevance.
82. See, for example, the report of the Birmingham and Midlands branch: "The Annual Dinner was held in November. Professor Betty Heimann was the principal Guest and talked to us about the fundamental philosophy and the peoples of India." Summary of Reports of Local Associations 1934–35, "BFUW Annual Report 1934–35," p. 17. The report notes that Dora Ilse captivated several meetings "with her beautiful butterfly films," "BFUW Annual Report 1938–39," p. 42, WL@LSE, printed collections.
83. See also the reports of the groups for the year 1939–40: "BFUW Annual Report 1939–40," pp. 24–8, WL@LSE, printed collections.
84. "Reports from Local Associations on Work for Refugees, 1938–39," "BFUW Annual Report 1938–39," pp. 22–4, WL@LSE, printed collections.
85. Gertrud Schlesinger to Hollitscher, May 14, 1939, WL@LSE, 5BFW (old Box 30), Refugees, Folder Schlesinger.
86. On this point, see also chapter 7, especially notes 35 and 36.
87. Minutes of a Meeting of the Committee on Refugees, November 22, 1938, WL@LSE, 5BFW (old Box 27), Sub-Committee on Refugees. In all likelihood, however, her sister's family did not come to Britain, so that in the end "only" Schlüchterer and her mother had to be looked after locally.
88. Folder Hinrichsen. On Hölzl, see BFUW, Minutes of the Meeting of the Emergency Sub-Committee on Refugees, September 23, 1939, WL@LSE, 5BFW (old Box 27), Sub-Committe on Refugees.
89. "BFUW Annual Report 1939–40," p. 22, WL@LSE, printed collections.
90. In early August, Kuranda had made contact with the BFUW, looking for the right people to help her set up a smoothly functioning cooperation between Vienna and London. She continued to pursue this goal in subsequent years, and her recommendations were given special credence in London: "Having been for several years chairman of the committee for international relations of the Austrian Federation, there are many of my colleagues from former days who come to me now for advice

and help, so it would be useful if I know just exactly who is the person to be approached either by me or them." Kuranda to BFUW, August 6, 1938, WL@LSE, 5BFW (old Box 27), Ad-hoc Committee on Refugees.

91. On Kuranda's activities in the Vienna Jewish Community (Israelitische Kultusgemeinde Wien), see Kuranda, Reference for Philipine Hannak, letter to Mrs. Omerod, August 3, 1938, WL@LSE, 5BFW (old Box 27), Sub-Committee on Refugees.
92. BFUW, Minutes of the Meeting of the Emergency Sub-Committee on Refugees, September 23, 1939, WL@LSE, 5BFW (old Box 27), Sub-Committee on Refugees.
93. Report of Emergency Sub-Committee for Refugees period May 15, 1939, to July 15, 1940, "BFUW Annual Report 1938–39," p. 21, WL@LSE, printed collections.
94. The International Federation of University Women, "BFUW Annual Report 1941–42," p. 28, WL@LSE, printed collections. Between March and June, 22 events were organized and 1,000 women attended them. Subsequently, the building was requisitioned by the Army until the end of the war. Parts of the old residential wing suffered bomb damage that made them unfit for habitation.
95. See Barnard College Archives, Gildersleeve papers, Folder 39–40, Correspondence 96 (IFUW), A–H and G–Z, Minutes of the Meeting of the International Federation Budget Sub-Commission in London, May 9, 1940. These minutes note: "Miss Bowie reported that, in view of the number of refugees in London, she had been in touch with Mrs. Wheeler, Hon. Sec. of the London branch, about the desirability of reviving the At Homes for foreign members which used to be held at Crosby Hall. Mrs. Wheeler had found enough hostesses for the summer term to make it possible to have an At Home once a fortnight in May, June and July. Prof. Cullis very kindly offered to give an informal tea-party at her house."
96. Hollitscher to Emma Reich, December 8, 1942, WL@LSE, 5BFW (old Box 27), Old Refugee Cases, Folder Reich.
97. The files reveal 27 cases in which the BFUW, mostly in cooperation with International Social Service, contributed to the costs of tuition fees and/or living expenses. Some of the women supported in this way joined the BFUW when they graduated, such as Brigitta Jacobs, whose studies were part-funded by the Glasgow Branch over many years. Minutes of a Meeting of the Sub-Committee on Refugees, April 5, 1941, WL@LSE, 5BFW (old Box 27), Sub-Committee on Refugees. The same goes for the economist Maria Kramer, a protégée of the Birmingham Branch who joined the BFUW out of gratitude for that support. Minutes of a Meeting of the Sub-Committee on Refugees, February 23, 1944, WL@LSE, 5BFW (old Box 27), Sub-Committee on Refugees.
98. IFUW, *Report of the Eighth Conference, Stockholm 1939*.
99. Ibid., 31–2. See also the following points.
100. On the negotiations in Budapest, see chapter 5.
101. IFUW, *Report of the Eighth Conference, Stockholm 1939*, 41–4. An additional, important reason for the American women to reject this amendment was the discriminatory policy that had kept most black women out of their own association. Susan Levine, *Degrees of Equality: The American Association of University Women and the Challenge of Twentieth-Century Feminism* (Philadelphia, PA: Temple University Press, 1995), 111.
102. IFUW, *Report of the Eighth Conference, Stockholm 1939*, 44.
103. The report on the Stockholm conference, which appeared in October 1939, already included a reference to this loss: "At the moment of writing, our sympathy goes to our colleagues of the Polish Federation in their appalling trial and especially to our newly-elected President, Lektor Adamovicz, who went back to Warsaw—to 'play her part', as she said—at the beginning of September. We have had no further news from her." Ibid., 16–17.

104. On Eder, see Bettina Vincenz, *Biederfrauen oder Vorkämpferinnen? Der Schweizerische Verband der Akademikerinnen (SVA) in der Zwischenkriegszeit* (Baden: hier + jetzt, 2011), 86.
105. Some associations, such as the Danish and the Belgian ones, paid their membership fees via members in exile, so as to be able to continue supporting the IFUW. Others, such as the French association, were active in refugee assistance despite their officially suspended status, especially in helping persecuted members and émigré women to escape Paris in June 1940 and reach the area outside Nazi occupation, in the south of France. See Edith C. Batho, *A Lamp of Friendship: A Short History of the International Federation of University Women, IFUW 1918–1968* (London: Sumfield & Day, 1969), 22.
106. "Report of the Special Committee for the Emergency Assistance of University Women," in IFUW, *Report of the Eighth Conference, Stockholm 1939*, 44–9, here 44.
107. Martha Forkl and Bertholda Plechl, "Elise Richter und der Verband der Akademikerinnen Österreichs," in *Frauenstudium und akademische Frauenarbeit in Österreich*, ed. Martha Forkl and Elisabeth Koffmahn (Vienna: Braumüller, 1968), 108–15, here 112; Robert Tanzmeister, *Elise Richter—Frau und Wissenschaftlerin. Vortrag, gehalten anläßlich der Buchpräsentation: Elise Richter, Summe des Lebens, am 17.10.1998 auf dem Campus der Universität Wien*, ed. Verband der Akademikerinnen Österreichs (Vienna: Wiener Universitätsverlag, 1998).
108. Her words were: "Alte Bäume verpflanzt man nicht." Adolf, "In Memoriam Elise Richter," 340.
109. Hollitscher to van der Kolf, October 6, 1939, Correspondence with IFUW, WL@LSE, 5BFW (old Box 33), IFUW Cases. The IFUW took action in a similar way to pay a modest pension to Helene Stöcker when she emigrated, via Sweden, to the United States at the age of 71. See AAUW Archives, RG Programs, International Relations, War Relief Case Files, Folder 48.
110. Tanzmeister, *Elise Richter*, 7.
111. It is not possible to tell now which women these were. The applicants included the Viennese philosopher Rose Rand. Vienna's first university lecturer in classical languages, Gertrud Herzog-Hauser, was offered a teaching position in Australia, though she turned it down when her husband received the opportunity to go to Switzerland. High school teacher Lilli Oppenheim had completed her application for Australia, but then decided against the move because her brother was emigrating to the United States and she wanted to be near him. See BFUW and AAUW Archives, War Relief Case Files.
112. IFUW, *Report of the Eighth Conference, Stockholm 1939*, 46, 48.
113. Vincenz, *Biederfrauen oder Vorkämpferinnen*, 108.
114. Schaetzel was the founder of the Swiss association, and was active on all levels of the association's work. She was responsible for the establishment of a Swiss fund to aid war refugees, which after 1945 took up the work that the IFUW had begun as part of its refugee assistance activities. Ibid., 128.
115. Martha Liefmann and Else Liefmann, *Helle Lichter auf dunklem Grund: Erinnerungen von Martha Liefmann und Else Liefmann* (Bern: Christliches Verlagshaus, 1966), 14, 81–2. Even if the French academic women as a body did not work for the German refugees, the president of the French association, Marie-Louise Puech, was an important contact person.
116. AAUW Archives, RG Programs, International Relations, War Relief Case Files, Folder 31.
117. AAUW Archives, RG Programs, International Relations, War Relief Case Files, Folder 48.
118. Esther Caukin-Brunauer, "The Women's Movement in Germany—Some Echos," *AAUW Journal* 28 (1935): 105.

119. Esther Caukin-Brunauer, "Notes on my experiences in Germany in 1933, presented very informally to the members of the board of directors of the AAUW," April 1934, AAUW Archives, Series I, Reel 1, Box 59.
120. See, for example, Esther Caukin-Brunauer, "The German Woman Marches—Back," *AAUW Journal* 27 (1934): 85–7; Esther Caukin-Brunauer, *The National Revolution in Germany, 1933* (Washington, DC: AAUW, 1933). Members could order the manuscript of the 1933 book for 50 cts. Personal communication from her daughter Kathryn Horvat, today the president of the AAUW's Utah branch; email of July 10, 2002.
121. "Speaking Engagements for Dr. Brunauer," *AAUW Journal* 27, no. 2 (1934): 89; "Dr. Brunauer Adresses Meetings," *AAUW Journal* 27, no. 3 (1934): 138.
122. This is demonstrated by numerous items in minutes and correspondence with the committee members of the International Relations Office.
123. On the restrictive immigration policy and its consequences, see Claus-Dieter Krohn, "Nobody Has the Right to Come to the United States," *Exilforschung* 3 (1985). A 1968 study that remains current is David Wyman, *Paper Walls: America and the Refugee Crisis 1938–1941* (Amherst: University of Massachusetts Press, 1968). For a useful overview, see Claus-Dieter Krohn, "Vereinigte Staaten von Amerika," in *Handbuch der deutschsprachigen Emigration 1933–1945*, ed. Claus-Dieter Krohn et al. (Darmstadt: Wissenschaftliche Buchgesellschaft, 1998), 446–66.
124. The AAUW did not have the resources to contribute to the funding of salaries for such posts, which would have been necessary in the case of, for example, the physicist Hertha Sponer from Göttingen. However, thanks to her own good connections and the mediation of her mentor and later husband, the physicist and Nobel laureate James Franck, in 1936 Sponer successfully completed lengthy negotiations to gain a position at Duke University in North Carolina without supplementary funding from aid organizations. See Marie-Ann Maushart, *"Um mich nicht zu vergessen": Hertha Sponer—Ein Frauenleben für die Physik im 20. Jahrhundert* (Bassum: Verlag für Geschichte der Naturwissenschaften und der Technik, 1997), 69–90. The AAUW played a direct part in finding jobs for physicist Hedwig Kohn, archaeologists Margarete Bieber and Elisabeth Jastrow (who, like Kohn, was offered a place at UNC Greensboro), and Germanists Melitta Gerhard and Susanne Engelmann. With the help of recommendations from the AAUW and the YMCA, Engelmann, a former high school principal, found a position in 1942–43 as "Refugee Scholar" at Wilson College in Philadelphia; this scholarship had been given to Erna Barschak the year before. A list of names regularly sent out to colleges and universities by the AAUW's International Relations Office from 1939 onward did not, however, yield any response.
125. Catherine Ashburner, "Shirley Farr," in *Those Intriguing Indomitable Vermont Women*, ed. Vermont State Division of the American Association of University Women (n.p., 1999), 44–5.
126. Brunauer to Roboz, February 6, 1940, AAUW Archives, RG Programs, International Relations, War Relief Case Files, Folder 28.
127. Bowie to Gildersleeve, May 29, 1940: "The position is that, as the only officer available, I am taking action which seems urgently necessary without waiting for replies from other Officers. I have the Budget Sub-Committee whose members are of course all in this country, whose advice and assistance I can call on. But as every day makes the position more difficult and as we do not know how long we shall be able to carry on our work here, I think it would be the best thing that could be done if you would organize an American Committee who will take over if circumstances make it necessary." Barnard Archives, Gildersleeve Papers, Box 45, IFUW Correspondence.

128. "Report of the National Committees at the Convention in Dallas, 1947," *AAUW Journal* 40/41 (1947): 245.
129. Disbursement from Reconstruction Aid Fund in the United States, March 8, 1946, AAUW Archives, Series V, Reel 130/38, Box 829: Fellowships, Funds, Reconstruction Aid and War Relief, 1940–1949. In November 1942, the policy for allocating money within the War Relief Committee was revised in order to allow an even more flexible response to a range of different emergency situations. See references in the file Fleischer: Brunauer to Lee, International Student Service, October 15, 1942.
130. Before Kohn, Hertha Sponer had escaped to the United States and Lise Meitner to Sweden; in the early period they also received support from the IFUW. On this and the following points, see AAUW Archives, RG Programs, International Relations, War Relief Case Files, Folder 22. On the stages of Kohn's emigration, see Brenda P. Winnewisser, "The Emigration of Hedwig Kohn, Physicist, 1940," *Mitteilungen der Österreichischen Gesellschaft für Wissenschaftsgeschichte* 18 (1998): 41–58.
131. In 1935, the German Scientist Relief Fund, an initiative launched in 1933 by physicists in the United States, had already provided the funding for Kohn to spend several months at the observatory in Arosa, Switzerland. Ibid., 44.
132. Meta Glass to Esther Brunauer, May 29, 1940, AAUW Archives, RG Programs, International Relations, War Relief Case Files, Folder 22.
133. McAfee to Brunauer, June 6, 1940, ibid. McAfee was another member active in the AAUW and the US "Movement for Democracy." In August 1942, she hit the headlines when she requested leave of absence in order to become the Navy's first female reserve officer; shortly afterward she was entrusted with heading the WAVES, the Navy's all-women volunteer corps, and commanding its soon-to-be eighty thousand members.
134. In 1940, Brunauer succeeded in getting the AAUW to speak out for the repeal of the Neutrality Act of 1935; see Levine, *Degrees of Equality*, 56.

7 Marked by Persecution

1. For gender history's critique of exile research, see Hiltrud Häntzschel's very informative survey "Geschlechterspezifische Aspekte," in *Handbuch der deutschsprachigen Emigration 1933–1945*, ed. Claus-Dieter Krohn (Darmstadt: Wissenschaftliche Buchgesellschaft, 1998), 101–16; Hiltrud Häntzschel, "Kritische Bemerkungen zur Erforschung der Wissenschaftsemigration unter geschlechterdiffenzierendem Blickwinkel," *Exilforschung* 14 (1996), 150–65. On paid employment among married women in exile, see also Christine Backhaus-Lautenschläger, *"...Und standen ihre Frau": Das Schicksal deutschsprachiger Emigrantinnen in den USA nach 1933* (Pfaffenweiler: Centaurus-Verlagsgesellschaft, 1991), 254–62.
2. In many such fields, no systematic research has yet been carried out on women university and college graduates or professionals in exile, but an important precedent was set by a collection arising from a German Historical Institute conference in Washington, which includes work on women physicians, psychologists and psychoanalysts, lawyers, and social workers: Sibylle Quack, ed., *Between Sorrow and Strength: Women Refugees of the Nazi Period* (Cambridge, MA: Cambridge University Press, 1995). See also Harriet Pass Freidenreich, *Female, Jewish, and Educated: The Lives of Central European University Women* (Bloomington: Indiana University Press, 2002), based on a selection of 460 biographies of Jewish graduate women from Europe (chapters 7 and 8 address persecution and exile). Apart from this, research on the exodus of the female educational elite depends primarily on biographical or autobiographical works on individual women; see, among many others, Ruth Lewin Sime, *Lise Meitner: A Life in Physics* (Berkeley: University of California

Press, 1996); Marie-Ann Maushart, *"Um mich nicht zu vergessen": Hertha Sponer—Ein Frauenleben für die Physik im 20. Jahrhundert* (Bassum: Verlag für Geschichte der Naturwissenschaften und der Technik, 1997); James W. Brewer and Martha K. Smith, eds., *Emmy Noether: A Tribute to Her Life and Work* (New York: Dekker, 1981).

3. Sibylle Quack, *Zuflucht Amerika: Zur Sozialgeschichte der Emigration deutsch-jüdischer Frauen in die USA, 1933–1945* (Bonn: Dietz, 1995), 200.
4. Emmy Klieneberger-Nobel, *Memoirs* (London: Academic Press, 1980), 74–5.
5. Ibid., 74.
6. Susanne Engelmann to Esther Brunauer, October 29, 1934, AAUW Archives, War Relief Committee, Individual Cases, Engelmann, Box 839, Folder 10.
7. Ibid.
8. On this, see also Freidenreich, *Female, Jewish, and Educated*, 173–4.
9. Maas to BFUW, March 26, 1939, WL@LSE, 5BFW (old Box 31), Refugee Cases, L–R, n.p.
10. Eduard Seidler, "Lucie Adelsberger, 1945–1971," in Lucie Adelsberger, *Auschwitz: Ein Tatsachenbericht. Das Vermächtnis der Opfer für uns Juden und für alle Menschen*, ed. Eduard Seidler (Bonn: Bouvier, 2001), 107–35.
11. Spiegel to Brunauer, February 17, 1940, AAUW Archives, RG Programs, International Relations, War Relief Case Files, Folder 38. In fact, Spiegel had been "temporarily suspended" since January 27, 1939. Gerhard Oberkofler, *Käthe Spiegel: Aus dem Leben einer altösterreichischen Historikerin und Frauenrechtlerin in Prag* (Innsbruck: Studienverlag, 2005), 105.
12. Maas to BFUW, March 26, 1939, WL@LSE, 5BFW (old Box 31), Refugee Cases, L–R, n.p. Atina Grossmann notes that for Jewish women physicians, losing their license to practice was the key factor triggering emigration: "Memoirs suggest that many women doctors remained in Germany for as long as they could practice and fulfill their vocation of helping others and making themselves useful." Atina Grossmann, "New Women in Exile: German Women Doctors and the Emigration," in *Between Sorrow and Strength: Women Refugees of the Nazi Period*, ed. Sibylle Quack (Cambridge, MA: Cambridge University Press, 1995), 215–38, here 228.
13. Seidler, "Lucie Adelsberger, 1945–1971," 115. See also chemist Else Hirschberg's appeal to the AAUW, February 15, 1939. Hirschberg asked about the possibility of reaching the United States outside the quota system, because she was about to lose her job in the medical laboratory of the "German-Israelite Community" hospital in Hamburg, which would shortly be closed down. See AAUW Archives, RG Programs, International Relations, War Relief Case Files, Folder 42.
14. Apt to Hollitscher, August 14, 1939, from London, WL@LSE, 5BFW (old Box 28), BFUW Cases, A–F, n.p. On the question of "domestic permits" in Britain, see Traude Bollauf, *Dienstmädchen-Emigration: Die Flucht jüdischer Frauen aus Österreich und Deutschland nach Großbritannien, 1938/39* (Vienna: LIT, 2010), 109–229.
15. Ehrmann to Brunauer, June 29 and October 21, 1938, from Berlin and May 22, 1939, from Bristol, AAUW Archives, RG Programs, International Relations, War Relief Cases, Folder 9. Georgette Schüler, Katherine Perl, and Rose Bluhm made similar decisions. Erna Barschak and Gertrud Schlesinger even returned to Germany, in 1936 and 1937 respectively, when the opportunity arose for them to find work in the Jewish education system. They emigrated again in 1939. See Christine von Oertzen, *Strategie Verständigung: Zur transnationalen Vernetzung von Akademikerinnen, 1917–1955* (Göttingen: Wallstein, 2012), 293–6.
16. It is very likely that Engelmann knew Heinrich Spiero, the president of the Paulusbund from 1935 to 1937, and that he found her this employment. Like Engelmann, Spiero had made his name as a literary scholar. He had converted to Protestantism and,

again like Engelmann, belonged to the Confessing Church. Werner Cohn, "Bearers of a Common Fate? The 'Non-Aryan' Christian 'Fate Comrades' of the Paulus-Bund, 1933–1939," *Leo Baeck Institute Year Book* 13 (1988): 327–68.

17. Interestingly, in a letter to Esther Brunauer, she stated that the Paulusbund was dissolved in 1937. Engelmann to Brunauer, November 29, 1938, AAUW Archives, War Relief Committee, Individual Cases, Susanne Engelmann, Box 839, Folder 10.
18. CV Engelmann, n.d. [1942], AAUW Archives, War Relief Committee, Individual Cases, Susanne Engelmann, Box 839, Folder 10.
19. Engelmann to Brunauer, November 29, 1938, ibid.
20. Ibid.
21. Statement by Susanne Engelmann on October 26, 1955, in her application for reparations: Erklärung zum Wiedergutmachungsantrag, LAB, B, Rep. 025, no. 8, WGA 1835/50, p. 17.
22. Engelmann to Brunauer, March 31, 1941, AAUW Archives, War Relief Committee, Individual Cases, Susanne Engelmann, Box 839, Folder 10.
23. Werner Röder and Herbert A. Strauss, eds., *Biographisches Handbuch der deutschsprachigen Emigration nach 1933: International Biographical Dictionary of Central European Emigrés 1933–1945* (Munich: Saur, 1983), vol. 2, 477.
24. "I live in Italy for five years now, and during this time I organized and directed independent speech, voice, and hearing services at the Universities of Milano, Genove, and Rome. In Rome I was appointed logopedist to the public schools for abnormal children. A textbook of logopedia which I wrote on request of the greatest italian publishers will be published soon." CV Jellinek n.d. [1938], AAUW Archives, RG Programs, International Relations, War Relief Cases, Folder 21.
25. Jellinek to Brunauer, September 7, 1938, ibid.
26. With Froeschels, Jellinek set up a night-school course for adult stammerers in St. Louis. Jellinek and Froeschels left St. Louis in 1940 to open a speech therapy practice in New York. Froeschels had also been offered a post as director of the speech and voice clinic at Mount Sinai Hospital. "[Obituary] Dr. Auguste Jellinek," *Folia Phoniatrica: International Journal of Phoniatry* 10 (1958): 58.
27. Thompson to Hollitscher, January 24, 1939. Hollitscher passed the letter on to Hinrichsen; Hollitscher to Hinrichsen, January 25, 1939, ibid.
28. Schlesinger to Hollitscher, June 20, 1939, WL@LSE, 5BFW (old Box 30), Folder Schlesinger. Schlesinger's unsatisfactory job lasted only three months, because her employer found the situation equally stressful and said she would prefer an English maid who would not need constant attention. Schlesinger to Hollitscher, August 5, 1939, ibid.
29. See especially Hollitscher's correspondence with the economist Hildegard Grünbaum-Sachs; Oertzen, *Strategie Verständigung*, 301.
30. Prevented by ill health from completing the course, Rosemann found a private placement with a BFUW member who promised her three months' instruction in housekeeping and English cooking. Minutes of the Meeting of the Committee held on Monday, September 23, 1939, WL@LSE, 5BFW (old Box 27), Sub-Committee on Refugees.
31. See the correspondence filed in WL@LSE, 5BFW (old Box 31), Refugee Cases, Folder Bleiberg.
32. Ibid.
33. Reich to Hollitscher, March 2, 1940, WL@LSE, 5BFW (old Box 27), War Refugees, Folder Reich.
34. Quack, ed., *Between Sorrow and Strength*, especially the contributions by Atina Grossmann, Mitchell G. Ash, Joachim Wieler, Frank Mecklenburg, and Catherine Epstein; also Backhaus-Lautenschl a ger, *Das Schicksal deutschsprachiger Emigrantinnen*, 90–7.

35. Atina Grossmann vividly describes the high degree of assimilation demanded by the US health system, so different from the system German women physicians had previously known. It was especially difficult for the women who, before 1933, had been professional pioneers in sexual medicine and social hygiene, and pioneers in private life as working wives. In the United States in the 1940s, such "modern" lifestyles were not yet accepted either on the employment market or in society. Atina Grossmann, "New Women in Exile," 238.
36. Bringing her young baby with her, the Viennese lawyer Friederike Fleischer had first emigrated to Britain, where she was cared for attentively by the Leicester Branch. BFUW, Minutes of the Meeting of the Emergency Sub-Committee on Refugees, September 23, 939, WL@LSE, 5BFW (old Box 27), Sub-Committee on Refugees. For many refugees, after arriving in the United States their contact with the AAUW was limited to a few letters. Appeals by the IFUW also went unanswered, such as the request to look after physician Anne Adler, who was sending severely depressed letters from New York to a friend in Britain. Holmes to Brunauer, n.d. [1941], AAUW Archives, RG Programs, International Relations, War Relief Cases, Folder 45.
37. Katharina Flesch to Brunauer, September 7, 1941, ibid.
38. Miss Shirley Farr, Chairman of the AAUW Committee on Refugee Aid, to Miss Sayre, March 28, 1942, ibid., Folder 48.
39. Erna Barschak, *My American Adventure* (New York: Washburn, 1945); for a more detailed analysis of this book, see Christine von Oertzen, "Rückblick aus der Emigration: Die Akademikerinnen Erna Barschak (1888–1958), Susanne Engelmann (1885–1963) und Lucie Adelsberger (1895–1971)," in *Erinnerungskartelle: Zur Konstruktion von Autobiografien nach 1945*, ed. Angelika Schaser (Bochum: Winkler, 2003), 169–95.
40. "Here, for the first time, I came across a serious problem which puzzles me even now, once in a while. What are the standards by which American employers or American employment agents measure the suitability on an applicant?...In Europe, there are, to be sure, well established standards in all the professions, such as grades in examinations. It makes a difference to have passed one's Ph.D. examination 'magna cum laude.' But a pleasant appearance, a slim figure, a charming personality are of little or no importance....Who cared? Careful grooming, on the other hand, might be considered by a prospective European employer as an indication of other than scholarly interests, of 'worldly inclinations,' interest in hunting for a husband instead of hunting for a job. What did count in the USA?" Barschak, *My American Adventure*, 17, see also 22–5.
41. Ibid., 19.
42. Ibid., 32.
43. Apart from her autobiographical account, Barschak published scholarly work in the mid-1950s. See Erna Barschak, *Today's Industrial Nurse and Her Job: A Study of Functions of Nurses and Their Relationship to Industry* (New York: G. P. Putnam, 1956).
44. See Fürst to Brunauer, October 17, 1940, AAUW Archives, RG Programs, International Relations, War Relief Cases, Folder 17. According to Grossmann, neither Fürst's husband nor the National Committee for Resettlement of Foreign Physicians showed any sympathy for her situation. Grossmann, "New Women in Exile," 215–38, especially 31–2.
45. Minutes of a Meeting of the Refugee Sub-Committtee, March 29, 1940, WL@LSE, 5BFW (old Box 27), Sub-Committee on Refugees.
46. See the Mühsam file, AAUW Archives, RG Programs (Series V), International Relations, War Relief Case Files, Folder 26.
47. Jellinek to Brunauer, October 13, 1939, ibid., Folder 21.

48. Since the mid-1920s, internationally renowned lecturers had been offered one year's accommodation at Hewitt Hall, an "attractive suite of rooms overlooking the Hudson." All the illustrious women professors from Britain, France, Spain, and Turkey who spent a year at Barnard were members of the IFUW and personally known to Gildersleeve. The service for the guest professors included a personal assistant, "familiar with Barnard, who attended to administrative details for her, and escorted her to social functions, interpreting if necessary." Virginia Crocheron Gildersleeve, *Many a Good Crusade: Memoirs* (New York: Macmillan, 1954), 80.
49. On Bieber's first impressions of New York, see Larissa Bonfante, "Margarete Bieber (1879–1978): An Archaeologist in Two Worlds," in *Women as Interpreters of the Visual Arts, 1820–1979*, ed. Claire Richter Sherman and Adele M. Holcomb (Westport, CT: Greenwood Press, 1981), 252.
50. AAUW San José, Mary Diugnan, March 13, 1941, AAUW Archives, War Relief Committee, Individual Cases, Susanne Engelmann, Box 839, Folder 10.
51. Reference from H. H. Fisher, History Department, Stanford University, to Brunauer, February 25, 1942, ibid.
52. Ibid.
53. Engelmann to Brunauer, April 9, 1942, ibid.
54. Engelmann to Sayre, AAUW, May 19, 1942, ibid.
55. Barschak quoted a colleague who had fled Germany: "We people from Germany are not so successful as the Viennese in meeting our new countrymen. When the worst comes to worst, when they are confronted with an emergency, they turn on the charm. It always works!" Barschak, *My American Adventure*, 25.
56. Friedmann to Brunauer, February 4, 1940, AAUW Archives, RG Programs, International Relations, War Relief Case Files, Folder 16.
57. Ibid.
58. Mrs. Paul M. Mitchell, Chicago, to Brunauer, March 3, 1942, ibid., Folder 11.
59. AAUW Chicago to Brunauer, March 3, 1942, ibid.
60. "Dr. Wunderlich Taught in Exile," *New York Times*, December 21, 1965, 21.
61. Drawing on the work of Pierre Bourdieu, Simone Lässig has incisively demonstrated the importance of cultural capital for upward mobility and integration into "majority society" through the example of the Jewish middle class in Germany. Simone Lässig, *Jüdische Wege ins Bürgertum. Kulturelles Kapital und sozialer Aufstieg im 19. Jahrhundert* (Göttingen: Vandenhoeck & Ruprecht, 2004).
62. Susan Levine, *Degrees of Equality: The American Association of University Women and the Challenge of Twentieth-Century Feminism* (Philadelphia, PA: Temple University Press, 1995), 74.
63. Quoted in Hollitscher to Hermes, February 3, 1943, WL@LSE, 5BFW (old Box 21), Hollitscher Personal Files.
64. Marguerite Bowie-Menzler, *Founders of Crosby Hall* (London: Privately published, 1981), 2.
65. Grossmann makes a similar observation for women émigré physicians in the United States. Atina Grossmann, *Reforming Sex: The German Movement for Birth Control and Abortion Reform, 1920–1950* (New York: Oxford University Press, 1995), 166–88. On the intellectual impact of academic immigrants, see Mitchell G. Ash, "Forced Migration and Scientific Change after 1933: Steps Towards a New Overview," in *Intellectual Migration and Cultural Transformation: Refugees from National Socialism in the English-Speaking World*, ed. Edward Timms and Jon Hughes (Vienna, New York: Springer, 2009), 241–63.
66. Klieneberger remained at the Lister Institute for 29 years, until she retired; Klieneberger-Nobel, *Memoirs*. For her scientific achievements, see Ruth Lemcke and L. H. Collier, "Obituary Notice, Emmy Klieneberger-Nobel, 1892–1985," *Journal of Medical Microbiology* 22 (1986): 183–95.

67. Bieber referred to her retirement as "enforced," and evidently experienced it as a blow comparable that of her dismissal in 1933, though it did prompt a surge of productivity. Bonfante, "Margarete Bieber," 257.
68. Röder and Strauss, eds., *Biographisches Handbuch*, vol. 2, 104.
69. C. M. Kauffmann, "Adelheid Heimann (1903–93)," *Burlington Magazine*, October 1993, 694.
70. Heimann died in 1961 during a trip to Italy. Röder and Strauss, *Biographisches Handbuch*, vol. 2, 477.
71. See the correspondence between the vice-chancellor of the University of Ceylon and the BFUW in spring 1949, WL@LSE, 5BFW (old Box 34), Refugees, Ringfolder, BFUW, Cases G–L.
72. See "Women of Attainment: Dr. Gertrude Kornfeld," *Museum Service* (ed. Rochester Museum of Arts and Sciences), March/April 1939: 65; also Angelika Timm, "Zur Biographie jüdischer Hochschullehrerinnen in Berlin bis 1933. Nach Materialien des Archivs der Humboldt-Universität zu Berlin," *Tel Aviver Jahrbuch für deutsche Geschichte* 21 (1992), 252–3; Annette Vogt, "Von Prag in die 'neue Welt': Die Wege der Chemikerin Gertrud Kornfeld," in *1933 Circuli 2003*, ed. Jana Nekvasilová (Prague: National Technical Museum, 2003), 281–97. These authors do not give details of Kornfeld's time in Rochester or her favorable research conditions at Kodak.
73. Minutes of a Meeting of the Sub-Committee on Refugees, July 27, 1944, WL@LSE, 5BFW (old Box 27), Sub-Committee on Refugees. In 1950, Ilse returned to Munich, before leaving again from 1952–55 to teach as a reader in zoology at the Institute of Biology, University in Poona, India. See Annette Vogt, *Wissenschaftlerinnen in Kaiser-Wilhelm-Instituten, A-Z* (Berlin: Archiv der Max-Planck-Gesellschaft, 2008), 62.
74. Rosenau to Hollitscher and Johnston, December 19, 1939, WL@LSE, 5BFW (old Box 31), Package Refugee Cases, L–R.
75. Gordon Higgott, "Helen Rosenau, 1900–1984," *Journal of Jewish Art* 11 (1985): 79–80. See also "Rosenau, Helen," in *Biographisches Handbuch deutschsprachiger Kunsthistoriker im Exil: Leben und Werk der unter dem Nationalsozialismus verfolgten und vertriebenen Wissenschaftler*, ed. Ulrike Wendland (Munich: Saur, 1999), 563–6.
76. Barschak, *My American Adventure*, 31–3.
77. CV Engelmann, n.d. [1945], and other correspondence from the AAUW folder, AAUW Archives, RG Programs, International Relations, War Relief Case Files, Folder 10. On Engelmann's scholarly publications, see Oertzen, "Rückblick aus der Emigration," 180–6, 192–5.
78. List of Engelmann's activities and residences: Auflistung der Tätigkeiten und Wohnorte von Susanne Engelmann, n.d. [1952], LAB, B Rep. 025, no. 81, WGA 3774/55, p. 9.
79. See Maushart, *Hertha Sponer*, 109; Brenda P. Winnewisser, "The Emigration of Hedwig Kohn, Physicist, 1940," *Mitteilungen der Österreichischen Gesellschaft für Wissenschaftsgeschichte* 18 (1998), 56–7. Kohn died in Durham, North Carolina, in 1964, aged 77. Ibid.
80. Leo Baeck Institute, Guide to the Papers of the Muehsam Family, 1828–1999, AR 25021, http://findingaids.cjh.org/?pID=121494 (accessed January 18, 2013).
81. Three other women found a place with the BBC in this way: Johanna Hinrichsen, Eva Reichmann, and Marianne Katz. Hinrichsen to Hollitscher, October 26, 1943, WL@LSE, 5BFW (old Box 28), File Dr. Johanna Hinrichsen, DAB; Reichmann to Hollitscher, June 2, 1939, WL@LSE, 5BFW (old Box 28), Refugee Cases, M–S.
82. Schlesinger to Hollitscher, August 23, October 3, and November 3, 1941, WL@LSE, 5BFW (old Box 30), War Refugees.

83. Minutes of a Meeting of the Sub-Committee on Refugees, July 27, 1944, WL@LSE, 5BFW (old Box 27), Sub-Committee on Refugees.
84. Schlesinger to Hollitscher, February 22, 1947, WL@LSE, 5BFW (old Box 29), Refugee Cases, Folder Schlesinger.
85. Flesch to Brunauer, September 7, 1941, AAUW Archives, RG Programs, International Relations, War Relief Case Files, Folder 15.
86. Emma Reich to Hollitscher, December 10, 1938, WL@LSE, 5BFW (old Box 27), War Refugees, Folder Reich.
87. To be sure, women were even less likely than men to receive invitations from Germany to return to the country. On remigration, see Irmela von der Lühe, Axel Schildt, and Stefanie Schüler-Springorum, eds., *"Auch in Deutschland waren wir nicht wirklich zu Hause": Jüdische Remigration nach 1945* (Göttingen: Wallstein, 2008); Irmela von der Lühe and Claus-Dieter Krohn, eds., *Fremdes Heimatland: Remigration und literarisches Leben nach 1945* (Göttingen: Wallstein, 2005); Alexander von der Borch Nitzling, *(Un)heimliche Heimat: Deutsche Juden nach 1945 zwischen Abkehr und Rückkehr* (Oldenburg: Paulo-Freire-Verlag, 2007).
88. Engelmann to Helen Reid, August 31, 1945, AAUW Archives, War Relief Committee, Individual Cases, Susanne Engelmann, Box 839, Folder 10.
89. Susanne Charlotte Engelmann, *German Education and Re-education* (New York: International University Press, 1945).
90. WL@LSE, 5BFW (old Box 27), Refugees, Folder Schirrmann, Folder Wreschner.
91. Freidenreich states that nearly 90 percent of the 460 women in her sample of Jewish university or college graduates succeeded in emigrating from Germany and other parts of Europe. However, the majority of these were women under 30; only around half of the over-fifty-fives managed to escape. Freidenreich, *Female, Jewish, and Educated*, 175. This is another clue to the effectiveness of the IFUW assistance, which was usually directed at the older women.
92. The book appeared in 1931: Käthe Spiegel, *Kulturgeschichtliche Grundlagen der amerikanischen Revolution* (Munich: Oldenbourg, 1931). Another book of Spiegel's was published in 1936: *Wilhelm Egon von Fürstenbergs Gefangenschaft und ihre Bedeutung für die Friedensfrage 1674–1679* (Bonn: Röhrscheid, 1936). See also Käthe Spiegel, "Die Prager Juden zur Zeit des dreißigjährigen Krieges," in *Die Juden in Prag: Bilder aus ihrer tausendjährigen Geschichte; Festausgabe der Loge Praga des Ordens B'nai B'rith zum Gedenktage ihres 25jährigen Bestandes*, ed. Samuel Steinherz (Prague: Die Bücherstube, 1927), and "Charakterzüge der amerikanischen Geschichte," *Historische Vierteljahreszeitschrift* 28 (1934): 119–41. On the German nationalist and anti-Semitic professors' blocking of her habilitation at Prague University, see Oberkofler, *Käthe Spiegel*, 81–8.
93. Because affidavits were only valid for six months, they could not be issued until it was certain that they would be recognized upon presentation to the American consulate. On US quota policies and the tightening of immigration regulations, see Krohn, "Vereinigte Staaten von Amerika," in *Handbuch der deutschsprachigen Emigration 1933–1945*, ed. Claus-Dieter Krohn et al. (Darmstadt: Wissenschaftliche Buchgesellschaft, 1998).
94. Spiegel to Brunauer, Halberg, and Cram, March 3, 1941, AAUW Archives, RG Programs, International Relations, War Relief Cases, Folder 38.
95. On Cuba as a "staging post," see especially Robert M. Levine, *Tropical Diaspora: The Jewish Experience in Cuba* (Gainesville: University Press of Florida, 1993).
96. Mrs. Woods to Brunauer, June 22, 1942. AAUW Archives, RG Programs, International Relations, War Relief Case Files, Folder 38. In this letter, Woods promised to give a verbal account of her visit to Spiegel; there is no documentation of that conversation. AAUW Archives, RG Programs, International Relations, War Relief Case Files, Folder 38.

97. Hamburg America Line to Brunauer, December 8, 1941, ibid.
98. Gerhard Oberkofler demonstrates that Spiegel was deported to the Łódź Ghetto on September 21, 1941, on one of the first death transports. The precise date and place of her death is unknown. Oberkofler, *Käthe Spiegel*, 110.
99. See Helene Richter to Holmes, February 13, 1939, and Hollitscher to Helene Richter, June 15, 1939, WL@LSE, 5BFW (old Box 28), Refugees, Folder M–S.
100. The manuscript was rediscovered in 1995 in the archives of the Vienna Provincial Library. It was then prepared and printed with funding from the Austrian association of university women. Elise Richter, *Summe des Lebens, hg. vom Verband der Akademikerinnen Österreichs* (Vienna: WUV Universitätsverlag, 1997).
101. Ibid., 221.
102. Helene Adolf, "In Memoriam Elise Richter," *Romance Philology* 1, no. 4 (1948): 339.
103. Hollitscher to Helene Richter, June 15, 1939, WL@LSE, 5BFW (old Box 28), Refugees, Folder M–S. On the Richter sisters' library, see Christiane Hoffrath, *Bücherspuren: Das Schicksal von Elise und Helene Richter und ihrer Bibliothek im Dritten Reich* (Vienna: Böhlau, 2009); Thierry Elsen and Robert Tanzmeister, "In Sachen Elise und Helene Richter: Die Chronologie eines 'Bibliotheksverkaufs,'" in *Geraubte Bücher: Die Österreichische Nationalbibliothek stellt sich ihrer NS-Vergangenheit*, ed. Murray G. Hall, Christina Köstner, and Margot Werner (Vienna: Österreichische Nationalbibliothek, 2004), 128–38.
104. Richter, *Summe des Lebens*, preface.
105. The others were Margot Melchior and Emma Blau-Balint, an insurance agent from Budapest. See Oertzen, *Strategie Verständigung*, 328–9.
106. On Johanna Maas, see Barbara Guttmann, *Den weiblichen Einfluss geltend machen…Karlsruher Frauen in der Nachkriegszeit, 1945–1955* (Karlsruhe: Badenia Verlag, 2000), 60; also Josef Werner, *Hakenkreuz und Judenstern: Das Schicksal der Karlsruher Juden im Dritten Reich* (Karlsruhe: Badenia-Verlag, 1988), 37–8, 154–5, 435. Olga Weiss, who worked at the Rothschild Hospital in Vienna until her deportation in 1942 and was held in Theresienstadt until summer 1945, survived even the last, particularly devastating typhus epidemic. Her husband, once a Viennese colleague of Hollitscher's, died in the camp. Olga Weiss to Hollitscher, November 14, 1950, WL@LSE, 5BFW (old Box 27), Package Old Refugee Cases.
107. Ibid.
108. BFUW to AAUW, International Relations Office, October 18, 1945, AAUW Archives, RG Programs, International Relations, War Relief Case Files, Folder 2.
109. While Adelsberger was in Amsterdam waiting for her US visa, she fought her way back to a sense of normality by writing. On her literary processing of her experiences, see Oertzen, "Rückblick aus der Emigration," 186–91.
110. Eduard Seidler, appendix to Lucie Adelsberger, *Auschwitz: Ein Tatsachenbericht. Das Vermächtnis der Opfer für uns Juden und für alle Menschen*, ed. Eduard Seidler (Bonn: Bouvier, 2001), 107–35. Lucie Adelsberger died in New York in 1971; ibid.

8 Continuity, Memory, and the Cold War

1. Edith C. Batho, *A Lamp of Friendship: A Short History of the International Federation of University Women, IFUW 1918–1968* (London: Sumfield & Day, 1969), 24, 25. Virginia Gildersleeve was the only female US delegate to the founding assembly of the United Nations in San Francisco in June 1945. See Justus D. Doenecke, "Virginia Crocheron Gildersleeve," in *Women in World History*, ed. Anne Commire and Deborah Klezmer (Waterford, CT: Yorkin Publications, 2000), 221–6. On the establishment of the UN and the role of international organisations in UNESCO after 1945, see Akira Iriye, *Global Community: The Role of International Organizations*

in the Making of the Contemporary World (Berkeley: University of California Press, 2002), especially 37–59.

2. The history of the IFUW after 1945 offers a fascinating, and hitherto underexplored, field for researching global networks and knowledge transfer. On the European discourse on equality after 1945, see Celia Donert, "Women's Rights in Postwar Europe: Disentangling Feminist Histories," *Past and Present* 218 (Supplement) (2013): 180–202.
3. In eastern Germany, the organized networking of university-trained women was not resumed until 1990. On the situation in the Soviet occupied zone and the GDR, see Gunilla Budde, *Frauen der Intelligenz: Akademikerinnen in der DDR 1945 bis 1975* (Göttingen: Vandenhoeck & Ruprecht, 2003).
4. Batho, *Lamp of Friendship*, 27.
5. Ibid., 22–3.
6. To this purpose, the AAUW published "names of Federation leaders in the liberated areas, suggesting that packages of food and clothing be sent to them for distribution among university women in their countries. Thousands of such packages were dispatched by our branches in response to this appeal," *AAUW Journal* 41, no. 3 (1947): 173. See also "For Those in Need," with precise instructions of what to include in the packages, *AAUW Journal* 40, no. 4 (1946): 234–5. The fall 1947 issue listed Berta Karlik of the Institute for Radium Research in Vienna under the heading "In Need of Food," *AAUW Journal* 41, no. 1 (1947): 61.
7. See Blanche Hegg-Hoffet, "University Women among the Displaced Persons," *AAUW Journal* 42, no. 2 (1948/49): 106–7. In "Where to Send Parcels," readers were informed: "A list is being prepared of displaced university women who are in camps in Austria, Germany, and Italy. As soon as it has been screened and compiled by the International Refugee Organization, ... it will be available upon request," ibid., 112.
8. On refugee assistance after 1945, see Batho, *Lamp of Friendship*, 24.
9. "AAUW International Students, 1946–1947: The Association Welcomes 37 Women from the Liberated Countries," *AAUW Journal* 40, no. 1 (1946): 25–30.
10. "The International AAUW Students," *AAUW Journal* 41, no. 1 (1947): 26–32; "Study Grants for 1948–49," *AAUW Journal* 42, no. 1 (1948): 35–50.
11. On the US policy of "re-education," see Konrad H. Jarausch, *After Hitler: Recivilizing Germans, 19451999* (Oxford: Oxford University Press, 2006); Ellen Latzin, *Lernen von Amerika? Das US-Kulturaustauschprogramm für Bayern und seine Absolventen* (Stuttgart: Steiner, 2005); Marianne Zepp, *Redefining Germany: Reeducation, Staatsbürgerschaft und Frauenpolitik im US-amerikanisch besetzten Nachkriegsdeutschland* (Göttingen: V&R Unipress, 2007), especially 77–92. See also Donna Harsch, "Public Continuity and Private Change? Women's Conciousness and Activity in Frankfurt, 1945–1955," *Journal of Social History* 27, no. 1 (1993): 29–58.
12. Major Mary S. Bell to AAUW Headquarters, August 13, 1945. AAUW Archives, AAUW War Relief Committee, Individual Cases, Susanne Engelmann, Box 839, Folder 10.
13. Helen Dwight Reid, Associate in International Education, to Major M. Bell, September 11, 1945, AAUW Archives, AAUW War Relief Committee, Individual Cases, Susanne Engelmann, Box 839, Folder 10. Reid compiled a study brochure on Germany for AAUW members. Helen Dwight Reid, *The Postwar Treatment of Germany: A Study Guide* (Washington, DC: American Association of University Women, 1945).
14. Helen Dwight Reid to Major M. Bell, September 11, 1945, AAUW Archives, AAUW War Relief Committee, Individual Cases, Susanne Engelmann, Box 839, Folder 10.
15. On the history of postwar Germany under Allied occupation, see Francis Graham-Dixon, *The Allied Occupation of Germany: The Refugee Crisis, Denazification, and the Path to Reconstruction* (London: Tauris, 2013); Mary Fulbrook, *A History of Germany, 1918–2008: The Divided Nation* (Chichester, West Sussex: Wiley-Blackwell, 2011), chapters 7 and 8.

16. Margaret Ross Jessup, Woodsmall, Ruth Frances, www.anb.org/articles/15/15–00773.html; *American National Biography Online* February 2000 (accessed January 18, 2013).
17. In addition, she arranged for funding of 10,000 deutschmarks to continue the women's bibliography that Agnes von Zahn-Harnack had published in 1934 on behalf of the Weimar DAB. See expression of thanks and obituary of Ruth Frances Woodsmall, *DAB Mitteilungsblatt* no. 20 (September 1963): 11.
18. On Zahn-Harnack's role as president of the BDF in 1933, see chapter 5. During the Nazi period, Zahn-Harnack kept up especially close contact with Anna von Gierke and Isa Gruner. With her sister Elisabet, she regularly invited opponents of the regime to discussions at her home. Her brother Ernst von Harnack was executed by the Nazis in the wake of the July 1944 plot to assassinate Hitler; her cousin Arvid Harnack was killed as a member of the "Rote Kapelle" resistance group. See Hans Mommsen, *Germans Against Hitler: The Stauffenberg Plot and Resistance under the Third Reich* (London: Tauris, 2009).
19. Irene Stoehr, "Traditionsbewußter Neuanfang: Zur Organisation der Frauenbewegung in Berlin, 1945–1950," in *Frauenpolitik und politisches Wirken von Frauen im Berlin der Nachkriegszeit, 1945–1949*, ed. Renate Genth et al. (Berlin: Trafo, 1996), 193–223.
20. Stoehr, "Traditionsbewußter Neuanfang," 200.
21. Agnes von Zahn-Harnack, "...einiges über die Lage der deutschen Frauen, insbesondere der Akademikerinnen," June 21, 1946, LAB, Rep. 235, BFB 3. The report appeared with two years' delay, in 1948, as the first German contribution to the *AAUW Journal*. By this time it no longer reflected the situation in Germany. Agnes von Zahn-Harnack, "Letter from Germany," *AAUW Journal* 42, no. 4 (1948/49): 205–8.
22. Agnes von Zahn-Harnack, "Eröffnungsansprache des Wilmersdorfer Frauenbundes 1945," quoted in Stoehr, "Traditionsbewußter Neuanfang," 200.
23. Agnes von Zahn-Harnack, "...einiges über die Lage der deutschen Frauen, insbesondere der Akademikerinnen," June 21, 1946, LAB, Rep. 235, BFB 3. The following quotations, ibid.
24. On the generation conflict among university women in the early Weimar Republic, see chapter 4.
25. Zahn-Harnack was alluding to Helene Lange's 1915 comments, to stress the special character of the first generation of female students. See Helene Lange, "Zur Bedeutung des Frauenstudiums (1915)," reproduced in *Die Frau* 44 (1937): 589.
26. Hollitscher to Lady Nunburnholme (Agnes von Zahn-Harnack's hostess in London), December 5, 1946, WL@LSE, 5BFW (old Box 28), German Correspondence, Zahn-Harnack.
27. Agnes von Zahn-Harnack to Miss Hensman, Headmistress from Glasgow, October 18, 1947, ibid.
28. Hollitscher to Mrs. Garton, BFUW, June 9, 1947, ibid.
29. This long-awaited meeting proved to be a difficult one; the two women continued to discuss it by letter. See Elvira Scheich, "Science, Politics, and Morality: The Relationship of Lise Meitner and Elisabeth Schiemann," in *Women, Gender, and Science: New Directions*, ed. Sally Gregory Kohlstedt and Helen E. Longino (Chicago, IL: University of Chicago Press, 1997), 143–68.
30. Schiemann to Hollitscher, September 28, 1947, WL@LSE, 5BFW (old Box 28), Schiemann.
31. Hollitscher to Mrs. Garton, BFUW, June 9, 1947, WL@LSE, 5BFW (old Box 28), German Correspondence, Zahn-Harnack. On the policy of the British occupation, see especially Ian D. Turner, ed., *Reconstruction in Post-War Germany: British Occupation Policy and the Western Zones, 1945–1955* (Oxford: Berg, 1989); Alan Brance, *The Cultural Legacy of the British Occupation in Germany* (Stuttgart: Heinz, 1997); Denise

Tscharntke, *Re-educating German Women: The Work of the Women's Affairs Section of British Military Government 1946–1951* (Frankfurt/Main: Lang, 2003).

32. Zahn-Harnack to Miss Hensman, October 18, 1947, WL@LSE, 5BFW (old Box 28), German Correspondence, Zahn-Harnack. Hollitscher passed the letter on to the AAUW; see the note in the bulletin's winter 1947 issue, "Berlin University Women Organize," *AAUW Journal* 41, no. 2 (1947/48), 116.
33. Minutes of the commission's meeting on July 21, 1947, LAB, HLA, B Rep. 236–07.
34. Zahn-Harnack to Hollitscher, Christmas 1949, WL@LSE, 5BFW (old Box 28), German Correspondence, Zahn-Harnack. The extremely misogynistic attempts to push women students out of western German universities after 1945 were part of academia's way of processing the past. In this period, male members of the academic elite tried to distance themselves from Nazism by citing the ideal of Humboldt's university reforms in the early nineteenth century: they blamed the rise of the Nazis on the advent of "mass" access to higher education during Weimar, an accusation that encompassed the opening of the universities to women. See Bernd Weisbrod, "Dem wandelbaren Geist: Akademisches Ideal und wissenschaftliche Transformation in der Nachkriegszeit," in *Akademische Vergangenheitspolitik*, ed. Bernd Weisbrod (Göttingen: Wallstein, 2002), 11–35.
35. Zahn-Harnack to Hollitscher, July 24, 1947, WL@LSE, 5BFW (old Box 28), German Correspondence, Zahn-Harnack.
36. Agnes von Zahn-Harnack, "From a Berlin Hospital," *AAUW Journal* 42, no. 2 (1948), 124–5.
37. Margarete Scherer to Hollitscher, August 16, 1949; December 1, 1949; April 5, 1950, WL@LSE, 5BFW (old Box 34), German Correspondence, Scherer.
38. Scherer to Hollitscher, April 5, 1950, ibid.
39. Hollitscher to Scherer, August 8, 1950, ibid.
40. I will turn to the Oosterbeek meeting of the IFUW, and the heated debate on whether to readmit the DAB, later in this chapter.
41. Kai Arne Linnemann, "Die Wiederkehr des akademischen Bürgers: Herrmann Nohl und die Pädagogik der Sittlichkeit," in *Akademische Vergangenheitspolitik: Beiträge zur Wissenschaftskultur der Nachkriegszeit*, ed. Bernd Weisbrod (Göttingen: Wallstein, 2002), 167–89. On the denazification of West Germany and East Germany more generally, see Dixon, *Allied Occupation of Germany*.
42. Thanks to his connections and recommendations, in 1952 Nohl succeeded in helping his former student Elisabeth Blochmann, who had emigrated to England in 1933, to obtain a full professorship in Marburg. She became one of only three female full professors in West Germany. Linnemann, "Wiederkehr," 175.
43. See the minutes of the constitutive assembly of the Deutscher Frauenbund 1945, LA Berlin, HLA/BFB, no. 10.
44. On the members of the academic commission and their very different experiences during the Nazi era, see Christine von Oertzen, *Strategie Verständigung: Zur transnationalen Vernetzung von Akademikerinnen, 1917–1955* (Göttingen: Wallstein, 2012), 348–51.
45. On Kohlund and her efforts to rebuild women's activism in the French occupation zone, see Grete Borgmann, "Dr. Phil. Johanna Kohlund. Die 'Grande Dame' der frühen Freiburger Frauenbewegung," in *"Nun gehen Sie hin und heiraten Sie!" Die Töchter der Alma Mater im 20. Jahrhundert*, ed. Isolde Tröndle-Weintritt and Petra Herkert (Freiburg: Kore, 1997), 15–43, here 31.
46. See Johanna Kohlund's spring 1946 statement as a defense witness for Maria Plum, StA Freiburg, M2/227, DNZ 21.836.
47. Personal statement by Maria Plum, appendix to her questionnaire of July 1945, StA Freiburg, M2/227.
48. Ibid.

49. Johanna Kohlund's spring 1946 statement as a defense witness for Maria Plum, StA Freiburg, M2/227, DNZ 21.836.
50. Carola Sachse, "Persilscheinkultur: Zum Umgang mit der NS-Vergangenheit in der Kaiser-Wilhelm-Gesellschaft," in *Akademische Vergangenheitspolitik: Beiträge zur Wissenschaftskultur der Nachkriegszeit,* ed. Bernd Weisbrod (Göttingen: Wallstein, 2002), 217–46. Plum's own statement mentions having legally represented several Jews during the Third Reich and enabled them to emigrate. Personal statement by Maria Plum, appendix to her questionnaire of July 1945; also defense witness statements, StA Freiburg, M2/227.
51. Chapter 5 describes the Freiburg group's expulsion of its Jewish members under Plum's presidency in July 1933.
52. Karl S. Bader, "In Memoriam Erica Sinauer," *Savigny: Zeitschrift für Rechtsgeschichte; Germanische Abteilung* 73 (1956), 556–7.
53. Isolde Tröndle-Weintritt, "Die Freiburger Gruppe des Deutschen Akademikerinnenbundes (DAB) und die Freiburger Akademikerinnen (FA)," in *"Nun gehen Sie hin und heiraten Sie!". Die Töchter der Alma Mater im 20. Jahrhundert,* ed. Isolde Tröndle-Weintritt and Petra Herkert (Freiburg: Kore, 1997), 374–85, here 380–4.
54. Barbara Guttmann, *Den Weiblichen Einfluss geltend machen... Karlsruher Frauen in der Nachkriegszeit, 1945–1955* (Karlsruhe: Badenia Verlag, 2000), 61.
55. Gertrud Carl's response to the indictment, December 7, 1946, GLA 465 a/57/7/2/21107, quoted ibid., 62.
56. Ibid., 62–3.
57. Quoted ibid., 63.
58. Ibid.
59. She said that in England she had been viewed as "a foreigner, a German!" and barred from paid employment. She returned to Heidelberg "to my house, my memories, and my friends who stood by me in times of need." Katharina Freifrau von Kuenssberg, *Vom langsamen Aufstieg der Frau: Dr. Katharina Freifrau von Kuenssberg erzählt aus ihrem Leben,* ed. Katharina Holger (Heidelberg: DAB, 1973), 22.
60. Käthe Kuenssberg, "Bericht über die Tätigkeit des deutsch-amerikanischen Frauenclubs," n.d. BArch Koblenz, B 232, no. 14 (Nachlass Kuenssberg).
61. On Strecker's visit to Crosby Hall, see Strecker to Hollitscher, October 20, 1949, WL@LSE, 5BFW (old Box 28), Correspondence with Germans, Strecker.
62. *Mitteilungsblatt des DAB,* no. 62 (1982): 87–9. Together with Nohl, Hakemeyer founded the private evening school Göttingen Institute for Teaching and Education, and set up an exchange program with British and French schools. Ibid.
63. Möller and Lorenz also lived together. List of DAB members working as university and college lecturers, n.d. [1966], BArch Koblenz, B 232, no. 92.
64. For several years Liebenam was also part of the Göttingen chapter's executive. "Lore Liebenam 85 Jahre," *Mitteilungsblatt des DAB,* no. 57 (1980), 100.
65. Christine von Oertzen, "Fräuleingeschichten," in *Geschichte in Geschichten: Ein historisches Lesebuch,* ed. Barbara Duden et al. (Frankfurt/Main: Campus, 2003), 70–7.
66. Anke Sawahn, "Sie half, wo sie konnte: Doktorin Else Wex (1884–1961). Fünf Jahrzehnte Fraueninteressen, Sozialengagement und Politik," *Cellesche Zeitung,* January 30, 1999.
67. "Festansprache von Anna Schönborn in Bremen anläßlich der Gründung des DAB Bremen," *Weser-Kurier,* February 3, 1948.
68. Johanna Lürssen, "Emmy Beckmann im Deutschen Akademikerinnenbund: Emmy Beckmann zum 75. Geburtstag," *Mädchenbildung und Frauenschaffen* 5, no. 4 (1955): 158.
69. "'Die Frau in der Wissenschaft'—Der Deutsche Akademikerinnenbund veranstaltete vom 28.-30. Juli in Stein bei Nürnberg eine Tagung für das gesamte Bundesgebiet und Westberlin," *Informationen für die Frau* 1 (1951): 9.

70. Irma Stoß, "Tagung des Deutschen Akademikerinnenbundes e.V. in Stein bei Nürnberg v. 28.-30.7.1951," *Mädchenbildung und Frauenschaffen* 1, no. 9 (1951): 2.
71. "Mitgliederversammlung des Deutschen Akademikerinnenbundes e.V. in Göttingen," *Mädchenbildung und Frauenschaffen* 2, no. 11 (1952): 5.
72. *Göttinger Tageblatt*, October 6, 1952, BArch Koblenz, B 232, no. 1.
73. For a detailed description of Auguste Hoffmann's difficulties as a professor in East Berlin, see Oertzen, *Strategie Verständigung*, 360–2.
74. *Göttinger Tageblatt*, October 6, 1952, BArch Koblenz, B 232, no. 1.
75. Ella Barowski, "Auguste Hoffmann 85 Jahre alt," *Konsens: Informationen des Deutschen Akademikerinnenbundes* 3, no. 4 (1987): 16.
76. Agnes von Zahn-Harnack, "Ein Jahrzehnt nach Kriegsausbruch," in *Agnes von Zahn-Harnack: Schriften und Reden, 1914–1950*, ed. Marga Anders and Ilse Reicke (Tübingen: Hopfer, 1964), 80–5, here 80.
77. Luise Berthold, *Erlebtes und Erkämpftes: Ein Rückblick* (Marburg: Privately published, 1969), 59.
78. "Rede der Alterspräsidentin des dritten Deutschen Bundestages über den Antisemitismus in der Bundesrepublik, 103. Sitzung am 18. Februar 1960," reproduced in Lüders, *Fürchte Dich nicht*, 227–30.
79. Lise Meitner to Gertrud Schiemann, January 21, 1947, Meitner papers, quoted in Elvira Scheich, "Elisabeth Schliemann (1881–1972): Patriotin im Zwiespalt," in *Autarkie und Ostexpansion: Pflanzenzucht und Agrarforschung im Nationalsozialismus*, ed. Susanne Heim (Göttingen: Wallstein, 2002), 250–79, here 277.
80. Norbert Frei, *Adenauer's Germany and the Nazi Past: The Politics of Amnesty and Integration*, translated by Joel Golb (New York: Columbia University Press, 2002), 305–6.
81. Frei refers to the "political deactualization" of the memory of the victims, which "considerably eased the life of earlier fellow travelers." Ibid., 308.
82. Minutes of the meeting of the academic commission, July 21, 1947, LAB, HLA, B Rep. 236–07. Marie Wreschner was the only one of the renowned Jewish women physicists not to escape from Germany in time. Threatened with deportation, she took her own life in Berlin on November 17, 1941. Annette Vogt, "Ehrendes Gedenken gegen das Vergessen," *Berlinische Monatsschrift* 9, no. 1 (2000): 22.
83. Lise Meitner, "Über einige Probleme in der Ausnützung der Kernenergie," *Mädchenbildung und Frauenschaffen* 8, no. 12 (1958).
84. Otto Hahn, "Mit Lise Meitner von der Holzwerkstatt in Berlin zum Kaiser-Wilhelm-Institut in Dahlem," BArch Koblenz, B 232, no. 1.
85. This applies to the Austrian organization as well as the German one. In the late 1940s, the Austrian university women's association VAÖ did not publish the memoirs of its former president Elise Richter, who had been killed in Theresienstadt, but left the manuscript to languish in the Vienna City and State Library. Elise Richter, *Summe des Lebens, hg. vom Verband der Akademikerinnen Österreichs* (Vienna: WUV Universitätsverlag, 1997), VAÖ's preface, n.p.
86. Kuenssberg, *Vom langsamen Aufstieg der Frau*, 21.
87. Unpublished memoirs of Katharina von Kuenssberg (Bad Wiessee, 1972), reproduced in "Transplant" (unpublished memoirs of Ekkehard von Kuenssberg, 1993), private archive of Dr. Dietlinde Raisig, Ruckersville, Virginia. I would like to thank Luise Hirsch of Heidelberg for forwarding me this document.
88. "Nachruf auf Marga Anders," DAB Gruppe Frankfurt n.d. [1984], Typescript, Papers Frau von Nidda.
89. Käthe Kuenssberg, "Bericht über die Tätigkeit des deutsch-amerikanischen Frauenclubs," n.d. BArch Koblenz, B 232, no. 14 (Nachlass Kuenssberg).
90. For a detailed analysis of this strategy, see Helen Laville, *Cold War Women: The International Activities of American Women's Organizations* (Manchester: Manchester University Press, 2002), especially 103–10.

91. On McCarthyism in Cold War America, see Caroline S. Emmons, ed., *Cold War and McCarthy Era: People and Perspectives* (Santa Barbara, CA: ABC-CLIO, 2010); Ellen Schrecker, *Many Are the Crimes: McCarthyism in America* (Princeton, NJ: Princeton University Press, 1999).
92. Brunauer's activities as secretary of the AAUW International Relations Committee are detailed in chapter 6.
93. Brunauer tried in vain to refute the absurd charges against her through extensive public statements. See "For the Press, Statement for Mrs. Esther Brunauer Before a Subcommittee of the Committee on Foreign Relations of the United States Senate," March 27, 1950, AAUW Archives, Series I, Reel 1/59, E. Brunauer, Old File Folder, Confidential Files. Esther Brunauer never recovered from the calumny to which she and her husband had been subjected, and died in Chicago aged only 57, in June 1959. See her obituary, *Washington Post*, June 28, 1959; also Elizabeth A. Collins, "Red-Baiting Public Women: Gender, Loyalty and Red Scare Politics," PhD diss., University of Illinois at Chicago, 2008, UMI No. 3345170, 155–213.
94. The file "Pink Ladies of the AAUW" is over 100 pages long. AAUW Archives, Series IV, Reel 89/39.
95. It was only the period of affirmative action in the early 1970s that halted this trend. See Margaret W. Rossiter, *Women Scientists in America*, vol. 3: *Forging a New World Since 1972* (Baltimore, MD: Johns Hopkins University Press, 2012).
96. See, for example, Latzin, *Lernen von Amerika?*; Zepp, *Redefining Germany*.
97. On the DAB's association politics after 1945 in general, see Bärbel Maul, *Akademikerinnen der Nachkriegszeit: Ein Vergleich zwischen der Bundesrepublik Deutschland und der DDR* (Frankfurt/Main: Campus, 2002). On the activities of the DAB's academic commission, see Christine von Oertzen, "Luise Berthold: Hochschulleben und Hochschulpolitik zwischen 1909 und 1957," *Feministische Studien: Zeitschrift für interdisziplinäre Frauen- und Geschlechterforschung* 20, no. 1 (2002); also Christine von Oertzen, "Was ist Diskriminierung? Westdeutsche Professorinnen ringen um ein hochschulpolitisches Konzept, 1949–1989," in *Zeitgeschichte als Geschlechtergeschichte: Perspektiven auf die Bundesrepublik (1945–1980)*, ed. Julia Paulus et al. (Frankfurt/Main: Campus, 2012), 101–17.
98. An example is Charlotte Möller, who tried unsuccessfully to find employment at West German universities after being dismissed from the University of Berlin in the eastern sector. Hans-Jürgen Brosin, "Charlotte Möller (1883–1973) und die gewässerkundlichen Arbeiten am Institut für Meereskunde Berlin," *Historisch-meereskundliches Jahrbuch* 6 (1999). The experiences of Charlotte Lorenz, Gertrud Savelsberger, and Else Esdorn were similar to Möller's. For further examples, see Heike Anke Berger, *Deutsche Historikerinnen, 1920–1970: Geschichte zwischen Wissenschaft und Politik* (Frankfurt/Main: Campus, 2007).
99. On the Nazification of the German universities, see Robert P. Ericksen, *Complicity in the Holocaust: Churches and Universities in Nazi Germany* (Cambridge: Cambridge University Press, 2012), 139–67.
100. One proponent of this line of argument was the jurist Wiltraud Krupp von Brünneck, who was appointed to the Federal Constitutional Court in the early 1960s as its second woman member. When a student and trainee, Brünneck had been in close contact with the Nazi Women's League, and called for a National Socialist legal system in which women would have a firm place (see chapter 5). After graduating, she taught at the University of Berlin and then moved to a post in the Ministry of Justice, where, as she put it after 1945, she "hibernated" for the duration of the dictatorship. Claudia Huerkamp reproduces this choice of term in *Bildungsbürgerinnen: Frauen im Studium und in akademischen Berufen 1900–1945* (Göttingen: Vandenhoeck & Ruprecht, 1996), 113.

101. Agnes von Zahn-Harnack, "Geschichte des Deutschen Akademikerinnenbundes, 1926–1933," in *Agnes von Zahn-Harnack: Schriften und Reden, 1914–1950*, ed. Marga Anders and Ilse Reicke (Tübingen: Hopfer, 1964), 6. The following points ibid.
102. This logic recalls the "right to political error" strikingly postulated by Eugen Kogon, "Das Recht auf den politischen Irrtum," *Frankfurter Hefte* 2 (1947): 641–55; on the reception of Kogon and the exoneration of former fellow travelers, see Frei, *Adenauer's Germany and the Nazi Past*, 29–30. Frei refers to Kogon's postulate as a fundamental principle or "Basic Law" of West Germany's politics of the past, ibid., 311.
103. The only exception to this rule is the formulation chosen by Ida Hakemeyer, high school teacher and founder of the Göttingen chapter. Asked about the DAB's past during a 1952 press conference on the association's third annual meeting, shortly to be convened in Göttingen, she answered: "In 1936 the association was dissolved; it had already been Nazified." "Deutsche Akademikerinnen tagen in Göttingen," *Göttinger Tageblatt*, October 6, 1952.
104. Emmy Beckmann to the local associations of university women, July 26, 1949, StA Karlsruhe, 8/StS 20/1181. See also Dorothee von Velsen and Elisabeth Serelmann-Küchler's call for the DAB to be refounded, December 19, 1948, ibid. The report of the DAB's second annual meeting, held in Nuremberg in 1951, notes that the federation "was founded in 1926 and until 1935, the year when it was forcibly dissolved, was presided over by Dr. Agnes von Zahn-Harnack and Dr. Marie-Elisabeth Lüders." Stoß, "Tagung des Deutschen Akademikerinnenbundes 1951," 1.
105. Marga Anders, "Geschichte des DAB: Gedenkrede zum 75; Geburtstag von Agnes von Zahn-Harnack," n.d. [1959], BArch Koblenz, N 1151, no. 268.
106. IFUW, *Report of the 52nd and 53rd Councils, Karlsruhe 1968* (London: The Federation, 1968), 59–60.
107. There is no record of the precise wording of Bowie's statements. According to Magda Staudinger, she said "even more—on this special and solemn occasion!" Staudinger to Maria Schlüter-Hermkes, August 27, 1968, Magda Staudinger papers, file "Zwischenfall Bowie" ("Bowie incident"). This file from the papers of the Freiburg biologist Magda Staudinger (1902–1992) is privately held by Isolde Tröndle-Weintritt. I thank her for allowing me to consult the documents.
108. Ibid.
109. Dr. M. Henrici, Zurich, to Staudinger, October 4, 1968, Magda Staudinger papers, file "Zwischenfall Bowie." Another Swiss participant, in contrast, believed that Bowie had "an extremely positive attitude to Germany," Erna Flamben, Lausanne, to Staudinger, August 18, 1968, ibid.
110. Staudinger to Schlüter-Hermkes, August 27, 1968, Magda Staudinger papers, file "Zwischenfall Bowie." A conversation with Erna Scheffler in London in 1969 also failed to yield results; see minutes of a DAB management meeting on January 25, 1969, Karlsruhe, BArch Koblenz, B 232/84.
111. The "Zwischenfall Bowie" file in Staudinger's papers includes a handwritten note: "This incident was a hate-filled speech against Germany by an English Jew, Mrs. Bowie, during a solemn ceremony! It was indescribably horrible!" Anonymous note, n.d. [c. 1990], Magda Staudinger papers, file "Zwischenfall Bowie."
112. IFUW, *Report of the 31st Council Meeting, Oosterbeek (Holland) 1951* (London: The Federation, 1951), 8.
113. Erna Scheffler to Prof. Steyn-Parvé, February 24, 1969, BArch Koblenz, N 1177, no. 77.
114. Staudinger to Dr. Henrici, Zurich, August 30, 1968, Magda Staudinger papers, file "Zwischenfall Bowie."

115. Isolde Tröndle-Weintritt, "Dr. Phil. Magda Staudinger, geb. 1902," in *"Nun gehen Sie hin und heiraten Sie!" Die Töchter der Alma Mater im 20. Jahrhundert*, ed. Isolde Tröndle-Weintritt and Petra Herkert (Freiburg: Kore, 1997), 78–94, here 86–7.
116. Staudinger to Dr. Henrici, Zurich, August 30, 1968, Magda Staudinger papers, file "Zwischenfall Bowie."
117. Schlüter-Hermkes to Staudinger, August 29, 1968, ibid. Agnes von Zahn-Harnack's one-time employee managed the few surviving items from her papers. Ibid.
118. Staudinger to Elisabeth von Schwartzkoppen, January 23, 1969, ibid.
119. Scheffler to Prof. Steyn-Parvé, February 24, 1969, BArch Koblenz, N 1171, no. 77.
120. Ibid. Staudinger's report was enclosed in Scheffler's letter.
121. Dorothea Frandsen, "Seit einem halben Jahrhundert aktiv: Der Deutsche Akademikerinnenbund," *Informationen für die Frau*, no. 1 (1978): 18–20.
122. Mohl was referring to the growing success of extreme right-wing parties in the Federal Republic. The "Socialist Reich Party" had entered Lower Saxony's regional parliament on May 6, 1951, as the fourth largest group, with 16 delegates. This development was watched with concern from abroad. See Frei, *Adenauer's Germany and the Nazi Past*, 256.
123. IFUW, *Report of the 31st Council Meeting, Oosterbeek (Holland) 1951*, 8.
124. Ibid.
125. Ibid., 9.
126. Kuenssberg to Staudinger, October 23, 1968, Magda Staudinger papers, file "Zwischenfall Bowie."
127. Magda Staudinger to Dorothea Frandsen, February 7, 1978, BArch Koblenz, N 1177, no. 81.
128. On the public discourse on the Holocaust in West Germany, see Wulf Kansteiner, *In Pursuit of German Memory: History, Television, and Politics after Auschwitz* (Athens: Ohio University Press, 2006); Philipp Gassert, *Coping With the Nazi Past: West German Debates on Nazism and Generational Conflict, 1955–1975* (New York: Berghahn, 2006).
129. Christine Eckelmann has made similar observations on the women physicians' association, showing how it retrospectively reinterpreted its Nazification. Eckelmann, *Ärztinnen in der Weimarer Zeit und im Nationalsozialismus: Eine Untersuchung über den Bund Deutscher Ärztinnen* (Wermelskirchen: WFT Verlag für Wissenschaft, Forschung und Technik, 1992), 46–7.
130. "40 Jahre Deutscher Akademikerinnenbund, Gruppe Frankfurt, 1949–1989," duplicated manuscript, n.d. [1989], 3. Privately owned by Gudrun von Nida.
131. Ella Barowski, "Sechzig Jahre Deutscher Akademikerinnenbund," in *Frauen in Wissenschaft und Politik*, ed. Dorothea Frandsen, Ursula Huffmann, and Annette Kuhn (Düsseldorf: Schwann, 1987), 15–22, here 19–20.
132. See Elisabeth Schwarzhaupt, "Der DAB in den 1920er Jahren." Manuscript marking the thirtieth anniversary of the Frankfurt chapter, November 16, 1979, BArch Koblenz, N 1177, no. 81.
133. Only when Barbara Guttmann found these documents in the Karlsruhe municipal archives did their explosive nature become clear, whereupon the Karlsruhe DAB group withdrew them from the archives. Fortunately, the papers are extensively cited by Guttmann, without DAB Karlsruhe's approval. Guttmann, *Weiblicher Einfluss*, 55–60.
134. "Geschichte" page on the DAB website, www.dab-ev.org/index.php?id=93 (accessed January 18, 2013).

Appendix: Biographies

This appendix provides information on 40 key figures mentioned in the book who were members of or associated with the IFUW network. If no other biographical sources are given, the information is drawn from the AAUW and BFUW archives. Notes on further IFUW members, currently a total of 240 people, can be found in the University Women's International Networks Database hosted by the Max Planck Institute for the History of Science, Berlin: http://uwind.mpiwg-berlin.mpg.de/en/home

Adamovicz, Stanislawa, Dr. (1888, ?–October 7, 1969, ?). Biologist, lecturer.

Education: University studies (medicine) in Kraków, Poland.

Career: Lecturer at the State School of Hygiene in Warsaw; editor of the epidemiological journal *Przegląd Epidemioliczny*.

Professional memberships: 1926 founding member of the Polish Federation of Academic Women; from 1930 responsible for the Polish federation's international contacts; 1932–36 second vice president of the IFUW; 1939–47 IFUW president.

Adelsberger, Lucie, Dr. med. (April 12, 1895, Nuremberg–November 2, 1970, New York). Physician.

Education: University studies (medicine) in Erlangen, Germany; 1920 state examination; 1923 doctorate.

Career: 1924–27 pediatrician in Berlin, Wedding district; 1927–33 research fellow at the Robert Koch Institute, Berlin; 1933 dismissed; from 1938 "Krankenbehandlerin," providing medical care for Jews; 1939 failed attempt to emigrate; 1943 deported to Auschwitz and put to work as camp physician; 1945 transferred to Ravensbrück, freed by US Army; 1945 emigration to the United States; 1947–49 tuberculosis specialist at the Montefiore Hospital, New York; from 1949 research fellow at the Montefiore Hospital and Medical Center; 1949 US medical license; from 1949 private allergological practice in New York.

Professional memberships: Until 1933 member of the German Society for Pediatrics (Deutsche Gesellschaft für Kinderheilkunde); 1926 Federation of German Women Physicians (BDÄ), resigned 1930; 1928 Berlin Medical Society (Berliner Medizinische Gesellschaft); Physicians' League of Greater Berlin (Groß-Berliner Ärztebund); 1930 Berlin Society for Internal Medicine and Pediatrics (Verein für innere Medizin und Kinderheilkunde in Berlin); 1956–65 Rudolf Virchow Medical Society.

Main works: Die Verdauungsleukozytose beim Säugling (diss., 1923); *Auschwitz. Ein Tatsachenbericht* (1956; English translation 1995).

Biographical literature: Gerhard Baader, "Lucie Adelsberger: A Forgotten Jewish Pioneer Allergist," *Korot* 12 (1996–97): 137–43; Eduard Seidler, "Lucie Adelsberger," in Lucie Adelsberger, *Auschwitz: Ein Tatsachenbericht. Das Vermächtnis der Opfer für uns Juden und für alle Menschen*, ed. Eduard Seidler (Bonn: Bouvier, 2001), 107–35.

Altmann-Gottheiner, Elisabeth, Dr., née Gottheiner (March 26, 1874, Berlin–October 21, 1930, Mannheim). Lecturer, women's rights advocate. Marries Sally Altmann 1906.

Education: University studies (economics) in London and Berlin, from 1902 in Zurich; 1903 doctorate.

Career: 1908 first woman lecturer at the Commercial College in Mannheim; from 1924 professor extraordinarius.

Professional memberships: 1924 council member, International Council of Women (ICW); 1925 founding member of the German Federation of University Women (DAB); 1926–30 German Association of Women University Lecturers (Verband der Hochschullehrerinnen Deutschlands); 1927 IFUW Fellowships Committee.

Main works: Studien über die Wuppertaler Textilindustrie und ihre Arbeiter in den letzten 20 Jahren (diss., 1903); editor of the periodical *Jahrbuch für Frauenforschung* (from 1912); *Die Berufsaussichten der deutschen Akademikerinnen* (1921).

Biographical literature: Alice Salomon, "Elisabeth Altmann-Gottheiner zum Gedächtnis," *Die Österreicherin* 9 (1930): 4.

Amieux, Anne-Léontine Nicolas (June 18, 1871, Lyon–1960 ?). School principal.

Education: 1883–89 girls' lycée in Lyon and École normale supérieure des jeunes filles de Sèvres; 1892 *agrégation* to teach science; 1898 Promotion Scientifique, Sorbonne, Paris, the ninth woman to achieve this in France; 1905–06 Albert Kahn "Autour du Monde" (Around the World) fellowship.

Career: 1893–98 teacher of natural sciences at a girls' lycée in Tournon-sur-Rhône, Ardèche; 1898–1905 teacher at the lycées Lamartine and Victor Hugo, Paris; 1905–06 Albert Kahn fellowship; 1906–13 teacher of mathematics at Lycée Victor Hugo, Paris; 1913–19 founding principal of the girls' school Lycée Jules Ferry, Paris; 1917 started the first classes preparing women to compete for entry to the École centrale des arts et manufactures; 1919–36 principal of the École normale supérieure des jeunes filles de Sèvres.

Professional memberships: 1906 "Autour du Monde" club; 1919 founding member of the French Association of University Women (AFFDU); IFUW Council member; 1933 Chevalière de la Légion d'honneur.

Biographical literature: "Amieux (Mlle) (Anne-Léontine-Nicolas)," in *Qui est-ce? Ceux dont on parle* (Paris: Lang, Blanchong et Cie, 1934), 20; Nicole Fouché et al., "Soixante-quinze ans d'histoire de l'AFFDU," *Diplômées* 180 (1997): 1–76, here 7; Evelyne Diebolt (ed.), *Dictionnaire biographique: Militer au XXe siècle: Femmes, féminismes, églises et société* (Paris: Michel Houdiard, 2009), 32.

Barschak, Erna, Dr. (1888, Berlin–October 12, 1958, Philadelphia). Vocational education teacher, psychologist, professor.

Education: Vocational training as typist and accountant; 1914 teaching diploma in commerce, Berlin; 1915 final high school examinations (*Abitur*), then university studies (economics, sociology, psychology) in Berlin and Tübingen; 1921 doctorate.

Career: 1908–14 typist and administrative accountant at a factory and a bank; 1914–21 night-school teacher; 1921 teacher at the municipal commercial college and the Pestalozzi-Fröbel-Haus in Berlin; 1922–31 editor of *Die deutsche Berufsschule*, the journal of the German Association for Vocational Education (Deutscher Verein für Berufsschulwesen); 1930–33 professor at the State Institute for Vocational Education (Staatliches Berufspädagogisches Institut), Berlin, dismissed 1933; until 1936 university studies (psychology) in London and Geneva; 1936–38 lecturer at the Jewish Teacher Training School (Jüdische Lehrerbildungsanstalt) in Berlin; 1939 emigration via Britain to the United States; 1941–42 refugee scholar at Wilson College, Pennsylvania, then counselor and consultant at YMCA camps; 1942–53 assistant professor of psychology at Miami University, Oxford, Ohio.

Professional memberships: 1926–33 DAB.

Main works: Die Schülerin der Berufsschule und ihre Umwelt (1926); *Die Idee der Berufsbildung* (1928); publications in *Soziale Praxis* and *Die Frau*; editor of *Wege zur Freude an Werk, Wissen und Welt*, a journal for young women workers published by the German Association for Social Welfare (Deutscher Verein für Soziale Fürsorge), 1922–30; *My American Adventure* (1945); *Today's Industrial Nurse and Her Job* (1956).

Biographical literature: Christine von Oertzen, "Rückblick aus der Emigration: Die Akademikerinnen Erna Barschak (1888–1958), Susanne Engelmann (1885–1963) und Lucie Adelsberger (1895–1971)," in *Erinnerungskartelle: Zur Konstruktion von Autobiografien nach 1945*, ed. Angelika Schaser (Bochum: Winkler, 2003); Martin Kipp, "'Wege zur Freude an Werk, Wissen und Welt'. Notizen zu einer Zeitschrift für weibliche Fortbildungsschuljugend—Zur Erinnerung an die exilierte Berufspädagogin Erna Barschak," in *Historische Berufsbildungsforschung: Beiträge zu einem gemeinsamen deutschen Wissensbereich der Berufs- und Wirtschaftpädagogik*, ed. Karlwilhelm Stratmann (Stuttgart: Steiner, 1992), 99–129, esp. 112–15.

Bäumer, Gertrud, Dr. (September 12, 1873, Hohenlimburg–March 25, 1954, Bethel). Women's rights advocate, politician, writer.

Education: 1888 girl's high school (*Höhere Töchterschule*) in Halle/Saale, then vocational seminar for women teachers (*Lehrerinnenseminar*) in Magdeburg; 1898–1900 seminar for women high school teachers (*Oberlehrerinnenseminar*) in Berlin; 1900–04 university studies (German, theology, philosophy, and sociology) in Berlin; 1904 doctorate.

Career: 1894–98 teacher in Halberstadt, Kamen, and Magdeburg; 1907–10 editor of the ADF journal *Die Neuen Bahnen*; 1914 founder of National Women's Service (Nationaler Frauendienst); 1916–20 director of Hamburg Institute for Social Education (Sozialpädagogisches Institut Hamburg); 1919–20 member of the Weimar National Assembly for the German Democratic Party (DDP); 1920–30 deputy in the German Reichstag and vice president of the DDP; 1922–33 chief executive of the cultural policy department at the Interior Ministry; 1930–32 Reichstag deputy for the German State Party (DStP); 1933 dismissed from civil service; 1933–52 writer of historical novels.

Professional memberships: 1910–19 chair of the Federation of German Women's Associations (BDF); 1918–30 DDP, then until 1933 DStP; 1923 initiator and from 1926 honorary member of the DAB; 1945 founding member of the Christian Social Union party (CSU).

Main works: Editor of *Handbuch der Frauenbewegung*, 5 vols. (1901–6, with Helene Lange), *Die Hilfe. Wochenzeitschrift für Politik, Literatur und Kunst* (1912–40, with Friedrich Naumann), and *Die Frau* (1916–36, with Helene Lange); *Lebensweg durch eine Zeitenwende* (1933); *Die Macht der Liebe: Der Weg des Dante Alighieri* (1941); *Der neue Weg der deutschen Frau* (1946).

Biographical literature: Angelika Schaser, *Helene Lange und Gertrud Bäumer: Eine politische Lebensgemeinschaft* (Cologne: Böhlau, 2000).

Beckmann, Emmy (April 12, 1880, Wandsbek–December 24, 1967, Hamburg). Teacher and schools superintendent (*Oberschulrätin*).

Education: 1906–10 university studies in Paris, Heidelberg, and Göttingen, examination for women high school teachers.

Career: From 1911 state-appointed teacher in Hamburg; 1926 principal of the high school Helene-Lange-Oberrealschule; 1927 superintendent for girl's secondary education, Hamburg; 1933 dismissed from all posts; 1945–49 reappointed as superintendent; 1957 awarded the title of professor by the Hamburg city government.

Professional memberships: 1915 founding member of the Municipal Federation of Hamburg Women's Associations (Stadtbund hamburgischer Frauenvereine); from 1914 member, 1921–33 chair of the General Federation of German Women Teachers; 1921–33 DDP deputy in the Hamburg city parliament; 1926–33 DAB; 1946 founding member of the Hamburg Council of Women (Hamburger Frauenring); 1948 founding member of the Hamburg Association of Academic Women (Hamburger Akademikerinnenverein); 1949–53 chair of the DAB's national council; 1949–57 Free Democratic Party (FDP) deputy in the Hamburg city parliament.

Main works: Helene Lange. Was ich hier geliebt: Briefe (1957).

Biographical literature: Walther Killy and Rudolf Vierhaus, *Deutsche Biographische Enzyklopädie*, vol. 1. (Munich: Saur, 1995), 479; Johanna Lürssen, "Emmy Beckmann im Deutschen Akademikerinnenbund. Emmy Beckmann zum 75. Geburtstag," *Mädchenbildung und Frauenschaffen* 5, no. 4 (1955).

Bowie-Menzler, Jessie Marguerite, née Bowie (? ?–? ?). Married to Frederick Menzler.

Education: 1913 BA (modern history), University of Bristol; 1915 Diploma in Education, Oxford University; 1934 called to the Bar by the Middle Temple, London.

Career: 1917–26 civil servant in the British Ministry of Labour, first implementing the minimum wage, then 1923–26 as trade inspector; 1926–34 insurance agent; from 1934 lawyer; 1939–45 works for the Ministry of Home Security, responsible for the evacuation of children to the United States and Canada, public welfare in London Underground air raid shelters, and welfare arrangements for fire wardens; 1945 works for UNRRA in the reconstruction of the formerly occupied countries and for the dissolution of Displaced Persons camps; 1945 heads British delegation to the UN Commission on Human Rights.

Professional memberships: From 1920 BFUW; founding member of IFUW and 1928–47 IFUW treasurer; 1933–45 active in the BFUW and IFUW's refugee assistance; 1944–76 member of the Crosby Hall Board of Directors; 1947–50 IFUW vice president; 1949 UN Human Rights Commission; 1955–58 BFUW president.

Main works: Founders of Crosby Hall (1981).

Biographical literature: Marguerite Bowie-Menzler, *Founders of Crosby Hall* (London: Privately published, 1981), 28–30; *The Lady's Who's Who* (London: Pallas, 1939), 90.

Brunauer (Caukin-Brunauer), Esther Delia, née Caukin, PhD (July 7, 1901, Jackson, CA–June 26, 1959, Evanston, IL). Marries Stephen Brunauer, chemist, in 1934. Two daughters.

Education: 1924 BA Mills College; 1925 MA Stanford University; 1926–27 AAUW Margaret Maltby scholarship for doctoral studies; 1927 doctorate (European history and international politics), Stanford University; 1932–33 Carl Schurz Foundation scholarship for studies in Berlin.

Career: 1927–44 research associate at the AAUW and secretary to the Committee for International Relations; in this role 1933–44 responsible for AAUW refugee aid; 1944 international security and international organizations consultant in the US State Department; 1946 US representative on the UNESCO preparatory commission; 1952 denounced and suspended by Senator McCarthy; 1953–59 associate director of the Film Council of America and editor of text books in Chicago.

Professional memberships: AAUW; American Historical Association (AHA).

Main works: The National Revolution in Germany 1933 (1933); *National Defense: Institutions, Concepts, and Policies* (1937); *Has America Forgotten? Myths and Facts about World Wars I and II* (1941).

Biographical literature: Durward Howes, *American Women. The Official Who's Who among the Women of the Nation* (Los Angeles: Richard Blank, 1939); Susan Levine, *Degrees of Equality: The American Association of University Women and the Challenge of Twentieth-Century Feminism* (Philadelphia, PA: Temple University Press, 1995), 55–61; Betty Miller Unterberger, "Esther Delia Caukin Brunauer," in *Notable American Women: The Modern Period; A Biographical Dictionary*, ed. Barbara Sicherman and Carol Hurd Green (Cambridge, MA: Harvard University Press, 1980).

Cullis, Winifred Clara, DSc (June 2, 1875, Gloucester–November 13, 1956, London). Physiologist, professor.

Education: 1896–1900 Natural Sciences Tripos as Sidgwick Scholar at Newnham College, Cambridge; 1908 DSc University of London.

Career: 1901 physiology demonstrator at the London School of Medicine for Women; 1903–8 co-lecturer; 1908 head of department; 1912 lecturer and university reader; 1916 visiting professor of physiology at the University of Toronto, Canada; 1919–39 full professor at the London School of Medicine for Women; 1939–45 head of the women's department of the British Information Service in New York City; honorary titles awarded by Vassar College, New York, in 1919, by Goucher College, Maryland, in 1931, and by the University of Birmingham, UK, in 1955.

Professional memberships: 1907 BFUW founding member; 1915 Physiological Society, London; 1919 IFUW founding member; 1924–29 BFUW president; 1929–32 IFUW president.

Main works: The Body and Its Health (1935, with Muriel Bond); *Your Body and the Way It Works* (1949).

Biographical literature: R. E. M. Bowden, "Cullis, Winifred Clara (1875–1956)," *Oxford Dictionary of National Biography* (Oxford University Press, 2004).

Engelmann, Susanne Charlotte, Dr. (September 26, 1886, Berlin–June 26, 1963, Berlin). Philologist, high school teacher, professor.

Education: 1900 completes girls' high school (Höhere Töchterschule) in Berlin, then the Helene Lange private grammar school; preparatory courses for admission to university; 1905 final high school examinations (*Abitur*) taken externally. University studies (German, English, psychology, and education) in Berlin and Heidelberg; 1909 doctorate in Heidelberg; 1910 trainee teacher at girls' schools in Berlin; 1912 state examination and Prussian license to teach in high schools; 1912–13 German Scholar at Bryn Mawr College, Pennsylvania.

Career: 1913–23 high school teacher at the Sophie Charlotte Schule, Berlin; 1923–25 high school teacher at the Municipal Center of Learning (Städtische Studienanstalt) and at a college in Berlin; 1925–28 principal of the Margarethen-Lyzeum and 1928–33 of the Viktoria-Oberlyzeum, both in Berlin; 1933 dismissed; until 1935 private tutor in English, psychology, and education; 1935–37 director of adult education for the St. Paul's League (Paulusbund), Berlin; 1939 emigration with her mother to Istanbul to her brother Konrad Engelmann; lecturer in educational psychology at the YMCA Social Service Center, Istanbul; 1940, after death of her mother, emigration to the United States; 1941–42 faculty member, Adult Education Center San Jose, California; archivist, Hoover Library, San Jose; 1941–42 AAUW scholarship to prepare for the Californian state teaching examination at the Stanford School of Education; 1942–43 visiting scholar at Wilson College, Pennsylvania; 1943 research fellowship of the Institute for International Education and lecturer at the Institute for International Relations in Wichita, Kansas; 1944 lecturer in international relations at Des Moines University, Iowa; 1943–45 visiting lecturer, German department, Smith College, Massachusetts; 1945 visiting lecturer, Weatherford College, Texas; 1946–47 visiting lecturer, foreign languages department, Texas Christian University, Dallas; 1947–51 associate professor, Mary Washington College of the University of Virginia, Fredericksburg; 1948 US citizenship; winter 1951 return to Berlin; 1952 restitution of retirement pension rights, which had been withdrawn in 1933.

Professional memberships: Until 1933 General Federation of Women Teachers and DAB; 1928 International League of Mothers and Women Teachers for the Promotion of Peace.

Main works: Der Einfluss des Volksliedes auf die Lyrik der Befreiungskriege (diss., 1909); *Die Literaturgeschichte im deutschen Unterricht* (1926); *Methodik des deutschen Unterrichts* (1926); *Die Krise der heutigen Mädchenerziehung* (1929); *Deutsche Sprachlehre für Kinder* (1930); *German Education and Re-education* (1945).

Biographical literature: Oertzen, "Rückblick aus der Emigration"; Renate Heuer, *Bibliographia Judaica. Verzeichnis jüdischer Autoren deutscher Sprache*, vol. 1 (Frankfurt/Main: Campus, 1981).

Gildersleeve, Virginia Crocheron, PhD (October 3, 1877, New York City–July 7, 1965, Centerville, MA). Professor, college dean.

Education: 1899 BA Barnard College; 1900 MA in medieval history, Columbia University; 1908 PhD in English studies and comparative literature.

Career: 1908–10 lecturer at Barnard College and graduate program in English at Columbia University, assistant professor of English at Barnard College; 1911–47 dean of Barnard College.

Professional memberships: Association of Collegiate Alumnae (ACA); AAUW founding member; 1924–26 and 1936–39 IFUW president; 1918 chair of the American Council on Education; 1945 US delegate to the San Francisco United Nations Conference on International Organization; 1946 Member of the United States Education Mission to Japan.

Main works: Many a Good Crusade. Memoirs (1954); *A Hoard for Winter* (1962).

Biographical literature: Justus D. Doenecke, "Virginia Crocheron Gildersleeve," in *Women in World History*, ed. Anne Commire and Deborah Klezmer (Waterford, CT: Yorkin, 2000), 221–6; Annette K. Blaxter, "Virginia Crocheron Gildersleeve," in *Notable American Women: The Modern Period; A Biographical Dictionary*, ed. Barbara Sicherman and Carol Hurd Green (Cambridge, MA: Harvard University Press, 1980), 273–5.

Gleditsch, Ellen, Dr. (December 29, 1879, Mandal, Norway–June 6, 1968, Oslo). Chemist, professor. One child (Nils Petter).

Education: 1897 apprenticeship as pharmacist; 1900–02 training in pharmacology and chemistry; 1905 examination in chemistry at the Trondheim Cathedral School; 1907–12 university studies (chemistry, mineralogy) in Paris; 1912 Licenciée des Sciences (Bachelor) in chemistry.

Career: 1903–07 assistant at the chemical laboratory of the University of Oslo; 1907–12 assistant of Marie Curie in Paris; 1912–16 director of a research group on radioactivity, University of Oslo; 1913–14 research under Bertram B. Boltwood, Yale University, where she discovers the radioactive half-life of radium; 1916–29 lecturer in chemistry; 1929–46 professor of inorganic chemistry.

Professional memberships: 1920 founding member and 1924 chair of the Norwegian Federation of University Women; 1926–29 IFUW president; 1962 honorary doctorates from University of Paris and University of Strasbourg; 1964 honorary member of the IFUW.

Main works: Papers on inorganic chemistry and radioactivity in national and international journals, including "Action de l'émanation du radium sur les solutions des sels de cuivre" (1908); "The Life of Radium" (1916).

Biographical literature: Anne-Marie Weidler Kubanek and Grete P. Grzegorek, "Ellen Gleditsch: Professor and Humanist," in *A Devotion to Their Science: Pioneer Women of Radioactivity*, ed. Marelene Rayner-Canham and Geoffrey Rayner-Canham (Montreal: McGill-Queen's University Press, 1997), 51–75; Annette Lykknes, Helge Kragh, and Lise Kvittingen, "Ellen Gleditsch: Pioneer Woman in Radiochemistry," *Physics in Perspective* 6 (2004): 126–55.

Hoffmann, Auguste, Dr. med. (December 2, 1902, Halle/Saale–October 24, 1989, Berlin). Physician, sports physician.

Education: 1922 final high school examinations (*Abitur*) in Berlin; university studies (medicine) in Halle, German College of Physical Education (Deutsche Hochschule für Leibesübungen), medical studies in Halle, Freiburg, and Vienna; 1927 state examination in Halle; 1929 doctorate and license to practice.

Career: 1930–31 German Center for Sport (Deutsches Sportforum Berlin); 1933–36 chief medical consultant, League of German Girls (BDM); 1934–36 physician at the German College of Physical Education; 1938 university lecturer qualification (*Habilitation*); from 1939 lecturer in women's sports medicine; assistant at the University of Berlin's Institute of Anatomy; from 1940 sports physician at the University of Berlin; 1946–51 full professor at Humboldt University, Berlin; 1948 head of anatomy, University of Greifswald, GDR; 1951 resignation of professorship; until 1952 research associate at the Max Planck Institute for Silicate Research; 1953–54 research associate at the Institute of Anatomy, Free University, Berlin; 1955 professor and 1965–69 chair at the Berlin College of Education; 1967 prorector of the Berlin College of Education.

Professional memberships: 1930–36 BDÄ; 1938 Association of German Physicians (Bund Deutscher Ärzte); 1938 Berlin Physiological Society (Berliner Physiologische Gesellschaft); 1946 Wilmerdorf Women's Union (Wilmersdorfer Frauenbund); 1947 Anatomical Society (Anatomische Gesellschaft); 1949 DAB; 1952–56 DAB president.

Main works: Zur Physiologie des Fettgewebes und der Fettablagerung (diss. med., 1929); *Die Bedeutung der Vereine für die sportliche Betätigung der Frau* (1971); *Sportmedizinische Grundlagen zum Leistungssport der Mädchen und Frauen* (1975).

Biographical literature: Ella Barowski, "Auguste Hoffmann 85 Jahre alt," *Konsens: Informationen des Deutschen Akademikerinnenbundes* 3, no. 4 (1987), 16.

Hollitscher, Erna, Dr. (May 1, 1897, Vienna–November 20, 1986, London). Anglicist, refugee worker.

Education: 1914 language studies in Britain; 1918–22 university studies (English, French) in Vienna; 1922 doctorate.

Career: 1922–26 Anglo-Austrian Bank, Vienna; 1926–30 legal department, Österreichische Creditanstalt; 1935–38 international department of the Wiener Automobil-Fabrik and adult education teacher at the Wiener Volksheim; 1938 emigration to Britain, initially guest of BFUW members in Manchester; 1938–54 secretary of the BFUW's Sub-Committee on Refugees.

Professional memberships: VAÖ; BFUW.

Karlik, Berta, Dr. (January 24, 1904, Vienna–February 4, 1990, Vienna). Physicist, professor.

Education: University studies (physics, mathematics) in Vienna; 1928 doctorate.

Career: Trainee teacher of math and physics; 1930–31 Junior Fellowship Holder, BFUW Crosby Hall Residential Fellowship; research in the laboratories of William Briggs, Ernest Rutherford, Marie Curie, the Pasteur Institute, and Louis de Broglie; from 1932 research assistant at the Institute for Radium Research in Vienna; 1934 research at the Oceanographic Commission in Bornö, Sweden; 1937 university lecturer qualification; 1940 assistant, then 1947 director of the Institute for Radium Research in Vienna and professor extraordinarius; 1956 full professor and thus Austria's first woman professor.

Professional memberships: VAÖ.

Main works: Über die Abhängigkeit der Szintillationen von der Beschaffenheit des Zinksulfids und das Wesen des Szintillationsvorganges (diss., 1927); *Tables of Cubic Crystal Structures* (1932, with I. Knaggs); "Uranium in SeaWater" (1935, with F. Hernegger); "Das Element 85 in den natürlichen Zerfallsreihen" (1944, with T. Bernert).

Biographical literature: Brigitte Bischoff, "Karlik, Berta," in *Wissenschafterinnen in und aus Österreich: Leben—Werk—Wirken*, ed. Brigitta Keintzel and Ilse Korotin (Vienna: Böhlau, 2002), 353–4.

Klieneberger-Nobel, Emmy, Dr., née Klieneberger (February 25, 1892, Frankfurt am Main–September 11, 1985, ?). Microbiologist. Marries Edmund Nobel, pediatrician, in 1944.

Education: 1911 women teachers' examination; 1913 final high school examinations (*Abitur*) and university studies (botany, zoology, mathematics, physics) in Göttingen; 1917 doctorate in Frankfurt; 1918 qualification for high school teaching.

Career: 1919–22 high school teacher in Dresden; 1922–30 microbiologist at the municipal Hygiene Institute in Frankfurt; 1930 university lecturer qualification (*Habilitation*); from 1930 lecturer; 1933 forced retirement without pension; 1933 emigration to London, stays at Crosby Hall; 1934 BFUW German Scholar Crosby Hall Residential Fellowship; 1933–62 research at the Lister Institute of Preventive Medicine, London.

Professional memberships: Until 1933 DAB; German Society for Hygiene and Microbiology (Deutsche Gesellschaft für Hygiene und Mikrobiologie); 1967 honorary member of the Robert Koch Institute, 1967 corresponding member of the German Society for Bacteriology and Hygiene (Deutsche Gesellschaft für Bakteriologie und Hygiene).

Main works: Über die Größe und Beschaffenheit der Zellkerne mit besonderer Berücksichtigung der Systematik (diss., 1917); *Pleuropneumonia-like Organisms (PPLO) Mycoplasmataceae* (1962); *Memoirs* (1980).

Biographical literature: Johanna Bleker and Sabine Schleiermacher, *Ärztinnen aus dem Kaiserreich: Lebensläufe einer Generation* (Weinheim: Deutscher Studien-Verlag, 2000) 266–7; Gary E. Rice, "Emmy Klieneberger-Nobel (1892–1985)," in *Women in the Biological Sciences. A Biobibliographic Sourcebook*, ed. Louise S. Grinstein, Carol A. Biermann, and Rose K. Rose (Westport, CN: Greenwood, 1997), 261–5.

Kock (Kock-Lindberg), Karin, Dr. (July 2, 1891, ?–July 28, 1976, Stockholm). Economist, professor; politician, married to Hugo Lindberg.

Education: University studies (economics) in Stockholm; 1929 doctorate.

Career: 1933–45 lecturer; 1945 professor of economics, Stockholm; 1947–49 member of the Swedish government; 1950–57 general director of the Swedish National Office of Statistics.

Professional memberships: Swedish Federation of University Women; 1916–32 chair of the Swedish University Women's Club; 1939–45 IFUW vice president; 1956 American Statistical Association; 1958 International Statistical Institute.

Main works: A Study of Interest Rates (1929); *International Trade and the GATT* (1969).

Biographical literature: Ingvar Ohlsson, "Karin Kock, 1891–1976," *International Statistical Review* 45, no. 1 (1977): 109; "Karin Kock," *Nationalencyklopedin* (1993), vol. 11, 145; Kirsti Niskanen, *Karriär i männens värld: Feministen och ekonomen Karin Kock* (Stockholm: SNS, 2007, with an English summary).

Kohn, Hedwig, Dr. (April 5, 1887, Breslau [Wrocław]–1964, Durham, NC). Physicist.

Education: 1906–13 university studies (physics) in Breslau; 1913 doctorate.

Career: 1914 research assistant, 1915–33 senior assistant in Breslau; 1920–21 scholarship, Kaiser Wilhelm Institute for Physics in Berlin; 1930 university lecturer qualification (*Habilitation*) in Breslau; 1933 banned from teaching; 1933–39 industrial physicist in Breslau (e.g., research for OSRAM); 1939 escape to the United States with the help of her colleagues and the IFUW; 1940–41 lecturer at Greensboro College, North Carolina; 1941–51 professor at Wellesley College, Massachusetts; 1946 US citizenship; 1952 retirement; until 1963 research associate at Hertha Sponer's laboratory, Duke University, Durham, North Carolina.

Professional memberships: DAB.

Main works: Über das Wesen der Emission der in Flammen leuchtenden Metalldämpfe (diss., 1913); *A Study of Optical Cross-Sections of Various Elements Based on Line Intensity and Temperature Measurements in a Flame Source* (1930, with E. Hinnov).

Biographical literature: Brenda P. Winnewisser, "The Emigration of Hedwig Kohn, Physicist, 1940," *Mitteilungen der Österreichischen Gesellschaft für Wissenschaftsgeschichte* 18 (1998); Brenda P. Winnewisser, "Hedwig Kohn—eine Physikerin des zwanzigsten Jahrhunderts," *Physik Journal* 2, no. 11 (2003): 51–5; Annette Vogt, "Hedwig Kohn," in *Einsteins Kolleginnen—Physikerinnen gestern und heute*, ed. Annette Vogt and Cornelia Denz (Bielefeld: TeDiC, 2005), 25–7.

Kornfeld, Gertrud, Dr. (July 25, 1891, Prague–July 4, 1955, Rochester, NY). Chemist.

Education: 1910–15 university studies (chemistry, physical chemistry) in Prague; 1915 doctorate.

Career: 1914–18 demonstrator, from 1915 assistant professor in Prague; 1919–25 unpaid researcher at the Hannover College of Technology; 1925–29 researcher at the Institute of Physical Chemistry, University of Berlin; 1927 Prussian citizenship; 1928 university lecturer qualification in chemistry; 1929 senior assistant at the Institute of Technology, University of Berlin; 1933 banned from teaching and dismissed; 1933 emigration to Britain; 1933–34 research fellow in the University of Nottingham physics department; 1934–35 BFUW German Scholar Residential Fellow at Crosby Hall and research fellow in the department of astronomy, Imperial College, University of London; 1935–36 AAUW scholarship for research in Vienna; 1937 emigration to the United States with AAUW support; 1938–55 scientist at the Research Laboratory Eastman Kodak Co. in Rochester, New York.

Professional memberships: DAB.

Main works: "Über Hydrate in Lösung" (diss., 1915); "Der Wirkungsquerschnitt von Gasmolekülen in der chemischen Kinetik" (1928), "Latent-image distribution by X-ray exposures" (1949).

Biographical literature: Annette Vogt, "Von Prag in die 'neue Welt': Die Wege der Chemikerin Gertrud Kornfeld," in *1933 Circuli 2003*, ed. Jana Nekvasilová (Prague: National Technical

Museum, 2003); "Women of Attainment: Dr. Gertrude Kornfeld," *Museum Service* (ed. Rochester Museum of Arts and Sciences), March/April 1939: 65.

Lüders, Marie-Elisabeth, Dr. (June 25, 1878, Adelsheim–March 23, 1966, Berlin). Economist, politician, women's rights advocate. Engaged until 1907. One child (Hans-Uwe, born 1922).

Education: 1910 final high school examinations (*Abitur*) taken externally, university studies (economics) in Berlin; 1912 doctorate.

Career: 1902–06 Center for Private Social Care (Zentrale für Private Fürsorge); 1914–16 establishes a department of social care in occupied Brussels; 1916–18 director of the division for women's work in the German War Ministry; from 1919 professional politician and until 1932 DDP deputy in the German Reichstag; councilwoman for social affairs, Berlin local government; 1934 banned from public speaking and writing; until 1936 historical research; 1937 short imprisonment by the Gestapo; 1938–44 research on international fisheries development for the Berlin Institute of Oceanography; 1949–51 senator for social affairs, West Berlin; 1953–61 Free Democratic Party deputy in the Bundestag; honorary doctorates from the Free University, Berlin, and University of Bonn.

Professional memberships: 1919 cofounder of the DDP, 1932 resignation; 1923 founder and president of the Union of Women Economists (VdN); 1926 founder and 1930–33 president of the DAB; initiator of the Berlin day home for women students; board member BDF; German delegate to the IFUW, 1932 delegate to the World Disarmament Conference in Geneva, IFUW Committee on Careers for Women, IFUW Committee on the Legal Status of Women; 1946 Liberal Democratic Party of Germany, later FDP, 1957 honorary president; 1947 academic commission of the Wilmersdorf Women's Union (Wilmersdorfer Frauenbund) and co-initiator of the DAB's reestablishment; 1952 Federal German Order of Merit; 1953–61 FDP deputy in the Bundestag; until 1957 honorary president of the Bundestag; 1958 honorary citizen of West Berlin.

Main works: Das unbekannte Heer: Frauen kämpfen für Deutschland 1914–1918 (1936); *Volksdienst der Frau* (1937); *Fürchte Dich nicht: Persönliches und Politisches aus mehr als 80 Jahren, 1878–1962* (1963).

Biographical literature: Helmut Stubbe-da-Luz, "Marie Elisabeth Lüders—erst Wohlfahrtspflegerin, dann Magistratsmitglied," *Das Rathaus* 12 (1984): 715–18; Irene Stoehr, "Marie Elisabeth Lüders," in *Frauenpolitik und politisches Wirken von Frauen im Berlin der Nachkriegszeit 1945–1949*, ed. Renate Genth, Reingard Jäkl, Rita Pawlowski, and Ingrid Schmidt-Harzbach (Berlin: Trafo, 1996): 289–301; Irene Stoehr, "Frauenerwerbsarbeit als Kriegsfall. Marie Elisabeth Lüders: Variationen eines Lebensthemas," in *Frauen arbeiten: Weibliche Erwerbstätigkeit in Ost- und Westdeutschland nach 1945*, ed. Gunilla-Friederike Budde (Göttingen: Vandenhoeck & Ruprecht, 1997); Bärbel Maul, *Akademikerinnen der Nachkriegszeit: Ein Vergleich zwischen der Bundesrepublik Deutschland und der DDR* (Frankfurt/Main: Campus, 2002), 423.

MacLean, Ida Smedley, DSc (June 14, 1877, Birmingham–March 2, 1944, London). Biochemist. Marries Hugh MacLean, professor of medicine, in 1913. Two children.

Education: 1896–99 BA (Natural Sciences Tripos), Newnham College, Cambridge; 1901–3 Bathurst scholarship; 1905 DSc, University of London.

Career: 1904 chemistry demonstrator, Newnham College; 1905 research fellow, Royal Institution (Davy Faraday Laboratory); 1906–10 assistant lecturer in chemistry, Manchester University; 1910–14 Lister Institute of Preventive Medicine (with Beit Memorial Fellowship); 1914–42 research chemist, from 1932 staff member, Lister Institute of Preventive Medicine.

Professional memberships: 1907 cofounder and from 1929–35 president of the BFUW; from 1920 fellow and 1931–34 council member, Chemical Society of London; 1923–36 chair, IFUW Fellowships Committee.

Main works: The Metabolism of Fat (1943).

Biographical literature: BFUW, *History of the British Federation of University Women 1907–1957* (London: BFUW, n.d., c. 1958); Mary R. S. Creese, "Maclean, Ida Smedley (1877–1944)," *Oxford Dictionary of National Biography* (Oxford University Press, 2004); Robin Mackie and Gerrylynn Roberts, *Biographical Database of the Chemical Community, 1880–1970* (Milton Keynes: Open University, 2006).

Maeztu, María de, Dr. (July 18, 1881, Vitoria, Spain–January 7, 1948, Buenos Aires). Educationist, professor.

Education: University studies (education, law) in Salamanca, Madrid, and Marburg, under José Ortega y Gasset, among others; 1919 LLD, Smith College, Massachusetts; 1936 PhD in philosophy and literature, University of Madrid.

Career: 1902 elementary school teacher in Madrid, introduces new methods; 1915 founds a women students' hall, the Residencia Internacional de Señoritas; 1927 visiting professor at Barnard College, New York; 1936 emigration to Argentina and professorship at the University of Buenos Aires.

Professional memberships: cofounder of the Instituto-Escuela de Segunda Enseñanza; 1919 cofounder and chair of the Spanish Federation of University Women; 1926 founder of the Lyceum Club in Spain.

Main works: El trabajo de la mujer. Nuevas perspectivas (1933); *El problema de la ética: la enseñanza de la moral* (1938); *Historia de la cultura europea* (1941); *Antología—siglo XX: prosistas españoles, semblanzas y comentarios* (1948).

Biographical literature: Diccionario de Literatura Española (Madrid: Revista de Occidente, 1949); Antonina Rodrigo, "La pedagoga María de Maeztu," *Tiempo de historia* 4, no. 47 (1978): 62–71; Isabel Pérez-Villanueva Tovar, *Maria de Maeztu. Una Mujer en el reformismo educativo español* (Madrid: Universidad Nacional de Educación a Distancia, 1989).

Matthias, Friederike (March 5, 1890–? ?). Senior high school teacher.

Career: 1933 high school teacher (*Studienrätin*) in Kiel; 1934–38 consultant on women's education for the National Socialist Teachers' League (NSLB); from 1935 senior high school teacher (*Oberstudiendirektorin*), Hindenburgschule girls' grammar school, Kiel.

Professional memberships: 1929 joins Nazi Party; May 18, 1933 elected to DAB executive; May 30, 1934 DAB acting president; from 1935 director of the Reich Federation of German Academic Women (Reichsbund Deutscher Akademikerinnen) until its dissolution in January 1936.

Meitner, Lise (Elise), Dr. (November 7, 1878, Vienna–October 27, 1968, Cambridge, UK). Nuclear physicist, professor.

Education: 1893–95 trains to teach French; 1898–1901 preparation for final high school examinations (*Matura*) taken externally; 1902–06 university studies (physics, mathematics, philosophy) in Vienna; 1906 doctorate.

Career: 1906–07 probationary teacher of mathematics and physics; 1907–12 work with Otto Hahn at the University of Berlin; 1912–15 assistant under Max Planck at the Physics Institute; 1912–38 Kaiser Wilhelm Institute for Chemistry, initially as visiting scholar, from 1914 as research associate and from 1917 as head of department; until 1938 director of the department of physical radioactivity at the Kaiser Wilhelm Institute, 1919 professorial title, 1922 university lecturer qualification (*Habilitation*);

1926 honorary professor, University of Berlin; 1933 banned from teaching; 1938 emigration via the Netherlands and Denmark to Sweden; 1938–47 research position at the Nobel Institute, Stockholm; 1947–53 research professor at the College of Technology, Stockholm; 1953–60 consultant at the Royal Swedish Academy of Engineering Sciences; 1960 moves to Cambridge, UK.

Professional memberships: 1926–38 corresponding member, Academy of Sciences, Göttingen; 1926–33 DAB, active in the Berlin chapter, and German Association of Women University Lecturers (Verband der Hochschullehrerinnen Deutschlands); 1926–32 IFUW Fellowships Committee; 1941 Academy of Sciences Gothenburg; 1945 Academy of Sciences, Stockholm; 1948 Max Planck Society; 1948 corresponding member, German Academy of Sciences (East Berlin); 1954 Otto von Hahn Prize; 1955 Royal Society, London; 1924–48 18 nominations for the Nobel Prize.

Main works: Der Aufbau der Atomkerne. Natürliche und künstliche Kernumwandlungen (1935, with M. Delbrück); "Einige Erinnerungen an das Kaiser-Wilhelm-Institut für Chemie in Berlin-Dahlem" (1954); "Looking Back" (1964).

Biographical literature: Patricia Rife, *Lise Meitner: Ein Leben für die Wissenschaft* (Hildesheim: Claassen, 1992); Patricia Rife, *Lise Meitner and the Dawn of the Nuclear Age* (Boston: Birkhäuser, 1999); Ruth Lewin Sime, *Lise Meitner: A Life in Physics* (Berkeley: University of California Press, 1996); "Lise Meitner," in Annette Vogt, *Wissenschaftlerinnen in Kaiser-Wilhelm-Instituten, A-Z* (Berlin: Archiv der Max-Planck-Gesellschaft, 2008), 97–100; Lore Sexl and Anne Hardy, *Lise Meitner* (Hamburg: Rowohlt, 2002); Ruth Lewin Sime, "Science, 'Race,' and Gender—The Forced Emigration of Lise Meitner and Marietta Blau," *Jahrbuch des Simon-Dubnow-Instituts* 3 (2004): 293–312; Annette Vogt, "Lise Meitner," in *Einsteins Kolleginnen—Physikerinnen gestern und heute*, ed. Annette Vogt and Cornelia Denz (Bielefeld: TeDiC, 2005), 28–33.

Mespoulet, Marguerite, Dr. (December 2, 1880, Paris–January 2, 1965, New York). Professor.

Education: University studies in Paris and London; 1905 *agrégation* to teach English; 1907 Albert Kahn "Autour du monde" fellowship.

Career: 1910 English teacher at the Lycée de Dijon; 1913, with Madeleine Mignon, produces a series of color photographs and a text about rural Celtic culture in Ireland for *Archives de la Planète* by philanthropist and utopian Albert Kahn; 1914–23 English teacher at the Lycée Victor Hugo in Paris; 1923 visiting professor of French literature, Wellesley College, Massachusetts; 1924 Barnard College, New York; 1924–34 associate professor, Wellesley College; 1934–50 professor, Barnard College.

Professional memberships: Founding member of the French Association of University Women (AFFDU); 1926–29 IFUW vice president.

Main works: Creators of Wonderland (1934).

Biographical literature: Nicole Fouché, "Soixante-quinze ans d'histoire de l'AFFDU," *Diplômées* 180 (1997): 1–76, esp. 55; Marguerite Mespoulet Papers (c. 1925–64), Rare Book and Manuscript Library, Columbia University.

Monod, Marie, née Chavannes (1875, Lyon–1967, ?). Marries Octave Monod, physician, in 1902. Two children (Noel, born 1911; Marie-Laure, born 1913).

Education: Lycée Adgar Quintet, Lyon; university studies (initially English, German, later geography, history) in Lyon and Paris; fails twice to achieve the *agrégation* to teach at high school level.

Career: Historical studies in Lyon; contributes to a bibliography of the history of Lyon and the publication of Lambert d'Harpigny's memoirs, specializes in the life and work of the

Comtesse d'Agoult; 1919 moves to Paris; until 1939 editor of the AFFDU bulletin; principal author of the *International Glossary of Academic Terms* issued by the IFUW in 1939.

Professional memberships: 1920 founding member, treasurer, and 1923–33 chair of the Société Nationale Féminine de Rapprochement Universitaire (from 1977 Association Française des Femmes Diplômées des Universités, AFFDU); IFUW delegate; 1932–36 IFUW vice president, IFUW representative at the World Disarmament Conference.

Biographical literature: Geneviève Poujol, *Un féminisme sous tutelle. Les protestantes françaises, 1810–1960* (Paris: Editions de Paris, 2003), 237; Evelyne Diebolt (ed.), *Dictionnaire biographique: Militer au XXe siècle: Femmes, féminismes, églises et société* (Paris: Michel Houdiard, 2009), 241–2.

Puech, Marie-Louise (July 6, 1876, Castres, Tarn–1966, Paris). Germanist. In 1908 marries Jules Puech (1879–1957: jurist, secretary of the European office of the Carnegie Foundation for International Peace, civil servant in the French Ministry of Foreign Affairs).

Education: High school, examination for women teachers (German) in Tarn; 1895 language studies in Britain.

Career: 1897–1900 four attempts to achieve the *agrégation* to teach German; 1900–1908 lecturer in French literature, McGill University, Montreal; 1908 returns to France, during World War I coeditor of the journal *La Paix par le Droit*; 1939 retreat to her estate La Borieblanque in Tarn, which she makes a center of academic refugee assistance; by 1945 has donated more than 274,000 francs to refugee relief for persecuted colleagues from German-occupied areas.

Professional memberships: From 1917 board member, French Council of Women (Conseil National des Femmes Françaises); 1919 French Union for Women's Suffrage (Union Française pour le Suffrage des Femmes), founding member of AFFDU; 1920 general secretary, later president, Women's Union for the League of Nations (Union féminine pour la Société des Nations); from 1924 chair of the IFUW Committee on Intellectual Cooperation; from 1929 responsible for AFFDU international affairs; from 1935 active in the ICW.

Biographical literature: Nicole Fouchet, "Marie-Louise Puech," in *Dictionnaire biographique: Militer au XXe siècle: Femmes, féminismes, Églises et société*, ed. Evelyne Diebolt (Paris: Houdiard, 2009), 266–9.

Reid, Helen Rogers, née Miles Rogers (November 23, 1882, Appleton, WI–July 27, 1970, New York City). Publisher, patron. In 1911 marries Ogden Mills Reid (private secretary to his father Whitelaw Reid, from 1911 owner and publisher of the daily *Herald Tribune*). Three children (including Ogden Rogers Reid, born 1925).

Education: 1903 BA Barnard College (Ancient Greek, Latin, zoology).

Career: 1903–11 private secretary to Mrs. Elizabeth Mills Reid; 1918 advertising solicitor at *Herald Tribune*, soon director of the advertising section; 1921–47 *Tribune*'s vice president and executive director, 1947–55 president; 1948 honorary doctorates from Smith College, Syracuse University, University of Wisconsin, Columbia University, Yale, and The New School for Social Research in New York City.

Professional memberships: AAUW; from 1914 trustee of Barnard College; president of New York Newspaper Women's Club; from 1947 director of Reid Foundation.

Biographical literature: Elizabeth V. Burt, "Helen Rogers Reid (1882–1970)," in *Women in Communication. A Biographical Sourcebook*, ed. Nancy Signorielli (Westport, CN: Greenwood, 1996), 312–20; Ruth Gruber, *Inside of Time. My Journey from Alaska to Israel* (New York: Carroll & Graf, 2003), esp. 146–57; Richard Kluger, *The Paper: The Life and Death of the New York Herald Tribune* (New York: Knopf, 1986).

Richter, Elise, Dr. (March 2, 1865, Vienna–June 21, 1943, Theresienstadt concentration camp). Lecturer.

Education: 1891 auditor at the University of Vienna; 1897 final high school examinations (*Matura*) taken externally; 1897 university studies (classical philology, Indo-European studies, Romance studies) in Vienna; 1901 doctorate.

Career: 1905 university lecturer qualification (*Habilitation*); 1907–21 lecturer in Romance languages and literature; 1921 temporary professor of linguistics and phonetics; 1935 refused full professorship by the University of Vienna; 1938 permission to lecture revoked; 1939 invitation from Dutch university women, which she refuses, deciding not to emigrate; until 1941 IFUW financial support in Vienna; 1942 deportation to Theresienstadt.

Professional memberships: Founding member and from 1922–30 president of the VAÖ; IFUW honorary member.

Main works: Zur Entwicklung der romanischen Wortstellung aus der lateinischen (diss., 1903); "Erziehung und Entwicklung" (1927); *Summe des Lebens* (1940, published 1997).

Biographical literature: Helene Adolf, "In Memoriam Elise Richter," *Romance Philology* 1, no. 4 (1948); Robert Tanzmeister, *Elise Richter—Frau und Wissenschaftlerin*, edited by Verband der Akademikerinnen Österreichs (Vienna: Wiener Universitätsverlag, 1998); Christiane Hoffrath, *Bücherspuren. Das Schicksal von Elise und Helene Richter und ihrer Bibliothek im Dritten Reich* (Vienna: Böhlau, 2009).

Rothbarth, Margarete, Dr. (June 7, 1887, Frankfurt am Main–September 7, 1953, Zurich). Historian.

Education: 1908 final high school examinations (*Abitur*) in Frankfurt, university studies (initially natural sciences, from 1909 history, German, and English) in Heidelberg, Munich, Berlin, and Freiburg; 1913 doctorate in medieval history.

Career: From 1913 teacher at girls' high school in Freiburg, simultaneously (1914–17) research assistant at the German Folk Song Archives, Freiburg (Deutsches Volksliedarchiv Freiburg); 1918 moves to Berlin, private research assistant to Friedrich Naumann, journalist on Naumann's weekly *Die Hilfe*, co-initiator of the German Federation for the League of Nations (Deutsche Liga für Völkerbund) and responsible for international communications, 1919 director of its library and archives; 1922 director of the Reich Ministry of Finance's international archives; 1926–39 director of the German office at the International Institute of Intellectual Cooperation in Paris; 1939 escape to Switzerland; 1946 successful claim against the Federal Foreign Office for compensation of income denied during the Nazi period.

Professional memberships: 1918 German Federation for the League of Nations; 1926 DAB.

Main works: Urban VI und Neapel (diss., 1913); *William Penns Völkerbundplan* (1920); *Die großen Vier am Werk: Beiträge zur Geschichte der Friedenskonferenz* (1921); *Archiv und Bibliothek der Deutschen Liga für Völkerbund* (1921); *Bibliographie zum Vertrag von Versailles* (1925); *Internationale geistige Zusammenarbeit* (1928); *Geistige Zusammenarbeit im Rahmen des Völkerbundes* (1931); *Die deutschen Gelehrten und die internationalen Wissenschaftsorganisationen* (1932); *Bibliographie internationale des travaux historiques publiés dans les volumes de "Mélanges," 1880–1939* (1955).

Biographical literature: Deutsche Biographische Enzyklopädie, 2nd ed., vol. 8 (Munich: Saur, 2007); Hans Manfred Bock, *Topographie deutscher Kulturvertretung im Paris des 20. Jahrhunderts* (Tübingen: Narr, 2010), 28–9; Peter Schöttler, "Margarethe Rothbarth," in *Historikerinnen. Eine biobibliographische Spurensuche im deutschen Sprachraum*, ed. Hiram Kümper (Kassel: Stiftung Archiv der deutschen Frauenbewegung, 2009), 182–3; Ute

Lemke, "Das Pariser Völkerbundinstitut für geistige Zusammenarbeit und die aus Deutschland geflüchteten Intellektuellen," in *Fluchtziel Paris. Die deutschsprachige Emigration 1933–1940*, ed. Anne Saint Sauveur-Henn (Berlin: Metropol, 2002), 51–9; Ute Lemke, "La femme, la clandestine de l'histoire. Margarete Rothbarth—ein Engagement für den Völkerbund," *Lendemains. Etudes comparées sur la France* 37, no. 146/147 (2012): 45–59.

Schaetzel, Mariette (1892, ?–1982, ?). Physician.

Professional memberships: 1923 founding member of the Association Genevoise de Femmes Universitaires and 1924 of the Swiss Federation of Women Academics (SVA); 1924 SVA secretary and 1924–35 president of the Geneva chapter; 1929 IFUW Conference Committee; 1938–41 president of the SVA, establishes an SVA assistance fund for academic refugees.

Biographical literature: Bettina Vincenz, *Biederfrauen oder Vorkämpferinnen? Der Schweizerische Verband der Akademikerinnen (SVA) in der Zwischenkriegszeit* (Baden: hier + jetzt, 2011), 128.

Spiegel, Käthe, Dr. (November 19, 1898, Prague–? ?). Historian.

Education: 1917 final high school examinations (*Abitur*) taken externally, then university studies (history) in Prague; 1919 language certificate in Greek; 1921 doctorate; 1924 international college classes in Vienna; 1926 summer school at the Commission on Intellectual Cooperation in Geneva; 1934–35 Czech language diploma and training in librarianship.

Career: Until 1926 secretary to her father, Prof. Ludwig Spiegel (1864–1926); 1927–29 Rockefeller Foundation Fellow at the Library of Congress, Washington, DC, including stays at the AAUW guest house; 1931, 1933, and 1936 unsuccessful attempts to attain university lecturer qualification in Prague; 1933–34 researcher at the State Archive of Bohemia; 1935–39 researcher at the Prague national and university library; 1939 dismissed; attempt to escape to the United States with AAUW support fails in fall 1941 when the United States enters the war; 1941 deportation to the Łódź ghetto; date of death unknown.

Professional memberships: Czech Federation of University Women, German group; 1935 delegate of the Prague-based German Association for the Advancement of Women (Deutscher Verein Frauenfortschritt) to the IAW congress, Istanbul.

Main works: Vom Karolinum: Ein Beitrag zur Geschichte der Prager Universität (diss., 1923); *Kulturgeschichtliche Grundlagen der amerikanischen Revolution* (1931).

Biographical literature: Guido Kisch, "Kaethe Spiegel 1898–1942," *Historia Judaica* 9 (1947): 193–4; Gerhard Oberkofler, *Käthe Spiegel: Aus dem Leben einer altösterreichischen Historikerin und Frauenrechtlerin in Prag* (Innsbruck: Studienverlag, 2005).

Spurgeon, Caroline Francis Eleanor (October 24, 1869, Punjab, India–October 24, 1942, Tucson, AZ). Professor.

Education: 1898 King's College and University College London; 1899 Oxford University; 1911 doctorate at the University of Paris.

Career: 1899 assistant, tutor, and lecturer at the Association for the Education of Women, Oxford; from 1901 Bedford College, University of London: until 1906 lecturer in English literature, 1906–13 Hildred Carlile Professor of English Literature, 1913–29 department chair; 1920–21 visiting professor, Barnard College, New York; 1929 honorary doctorates from University of London and University of Michigan.

Professional memberships: 1907 founding member and long-serving chair of the BFUW; 1916 Fellow of the Royal Society of Literature; 1918 British Educational Mission to the United States; initiator and 1919–22 president of the IFUW.

Main works: Chaucer devant la critique en Angleterre et en France depuis son temps jusqu'à nos jours (diss., 1911); *Mysticism in English Literature* (1913); *The Privilege of Living in War-Time: An Inaugural Address to King's College for Women* (1914); *Five Hundred Years of Chaucer Criticism and Allusion* (1929); *Shakespeare's Imagery and What It Tells Us* (1935).

Biographical literature: Caroline Spurgeon, "Dr. phil. Caroline F.E. Spurgeon, o. Professor des Bedford College der Universität London (Final Honours English, Oxford; D. Litt., London; Docteur de l'Université Paris; Hon. Litt. D., Michigan)," in *Führende Frauen Europas. Neue Folge*, ed. Elga Kern (Munich: E. Reinhardt, 1930), 37–40; John H. Schwarz, "Spurgeon, Caroline Frances Eleanor (1869–1942)," *Oxford Dictionary of National Biography* (Oxford University Press, 2004); Renate Haas, "Caroline Spurgeon: English Studies, the United States, and Internationalism," *Studia Anglica Posnaniensia: International Review of English Studies* (2002): 1–15.

Staudinger, Magda Mathilde Jenny, Dr., née Woit (August 12, 1902, Uderna, Estonia–April 21, 1997, Freiburg, Germany). Biologist, chemist. Marries Hermann Staudinger, chemist, in 1928. Two children (from Hermann's first marriage).

Education: Final high school examinations (*Abitur*) at the Russian girls' grammar school in Yeysk; 1921–25 university studies (biology) in Berlin; 1925 doctorate and state examination in Riga.

Career: Kindergarten teacher; after completing the state examination, appointed as assistant at the botanical laboratory, University of Riga; 1927–28 visiting scholar at the Biological Institute Helgoland; 1937–44 guest researcher at Hermann Staudinger's chemical laboratory; from 1945 lecture tours.

Professional memberships: 1949 founding member and until 1957 chair of the Freiburg University Women (Freiburger Akademikerinnen); 1959–68 IFUW Fellowships Committee; 1961 DAB representative on the German Commission for UNESCO; 1966–80 German delegate to the UNESCO General Conferences; 1968 committee investigating the DAB's history in the Nazi period.

Main works: Umgestaltungen an Blattgeweben infolge des Wundreizes (diss., 1925); *Das wissenschaftliche Werk von Hermann Staudinger*, 7 vols. (1969–70).

Biographical literature: Isolde Tröndle-Weintritt, "Dr. phil. Magda Staudinger, geb. 1902," in *"Nun gehen Sie hin und heiraten Sie!" Die Töchter der Alma Mater im 20. Jahrhundert*, ed. Isolde Tröndle-Weintritt and Petra Herkert (Freiburg: Kore, 1997).

Szagunn, Ilse, Dr. med. (September 16, 1887, Berlin–March 10, 1971, Berlin). Physician. Marries Walter Szagunn, jurist and bank manager, in 1914. Two children (Volkhard, born 1923; Helfried, 1925–45).

Education: 1912 state examination in medicine in Berlin; 1913 license to practice and doctorate.

Career: 1914–27 director of the mother-and-baby advisory office in Charlottenburg, Berlin; school physician at municipal high schools in Charlottenburg, physician at vocational schools; 1917–25 private practice; 1919–31 lecturer at Alice Salomon's Women's School for Social Work (Soziale Frauenschule), the Charlottenburg Women's School (Frauenschule Charlottenburg), and the School for Public Welfare of the Protestant Home Mission (Wohlfahrtsschule der Inneren Mission), Berlin; 1927–43 director of the Protestant marriage counseling center in Friedenau, Berlin; 1940 admitted to the Nazi Office of Public Health (Amt für Volksgesundheit); 1941–44 editor of the journal *Die Ärztin*; 1945–67 private practice in Lichterfelde, Berlin.

Professional memberships: From 1909 German Academic Women's Union (Deutsch-Akademischer Frauenbund); 1926 president of the German League of University Women's Associations (DVAF); 1924–36 BDÄ; 1925–28 German People's Party (DVP) deputy in the district assembly; 1926–33 DAB vice president; 1953 Federal German Order of Merit, honorary member of the reestablished BDÄ; 1961 honorary member of the Berlin Medical Society (Berliner Medizinische Gesellschaft).

Main works: "Wandlungen in den geistigen Grundlagen der Familie" (1929); "Psychotherapie und Religion" (1964); "Mensch und Automation" (1965).

Biographical literature: Bleker and Schleiermacher, *Ärztinnen aus dem Kaiserreich*, 296–7; Louisa Sach, "'Gedenke, daß du eine deutsche Frau bist': Die Ärztin und Bevölkerungspolitikerin Ilse Szagunn (1887–1971)," PhD diss., University of Berlin (Charité Universitätsmedizin), 2006.

Thomas, Martha Carey, Dr. (January 2, 1857, Baltimore–December 2, 1935, Philadelphia). College president, professor.

Education: 1877 BA Cornell University; Johns Hopkins University; 1879–81 university studies (philology) in Leipzig; 1882 doctorate in Zurich.

Career: 1884–94 dean and professor of English language at the newly established Bryn Mawr College, 1894–22 president; 1919–20 sabbatical and travels through France and Spain to recruit members for the IFUW.

Professional memberships: Founding member, Naples Table Association for Promoting Scientific Research by Women; president, National College Equal Suffrage League; 1917–18 chair, ACA War Service Committee; IFUW cofounder, member of the Crosby Hall fund-raising committee.

Main works: Sir Gawayne and the Green Knight (1887); *Education of Women* (1900).

Biographical literature: Helen Lefkowitz Horowitz, *The Power and Passion of M. Carey Thomas* (New York: Knopf, 1994).

Westerdijk, Johanna, Dr. (January 4, 1883, Amsterdam–November 15, 1961, ?). Botanist, professor.

Education: 1900–05 university studies (botany) in Amsterdam, Munich, and Zurich; 1905 doctorate.

Career: 1906 director of the W. C. Scholten Laboratory; 1907 receives collection of 80 fungus cultures, establishes Central Bureau of Fungus Cultures with more than 8,000 species; 1917 professor extraordinarius of plant pathology, University of Utrecht, 1930 appointment to the University of Amsterdam; 1957 honorary doctorate, University of Utrecht; 1958 honorary doctorate, University of Giessen.

Professional memberships: 1933–36 IFUW president; Royal Netherlands Academy of Sciences.

Main works: Zur Regeneration der Laubmoose (diss., 1905).

Biographical literature: L. C. P. Kerling and J. G. ten Houten, "Johanna Westerdijk: Pioneer Leader in Plant Pathology," *Annual Review of Phytopathology* 24 (1986): 33–41; Marie P. Löhnis, "Professor Dr Johanna Westerdijk 1917–1942," *Antonie van Leeuwenhoek* 8, no. 1 (1942): 1–9.

Willich, Johanna (? ?–? ?). High school teacher.

Career: High school teacher (*Studienrätin*) in Berlin; 1928–29 teaches in St. Louis, Missouri, as part of an IFUW international teacher exchange program.

Professional memberships: German National People's Party (DNVP); until 1934 executive of DAB's Berlin chapter; DAB president following Nazification in May 1933.

Zahn-Harnack, Agnes von, Dr., née von Harnack (June 19, 1884, Giessen–May 22, 1950, Berlin). Historian, writer. Married to Dr. Karl von Zahn (1877–1944), jurist and senior civil servant. Two children (Edward, born 1921; Margarete, born 1924).

Education: 1903 women teachers' examination; 1908 university studies (theology, German, English) in Berlin; 1912 doctorate in Greifswald.

Career: 1903–14 schoolteacher; from September 1914 National Women's Service (Nationaler Frauendienst); 1916 directs women's division of the Central War Office; 1918–24 returns to teaching; 1927–32 historical studies; editor of the *Bibliographie zur Frauenfrage* in cooperation with the Prussian State Library; 1933 resignation from all posts; writer; 1945 initiates reestablishment of Berlin women's associations; 1949 honorary doctorate, University of Marburg.

Professional memberships: DAB founding member, 1926–30 DAB president; from 1928 German National Academic Foundation (Studienstiftung des Deutschen Volkes); 1931–33 chair of BDF; Protestant Social Association (Evangelisch-Soziale Vereinigung); 1919–33 DDP; until 1934 IFUW delegate and member of the IFUW Conference Committee; 1945 founding member of the Wilmersdorf Women's Union (Wilmersdorfer Frauenbund, later Berlin Women's Union 1945); initiates academic commission of the Wilmersdorfer Frauenbund and reestablishment of the DAB; 1949 founding member of the Working Group of Berlin Women's Associations (Arbeitsgemeinschaft Berliner Frauenverbände).

Main works: Die arbeitende Frau (1924); *Die Frauenbewegung. Geschichte, Probleme, Ziele* (1928); *Die Frauenfrage in Deutschland* (1934); *Adolf von Harnack* (1936); "Geschichte des Deutschen Akademikerinnenbundes 1926–1933" (1948).

Biographical literature: Irene Stoehr, "Agnes von Zahn-Harnack," in *Frauenpolitik und politisches Wirken von Frauen im Berlin der Nachkriegszeit 1945–1949*, ed. Renate Genth, Reingard Jäkl, Rita Pawlowski, and Ingrid Schmidt-Harzbach (Berlin: Trafo, 1996), 348–58; Hans Cymorek and Friedrich Wilhelm Graf, "Agnes von Zahn-Harnack (1884–1950)," in *Frauenprofile des Luthertums. Lebensgeschichten im 20. Jahrhundert*, ed. Inge Mager (Gütersloh: Gütersloher Verlags-Haus, 2005), 202–51; Gisa Bauer, *Kulturprotestantismus und frühe bürgerliche Frauenbewegung in Deutschland: Agnes von Zahn-Harnack (1884–1950)* (Leipzig: Evangelische Verlagsanstalt, 2006).

Sources and Bibliography

Archival Sources

AAUW Archives, Washington DC
Series IX, Reel 150/35

Archiv der Humboldt-Universität Berlin (University Archives, Humboldt University Berlin)
Berufungsakten
Personalakten

Archiv der Universität Marburg (University Archives, Marburg University)
Personalakten Luise Berthold

Archives nationales de France, Site de Fontainebleau
Archives AFFDU

BFUW Archives, The Women's Library @ LSE, London
5BFW: Records of the British Federation of University Women

BFUW News Sheet: WL@LSE, printed collections

Bundesarchiv Berlin (German Federal Archives, Berlin)
BDC Mitgliederkartei der NSDAP
NS 12 NS Lehrerbund
NS 44 Deutsches Frauenwerk

Bundesarchiv, Abteilung Koblenz (BArch Koblenz) (German Federal Archives, Koblenz)
B 232 Geschäftsakten Deutscher Akademikerinnenbund
B 232/14 Nachlass Katharina von Kuenssberg
N 1151 Nachlass Marie-Elisabeth Lüders
N 1177 Nachlass Elisabeth Schwarzhaupt

International Information Centre and Archives for the Women's Movement (IIAV), Amsterdam
IFUW Archives

Landesarchiv Berlin (LAB) (Berlin State Archive)
HLA Helene-Lange-Archiv der deutschen Frauenbewegung
Entschädigungsakten

League of Nations Archives, Geneva
IFUW
Répertoire general, 1919–1946, 2ième volume
Section des bureaux internationales et de la coopération intellectuelle

Politisches Archiv des Auswärtigen Amtes (Political Archive, Federal Foreign Office)
Gz 117–251.070

Stadtarchiv Freiburg im Breisgau (Freiburg municipal archives)
M 2/227 Entnazifizierungsakten

Bryn Mawr College Archives
Carey M. Thomas Papers

Columbia University, Rare Book Collection
Virginia Gildersleeve Collection

Barnard College Archives
Virginia Gildersleeve Papers

RHUL Archives, Royal Holloway, University of London
Caroline Spurgeon Papers

Papers held privately
Magda Staudinger papers, privately held by Isolde Tröndle-Weintritt

Bibliography

Adams, Pauline. *Somerville for Women: An Oxford College, 1879–1993* (Oxford: Oxford University Press, 1996).

Adelsberger, Lucie. *Auschwitz: Ein Tatsachenbericht. Das Vermächtnis der Opfer für uns Juden und für alle Menschen*. Edited by Eduard Seidler (Bonn: Bouvier, 2001).

Adolf, Helene. *Gesammelte Schriften*, ed. Renate Heuer and Michael Dallapiazza (Triest: Edizioni Parnaso, 2004).

Adolf, Helene. "In Memoriam Elise Richter," *Romance Philology* 1, no. 4 (1948): 338–41.

Albisetti, James C. "American Women's Colleges through European Eyes, 1865–1914," *History of Education Quarterly* 32, no. 4 (1992): 439–58.

Albisetti, James C. *Schooling German Girls and Women: Secondary and Higher Education in the Nineteenth Century* (Princeton, NJ: Princeton University Press, 1988).

Albrecht, Nele. "Gender History in a Transnational Perspective." http://h-net.msu.edu/cgi-bin/logbrowse.pl?trx=vx&list=h-soz-u-kult&month=0712&week=b&msg=LnsuGFwJSC4xvTgbAez37w&user=&pw= (accessed January 18, 2013).

Allen, Keith R. *Überprüfung, Befragung, Kontrolle: Die Aufnahme von DDR-Flüchtlingen in West-Berlin bis 1961* (Berlin: Christoph Links, 2013).

Alp, Emma. "Meinungsaustausch. Ein internationaler Studentinnenkongress in Kristiania," *Die Studentin* 1, no. 1 (1924): 14–15.

Alp, Emma. "Meinungsaustausch. Noch etwas über den Internationalen Verband Akademischer Frauen," *Die Studentin* 1, no. 12 (1924): 65.

Alp, Emma. "Völkerbund und Pazifismus," *Die Studentin* 2, nos. 1/2 (1924): 3–6.

Alter, Peter. *Wissenschaft, Staat, Mäzene: Anfänge moderner Wissenschaftspolitik in Großbritannien 1850–1920* (Stuttgart: Klett-Cotta, 1982).

Altmann-Gottheiner, Elisabeth. *Die Berufsaussichten der deutschen Akademikerinnen* (Halle: Niemeyer, 1921).

Altmann-Gottheiner, Elisabeth. "Die Krisis im Vereinsleben. Gedanken einer Schatzmeisterin," *Die Frau* 29 (1921/22): 231–5.

Anheier, Helmut, Marlies Glasius, and Mary Kaldor, eds. *Global Civil Society* (Oxford: Oxford University Press, 2001).

Arwill-Nordbladh, Elisabeth. "Hanna Rydh. Eine Forscherin mit vielen Dimensionen." In *Eine Dame zwischen 500 Herren: Johanna Mestorf, Werk und Wirkung*, edited by Julia K. Koch and Eva-Maria Mertens (Münster: Waxmann, 2002), 281–94.

Ash, Mitchell G. "Forced Migration and Scientific Change after 1933: Steps towards a New Overview." In *Intellectual Migration and Cultural Transformation: Refugees from National Socialism in the English-Speaking World*, edited by Edward Timms and Jon Hughes (Vienna: Springer, 2009), 241–63.

Ashburner, Catherine. "Shirley Farr." In *Those Intriguing Indomitable Vermont Women*, edited by Vermont State Division of the American Association of University Women (n.p., 1999), 44–5.

Auswärtiges Amt, ed. *Biographisches Handbuch des Auswärtigen Dienstes 1871–1945* (Paderborn: Schöningh, 2000).

Backhaus-Lautenschläger, Christine. *"...Und standen ihre Frau": Das Schicksal deutschsprachiger Emigrantinnen in den USA nach 1933* (Pfaffenweiler: Centaurus-Verlagsgesellschaft, 1991).

Bader, Karl S. "In Memoriam Erica Sinauer," *Savigny. Zeitschrift für Rechtsgeschichte. Germanische Abteilung* 73 (1956): 556–7.

Baecker-Vowinkel, Elisabeth. "8. Schulungslehrgang für Ärztinnen in Alt-Rehse v. 5.-15.7.1939," *Die Ärztin* 15 (1939): 260–3.

Banaszak, Lee Ann. *The U.S. Women's Movement in Global Perspective* (Lanham, MD: Rowman & Littlefield, 2006).

Bard, Christine. "Feministinnen in Frankreich. Frauenstimmrecht und Frieden, 1914–1940." In *Feminismus und Demokratie: Europäische Frauenbewegungen der 1920er Jahre*, edited by Ute Gerhard (Königstein: Helmer, 2001), 84–103.

Bard, Christine. *Les femmes dans la société francaise au 20e siècle* (Paris: Colin, 2001).

Barowski, Ella. "Auguste Hoffmann 85 Jahre alt," *Konsens: Informationen des Deutschen Akademikerinnenbundes* 3, no. 4 (1987): 16.

Barowski, Ella. "Sechzig Jahre Deutscher Akademikerinnenbund." In *Frauen in Wissenschaft und Politik*, edited by Dorothea Frandsen, Ursula Huffmann, and Annette Kuhn (Düsseldorf: Schwann, 1987), 15–22.

Barschak, Erna. *My American Adventure* (New York: Washburn, 1945).

Barschak, Erna. *Today's Industrial Nurse and Her Job: A Study of Functions of Nurses and Their Relationship to Industry* (New York: G. P. Putnam, 1956).

Baten, Jörg, and Andrea Wagner. *Autarchy, Market Disintegration, and Health: The Mortality and Nutritional Crisis in Nazi Germany, 1933–1937* (Munich: CES, 2002).

Batho, Edith C. *A Lamp of Friendship: A Short History of the International Federation of University Women, IFUW 1918–1968* (London: Sumfield & Day, 1969).

Bauer, Gisa. *Kulturprotestantismus und frühe bürgerliche Frauenbewegung in Deutschland: Agnes von Zahn-Harnack (1884–1950)* (Leipzig: Evangelische Verlagsanstalt, 2006).

Bäumer, Gertrud. "Die Akademikerin und die Volkskultur. Zur Gründung des Verbandes Deutscher Akademikerinnen," *Die Frau* 33 (1926): 513–17.

Bäumer, Gertrud. "Berufsorganisationen und Frauenbewegung," *Die Frau* 32 (1925): 97–100.

Bäumer, Gertrud. "Der Frauenweltbund im Haag," *Die Frau* 29 (1921/22): 264–70.

Bäumer, Gertrud. "Die Not der geistigen Arbeiterin," *Die Frau* 30 (1923): 204–10.

Bäumer, Gertrud. "Prinzipienfragen des Frauenweltbundes," *Die Frau* 28 (1920): 1–4.

Behr, Tatjana. "Der Deutsche Verband Akademischer Frauenvereine," *Die Studentin* 2, nos. 9/10 (1925): 67–70.

Belser, Katharina. *Ebenso neu als kühn: 120 Jahre Frauenstudium an der Universität Zürich* (Zurich: eFeF-Verlag, 1988).

Bender, Thomas. *Rethinking American History in a Global Age* (Berkeley: University of California Press, 2002).

Berger, Heike Anke. *Deutsche Historikerinnen, 1920–1970: Geschichte zwischen Wissenschaft und Politik* (Frankfurt/Main: Campus, 2007).

Berghahn, Marion. *German-Jewish Refugees in England: The Ambiguities of Assimilation* (New York: St. Martin's Press, 1984).

Berthold, Luise. *Erlebtes und Erkämpftes: Ein Rückblick* (Marburg: Privately published, 1969).

Bieler, Katharina. *Im preußischen Schuldienst: Arbeitsverhältnisse und Berufsbiographien von Lehrerinnen und Lehrern in Berlin-Schöneberg 1871–1933* (Cologne: Böhlau, 2007).

Bischof, Brigitte. "*...Junge Wienerinnen zertrümmern Atome....*" *Physikerinnen am Wiener Institut für Radiumforschung* (Mössingen: Talheimer-Verlag, 2004).
Blaxter, Annette K. "Virginia Crocheron Gildersleeve." In *Notable American Women: The Modern Period; A Biographical Dictionary*. Edited by Barbara Sicherman and Carol Hurd Green (Cambridge, MA: Harvard University Press, 1980), 273–5.
Bleker, Johanna. "Anerkennung durch Unterordnung? Ärztinnen und Nationalsozialismus." In *Weibliche Ärzte. Zur Durchsetzung eines Berufsbildes*, edited by Eva Brinkschulte (Berlin: Edition Hentrich, 1995), 126–35.
Bleker, Johanna, and Christine Eckelmann. "'Der Erfolg der Gleichschaltungsaktion kann als durchschlagend bezeichnet werden.' Der 'Bund Deutscher Ärztinnen' 1933–1936." In *Medizin im Dritten Reich*, edited by Johanna Bleker and Norbert Jachertz (Cologne: Deutscher Ärzte-Verlag, 1989), 87–96.
Bleker, Johanna, and Sabine Schleiermacher. *Ärztinnen aus dem Kaiserreich: Lebensläufe einer Generation* (Weinheim: Deutscher Studien-Verlag, 2000).
Bolbecher, Siglinde, and Konstantin Kaiser. *Lexikon der österreichischen Exilliteratur* (Vienna: Deuticke, 2000).
Bollauf, Traude. *Dienstmädchen-Emigration: Die Flucht jüdischer Frauen aus Österreich und Deutschland nach Großbritannien, 1938/39* (Vienna: LIT, 2010).
Bonfante, Larissa. "Margarete Bieber (1879–1978). An Archaeologist in Two Worlds." In *Women as Interpreters of the Visual Arts, 1820–1979*, edited by Claire Richter Sherman and Adele M. Holcomb (Westport, CT: Greenwood Press, 1981), 239–74.
Borch-Nitzling, Alexander von der. *(Un)heimliche Heimat: Deutsche Juden nach 1945 zwischen Abkehr und Rückkehr* (Oldenburg: Paulo-Freire-Verlag, 2007).
Borgmann, Grete. "Dr. Phil. Johanna Kohlund. Die 'Grande Dame' der frühen Freiburger Frauenbewegung." In *"Nun gehen Sie hin und heiraten Sie!" Die Töchter der Alma Mater im 20. Jahrhundert*, edited by Isolde Tröndle-Weintritt and Petra Herkert (Freiburg: Kore, 1997), 15–43.
Bosanquet, Theodora. "Der Internationale Akademikerinnen-Verband," *Die Studentin* 2, nos. 1/2 (1925): 7.
Bosch, Mineke. *Het Geslacht van het Wetenschap* (Amsterdam: Sua, 1994).
Bosch, Mineke, and Annemarie Kloosterman, eds. *Politics and Friendship: Letters from the International Woman Suffrage Alliance, 1902–1942* (Columbus: Ohio State University Press, 1990).
Boustedt, Olaf. "Prof. Charlotte Lorenz zum 65. Geburtstag." In *Allgemeines Statistisches Archiv*, edited by Karl Wagner (Heidelberg: Physica, 1960), 319–21.
Bowie-Menzler, Marguerite. *Founders of Crosby Hall* (London: Privately published, 1981).
Brance, Alan. *The Cultural Legacy of the British Occupation in Germany* (Stuttgart: Heinz, 1997).
Braun, Lily. *Die Frauenfrage, ihre geschichtliche Entwicklung und wirtschaftliche Seite* (Leipzig: Europäischer Hochschulverlag, 1901).
Brehmer, Ilse, ed. *Frauen in der Geschichte IV:. "Wissen heisst Leben...": Beiträge zur Bildungsgeschichte von Frauen im 18. und 19. Jahrhundert* (Düsseldorf: Schwann, 1983).
Breuer, Stefan. "Der Neue Nationalismus in Weimar und seine Wurzeln." In *Mythos und Nation*, edited by Helmut Berding (Frankfurt/Main: Suhrkamp, 1996), 257–74.
Brewer, James W., and Martha K. Smith, eds. *Emmy Noether: A Tribute to Her Life and Work* (New York: Dekker, 1981).
Bridenthal, Renate. "Professional Housewives: Stepsisters of the Women's Movement." In *When Biology Became Destiny: Women in Weimar and Nazi Germany*, edited by Renate Bridenthal, Atina Grossmann, and Marion A. Kaplan (New York: Monthly Review Press, 1984), 153–73.
Brosin, Hans-Jürgen. "Charlotte Möller (1883–1973) und die gewässerkundlichen Arbeiten am Institut für Meereskunde Berlin," *Historisch-meereskundliches Jahrbuch* 6 (1999): 19–34.

Brünneck, Wiltraud von. "Die Aufgaben der Frau im Recht," *Frauenkultur im deutschen Frauenwerk* 3, no. 11 (1937): 9–10.

Bubenicek, Hanna. "Wissenschaftlerin auf Umwegen. Christine Touaillon, geb. Auspitz (1878–1928), Versuch einer Annäherung." In *Über Frauenleben, Männerwelt und Wissenschaft: Österreichische Texte zur Frauenforschung*, edited by Beate Frakele (Vienna: Verlag für Gesellschaftskritik, 1987), 5–17.

Buchheim, Christoph, ed. *German Industry in the Nazi Period* (Stuttgart: Steiner, 2008).

Budde, Gunilla. *Frauen der Intelligenz: Akademikerinnen in der DDR 1945 bis 1975* (Göttingen: Vandenhoeck & Ruprecht, 2003).

Budde, Gunilla. *Transnationale Geschichte: Themen, Tendenzen und Theorien* (Göttingen: Vandenhoeck & Ruprecht, 2006).

Burchardt, Anja. *Blaustrumpf, Modestudentin, Anarchistin? Deutsche und russische Medizinstudentinnen in Berlin, 1896–1918* (Stuttgart: Metzler, 1997).

Burr Margadant, Jo. *Madame le Professeur: Women Educators in the Third Republic* (Princeton, NJ: Princeton University Press, 1990).

Burstall, Sara Annie. *Impressions of American Education in 1908* (London: Longmans, 1909).

Bussemer, Ulrike. *Frauenemanzipation und Bildungsbürgertum: Sozialgeschichte der Frauenbewegung in der Reichsgründungszeit* (Weinheim: Beltz, 1985).

Caukin-Brunauer, Esther. "The German Frauenfront. 'Kinder, Küche, Kirche' and the Women's Organizations," *AAUW Journal* 27 (1934): 131–7.

Caukin-Brunauer, Esther. "The German Woman Marches—Back," *AAUW Journal* 27 (1934): 85–7.

Caukin-Brunauer, Esther. "The Women's Movement in Germany—Some Echos," *AAUW Journal* 29 (1935): 105–6.

Caukin-Brunauer, Esther. *The National Revolution in Germany, 1933* (Washington, DC: AAUW International Relations Office, 1933).

Charle, Christophe, Jürgen Schriewer, and Peter Wagner, eds. *Transnational Intellectual Networks: Forms of Academic Knowledge and the Search for Cultural Identities* (Frankfurt/Main: Campus, 2004).

Churchill White, Marian, *A History of Barnard College* (New York: Columbia University Press, 1954).

Cofer, Rebecca H. *The Straight Story: An Informal History of Willard Straight Hall, 1925–1990* (Ithaca, NY: Cornell University Press, 1990).

Cohen, Susan. "How the British Federation of University Women Helped Rebuild the Lives of Academic Refugee Women in the 1930s and 40s," presented on February 26, 2008, at the Centre for the History of Women's Education, University of Winchester. http://sybilcampbellcollection.org.uk/scc2/images/stories/downloads/BFUW-and-AcademicRefugeeWomen.pdf (accessed January 18, 2013).

Cohen, Susan. "'Now You See Them, Now You Don't': The Archives of the Refugee Committee of the British Federation of University Women." In *Refugee Archives: Theory and Practice*, edited by Andrea Hammel, Anthony Grenville, and Sharon Krummel (Amsterdam: Rodopi, 2008), 109–22.

Cohen, Willy. "Crosby Hall," *New York Times*, February 3, 1986.

Cohn, Werner. "Bearers of a Common Fate? The 'Non-Aryan' Christian 'Fate Comrades' of the Paulus-Bund, 1933–1939," *Leo Baeck Institute Year Book* 13 (1988): 327–68.

Collins, Elizabeth A. "Red-Baiting Public Women: Gender, Loyalty and Red Scare Politics." Diss., University of Illinois at Chicago, 2008, UMI No. 3345170.

Complete Report of the Chairman of the Committee on Public Information. 1917–1918–1919 (Washington, DC: Public Information Committee, 1920).

Conrad, Sebastian, and Jürgen Osterhammel, eds. *Das Kaiserreich transnational: Deutschland in der Welt 1871–1914* (Göttingen: Vandenhoeck & Ruprecht, 2006).

Conze, Vanessa. "'Unverheilte Brandwunden in der Außenhaut des deutschen Volkskörpers.' Der deutsche Grenzdiskurs in der Zwischenkriegszeit." In *Ordnungen in der Krise: Zur Politischen Kulturgeschichte Deutschlands, 1900–1933*, edited by Wolfgang Hardtwig (Munich: Oldenbourg, 2007), 21–48.

Cook, Blanche Wiesen. "Female Support Networks and Political Activism. Lillian Wald, Crystal Eastman, Emma Goldman." In *A Heritage of Her Own: Toward a New Social History of American Women*, edited by Nancy F. Cott and Elizabeth H. Pleck (New York: Simon and Schuster, 1979), 412–44.

Crawford, Elizabeth. "The Universe of International Science, 1880–1939." In *Solomon's House Revisited: The Organization and Institutionalization of Science*, edited by Tore Frängsmyr (Canton, MA: Science History Publications, 1990), 251–69.

Crawford, Elizabeth. *The Women's Suffrage Movement in Britain and Ireland: A Regional Survey* (London: Routledge, 2006).

Crawford Mitchell, Ruth. *Alice Garrigue Masaryk, 1879–1966: Her Life as Recorded in Her Own Words and by Her Friends* (Pittsburgh, PA: University Center for International Studies, 1980).

Dabernat, Nicole. "Marie-Louise Puech-Milhau, une femme engagée, 1876–1966," *Diplômées* 204 (March 2003): 17–30.

Dageförde, Astrid. "Frauen an der Hamburger Universität 1933–1945. Emanzipation oder Repression?" In *Hochschulalltag im "Dritten Reich": Die Hamburger Universität 1933–1945*, edited by Eckart Krause (Berlin: Reimer, 1991), 255–70.

Davies, Thomas Richard. "The Rise and Fall of Transnational Civil Society: The Evolution of International Non-Governmental Organizations Since 1839." City University Working Papers on Transnational Politics 003 (2008). www.staff.city.ac.uk/tom.davies/CUWPTP003.pdf (accessed January 18, 2013).

Deichmann, Ute. *Biologen unter Hitler: Porträt einer Wissenschaft im NS-Staat* (Frankfurt/Main: Fischer, 1995).

Deichmann, Ute. "Frauen in der Genetik. Forschung und Karrieren bis 1950." In *"Aller Männerkultur zum Trotz": Frauen in Mathematik und Naturwissenschaften*, edited by Renate Tobies (Frankfurt/Main: Campus, 1996): 221–51.

Delamont, Sara, ed. *A Women's Place in Education: Historical and Sociological Perspectives in Gender and Education* (Aldershot: Avebury, 1996).

Dennison, Mariea Caudill. "The American Girls' Club in Paris. The Propriety and Imprudence of Art Students, 1890–1914," *Woman's Art Journal* 26, no. 1 (2005): 32–7.

Desser, Michael. *Zwischen Skylla und Charybdis. Die "scientific community" der Physiker, 1919–1939* (Vienna: Böhlau, 1991).

Deutscher Akademikerinnenbund, ed. *Die Frauenfrage in Deutschland 1951–1960* (Cologne: Deutscher Akademikerinnenbund, 1961).

Deutscher Akademikerinnenbund, ed. *Die Frauenfrage in Deutschland, Bd. IV, 1961–65* (Cologne: Deutscher Akademikerinnenbund, 1966).

Deutscher Akademikerinnenbund, ed. *Die Frauenfrage in Deutschland, Bd. V, 1966–70* (Cologne: Deutscher Akademikerinnenbund, 1972).

Diethe, Carol. *The Life and Work of Germany's Founding Feminist, Louise Otto-Peters (1819–1895)*. (Lewiston, NY: Edwin Mellen Press, 2002).

Doenecke, Justus D. "Virginia Crocheron Gildersleeve." In *Women in World History*, edited by Anne Commire and Deborah Klezmer (Waterford, CT: Yorkin, 2000), 221–6.

Dohm, Hedwig. *Die wissenschaftliche Emancipation der Frau* (Berlin: Wedekind und Schwieger, 1874).

Donert, Celia. "Women's Rights in Postwar Europe: Disentangling Feminist Histories," *Past and Present* 218 (Supplement) (2013): 180–202.

Drummer, Heike, and Jutta Zwilling. "Elisabeth Schwarzhaupt. Eine Biographie." In *Elisabeth Schwarzhaupt (1901–1986). Portrait einer streitbaren Politikerin und Christin*, edited by Hessische Landesregierung (Freiburg: Herder, 2001), 14–115.

DuBois, Ellen Carol. *Feminism and Suffrage: The Emergence of an Independent Women's Movement in America, 1848–1869* (Ithaca, NY: Cornell University Press, 1999).

Duggan, Stephen, and Betty Drury. *The Rescue of Science and Learning: The Story of the Emergency Committee in Aid of Displaced Foreign Scholars* (New York: Macmillan, 1948).

Düwell, Kurt. "Hilfsorganisationen für deutsche Wissenschaftler im Ausland." In *Emigration: Deutsche Wissenschaftler nach 1933; Entlassung und Vertreibung*, edited by Herbert Arthur Strauss, Tilmann Buddensieg, and Kurt Düwell (Berlin: Technische Universität, 1987), viii–xxii.

Dyhouse, Carol. *No Distinction of Sex? Women in British Universities, 1870–1939* (London: UCL Press, 1995).

Eben-Servaes, Ilse. "Wissen ist uns Verpflichtung," *Frauenkultur im Deutschen Frauenwerk* 4, no. 2 (1938): 13.

Eckelmann, Christine. *Ärztinnen in der Weimarer Zeit und im Nationalsozialismus: Eine Untersuchung über den Bund Deutscher Ärztinnen* (Wermelskirchen: WFT Verlag für Wissenschaft, Forschung und Technik, 1992).

Elsen, Thierry, and Robert Tanzmeister. "In Sachen Elise und Helene Richter. Die Chronologie eines 'Bibliotheksverkaufs.'" In *Geraubte Bücher: Die Österreichische Nationalbibliothek stellt sich ihrer NS-Vergangenheit*, edited by Murray G. Hall, Christina Köstner, and Margot Werner (Vienna: Österreichische Nationalbibliothek, 2004), 128–38.

Emmons, Caroline S., ed. *Cold War and McCarthy Era: People and Perspectives* (Santa Barbara, CA: ABC-CLIO, 2010).

Engelmann, Susanne Charlotte. *German Education and Re-education* (New York: International University Press, 1945).

Eoff, Shirley M. *Viscountess Rhondda: Equalitarian Feminist* (Columbus: Ohio State University Press, 1991).

Erdmann, Rhoda. "Typ eines Ausbildungsgangs weiblicher Forscher." In *Führende Frauen Europas: In sechzehn Selbstschilderungen*, edited by Elga Kern (Munich: E. Reinhardt, 1928), 35–54.

Erhart, Hannelore. "Hanna Marie Margarete Jursch (1902–1972)." In *Biographisch-Bibliographisches Kirchenlexikon*, edited by Traugott Bautz (Herzberg: Verlag Traugott Bautz, 2000), 733–93.

Erichsen, Regine. "Vom Nationalsozialismus vertriebene Wissenschaftler auf dem Markt. Die Arbeitsvermittlung des englischen Academic Assistance Council (SPSL) am Beispiel von Türkeiemigranten," *Berichte zur Wissenschaftsgeschichte* 19, no. 4 (1997): 219–45.

Ericksen, Robert P. *Complicity in the Holocaust: Churches and Universities in Nazi Germany* (Cambridge: Cambridge University Press, 2012).

Evans, Richard J. *Comrades and Sisters: Feminism, Socialism, and Pacifism in Europe, 1870–1945* (Brighton, Sussex: Wheatsheaf Books, 1987).

Evans, Richard J. *The Feminist Movement in Germany 1894–1933* (London: Sage, 1976).

Falk, G. "The Reaction of Americans to the Persecution of the European Scholars during the Nazi Era," *International Review of History and Political Science* 9 (1972): 78–98.

Filla, Wilhelm, "Miss A. Levetus—Kunsthistorikerin und Volksbildnerin. Portrait einer grenzüberschreitenden Pionierin," *Spurensuche. Zeitschrift für Geschichte der Erwachsenenbildung und Wissenschaftspopularisierung* 12, no. 1 (2001): 24–39.

Finkelman, Paul. "The War on German Culture, 1917–1925." In *Confrontation and Cooperation: Germany and the United States in the Era of World War I, 1900–1924*, edited by Hans-Jürgen Schröder (Providence, RI: Berg, 1993), 177–206.

Fischer, Conan. *The Ruhr Crisis, 1923–1924* (Oxford: Oxford University Press, 2003).

Flexner, Eleanor, and Ellen F. Fitzpatrick. *Century of Struggle: The Woman's Rights Movement in the United States* (Cambridge, MA: Belknap Press of Harvard University Press, 1996).

Forgan, Sophie. "Bricks and Bones: Architecture and Science in Victorian Britain." In *The Architecture of Science*, edited by Peter Galison and Emily Thompson (Cambridge, MA: MIT Press, 1999), 181–212.

Forkl, Martha, and Bertholda Plechl. "Elise Richter und der Verband der Akademikerinnen Österreichs." In *Frauenstudium und akademische Frauenarbeit in Österreich*, edited by Martha Forkl and Elisabeth Koffmahn (Vienna: Braumüller, 1968), 108–15.

Fouché, Nicole. "Des Américaines protestantes à l'origine des 'University Women' françaises, 1919–1964," *Bulletin de la Société d'histoire du protestantisme français* 146 (2000): 133–52.

Fouché, Nicole. "Reid Hall, l'Association française des femmes diplômées des universités et la Fédération internationale des femmes diplômées des universités, 1919–1993," *Diplômées* 178 (1994): 190–200.

Frandsen, Dorothea. "Seit einem halben Jahrhundert aktiv. Der Deutsche Akademikerinnenbund," *Informationen für die Frau*, no. 1 (1978): 18–20.

Frei, Norbert. *Adenauer's Germany and the Nazi Past: The Politics of Amnesty and Integration*. Translated by Joel Golb (New York: Columbia University Press, 2002).

Freidenreich, Harriet Pass. *Female, Jewish, Educated: The Lives of Central European University Women* (Bloomington: Indiana University Press, 2002).

Fritz, Jan-Stefan. "Internationalism and the Promise of Science." In *Imperialism and Internationalism in the Discipline of International Relations*, edited by David Long and Brian C. Schmidt (Albany: State University of New York Press, 2005), 141–59.

Fuchs, Eckhardt. "Wissenschaftsinternationalismus in Kriegs- und Krisenzeiten. Zur Rolle der USA bei der Reorganisation der internationalen 'Scientific Community,' 1914–1925." In *Wissenschaft und Nation in der europäischen Geschichte*, edited by Ralph Jessen and Jakob Vogel (Frankfurt/Main: Campus, 2002), 263–84.

Fuchs, Eckhardt, and Matthias Schulze. "Globalisierung und transnationale Zivilgesellschaft in der Ära des Völkerbundes," *Zeitschrift für Geschichtswissenschaft* 54, no. 10 (2006): 837–40.

Fulbrook, Mary. *A History of Germany, 1918–2008: The Divided Nation* (Chichester, West Sussex: Wiley-Blackwell, 2011).

Gallmeister, Hildegard. "Die Studentin im akademischen Leben," *Die Frau* 37 (1929/30): 623–30.

Gandal, Keith. *The Gun and the Pen: Hemingway, Fitzgerald, Faulkner, and the Fiction of Mobilization* (Oxford: Oxford University Press, 2008).

Gassert, Philipp. *Coping With the Nazi Past: West German Debates on Nazism and Generational Conflict, 1955–1975* (New York: Berghahn, 2006).

Gerhard, Ute, and Ulla Wischermann. *Unerhört: Die Geschichte der deutschen Frauenbewegung* (Reinbek: Rowohlt, 1996).

Gienow-Hecht, Jessica C. E. *Sound Diplomacy: Music and Emotions in Transatlantic Relations, 1850–1920* (Chicago, IL: University of Chicago Press, 2009).

Gildersleeve, Virginia Crocheron. *Many a Good Crusade: Memoirs* (New York: Macmillan, 1954).

Glass, Meta. "Some Highlights of the Council Meeting," *AAUW Journal* 28, no. 2 (Winter 1934/35): 121–2.

Glauser, Andrea. *More Than a Watchdog: Marion Talbot und die Chicago Sociology* (Bern: Institut für Soziologie, 2003).

Goodman, Joyce. "International Citizenship and the International Federation of University Women before 1939," *History of Education* 40, no. 6 (2011): 701–21.

Goodman, Joyce. "Women and International Intellectual Co-operation," *Paedagogica Historica* 48, no. 3 (2012): 357–68.

Graham-Dixon, Francis. *The Allied Occupation of Germany: The Refugee Crisis, Denazification, and the Path to Reconstruction* (London: Tauris, 2013).

Greven-Aschoff, Barbara. *Die bürgerliche Frauenbewegung in Deutschland 1894–1933* (Göttingen: Vandenhoeck & Ruprecht, 1981).

Grossmann, Atina. "New Women in Exile. German Women Doctors and the Emigration." In *Between Sorrow and Strength: Women Refugees of the Nazi Period*, edited by Sibylle Quack (Cambridge, MA: Cambridge University Press, 1995), 215–38.
Grossmann, Atina. *Reforming Sex: The German Movement for Birth Control and Abortion Reform, 1920–1950* (New York: Oxford University Press, 1995).
Grüttner, Michael. *Studenten im Dritten Reich* (Paderborn: Schöningh, 1995).
Guttmann, Barbara. *Den weiblichen Einfluss geltend machen…Karlsruher Frauen in der Nachkriegszeit, 1945–1955* (Karlsruhe: Badenia Verlag, 2000).
Haas, Renate. "Caroline Spurgeon: English Studies, the United States, and Internationalism," *Studia Anglica Posnaniensia: International Review of English Studies* (2002): 1–15.
Hanna, Martha. "French Women and American Men. 'Foreign' Students at the University of Paris, 1915–1925," *French Historical Studies* 22, no. 1 (1999): 87–112.
Häntzschel, Hiltrud. "Geschlechterspezifische Aspekte." In *Handbuch der deutschsprachigen Emigration 1933–1945*, edited by Claus-Dieter Krohn (Darmstadt: Wissenschaftliche Buchgesellschaft, 1998), 101–16.
Häntzschel, Hiltrud. "Kritische Bemerkungen zur Erforschung der Wissenschaftsemigration unter geschlechterdiffenzierendem Blickwinkel," *Exilforschung* 14 (1996): 150–65.
Harrison, Brian. *Prudent Revolutionaries: Portraits of British Feminists between the Wars* (Oxford: Clarendon Press, 1987).
Harsch, Donna. "Public Continuity and Private Change? Women's Consciousness and Activity in Frankfurt, 1945–1955," *Journal of Social History* 27, no. 1 (1993): 29–58.
Harvey, Elizabeth. "The Failure of Feminism? Young Women and the Bourgeois Feminist Movement in Weimar Germany 1918–1933," *Central European History* 28, no. 1 (1995): 1–28.
Harvey, Elizabeth. "Seeing the World: Photography, Photojournalism, and Visual Pleasure in the Third Reich." In *Pleasure and Power in Nazi Germany*, edited by Pamela E. Swett, Corey Ross, and Fabrice d'Almeida (Basingstoke: Palgrave Macmillan, 2011), 177–204.
Heim, Susanne. *Plant Breeding and Agrarian Research in Kaiser-Wilhelm-Institutes, 1933–1945. Calories, Caoutchouc, Careers* (Dordrecht: Springer, 2004).
Heinsohn, Kirsten. *Politik und Geschlecht: Zur politischen Kultur bürgerlicher Frauenvereine in Hamburg* (Hamburg: Verein für Hamburgische Geschichte, 1997).
Hellpach, Willy. *Die Wesensgestalt der deutschen Schule* (Leipzig: Quelle & Meyer, 1925).
Herbst, Jürgen. *The German Historical School in American Scholarship: A Study in the Transfer of Culture* (Ithaca, NY: Cornell University Press, 1965).
Hernegger, Friedrich, and Berta Karlik. "Uranium in Sea-Water." *Meddelanden från Göteborgs Högskolas Oceanografiska Institution* 12 (1935).
Higgott, Gordon. "Helen Rosenau, 1900–1984," *Journal of Jewish Art* 11 (1985): 79–80.
Hillgruber, Andreas. "Unter dem Schatten von Versailles—Die außenpolitische Belastung der Weimarer Republik. Realität und Perzeption bei den Deutschen." In *Weimar: Selbstpreisgabe einer Republik. Eine Bilanz heute*, edited by Karl-Dietrich Erdmann and Hagen Schulze (Düsseldorf: Droste, 1980), 51–67.
Hirschfeld, Gerhard. "'The Defense of Learning and Science….' Der Academic Assistance Council in Großbritannien und die wissenschaftliche Emigration aus Nazi-Deutschland," *Exilforschung* 6 (1988): 28–45.
Hoffmann, Auguste. "Die Aufgaben der Ärztin im BDM," *Die Ärztin* 10, no. 2 (1934): 31–3.
Hoffmann, Auguste. "Olympiastreiflichter," *Die Ärztin* 12 (1936): 188–92.
Hoffrath, Christiane. *Bücherspuren. Das Schicksal von Elise und Helene Richter und ihrer Bibliothek im Dritten Reich* (Vienna: Böhlau, 2009).
Hollihan, Thomas A. "Propagandizing in the Interest of War. A Rhetorical Study of the Committee on Public Information," *Southern Speech Communication Journal* 49 (1984): 241–57.

Hoppe, Brigitte. "Die Institutionalisierung der Zellforschung durch Rhoda Erdmann (1870–1935)," *Biologie heute: Beilagen zur Zeitschrift Naturwissenschaftliche Rundschau* 366, no. 7 (1989): 2–4, 9.
Horne, John, and Alan Kramer. *German Atrocities, 1914: A History of Denial* (New Haven, CT: Yale University Press, 2001).
Horowitz, Helen Lefkowitz. *Alma Mater: Design and Experience in the Women's Colleges from Their Nineteenth-Century Beginnings to the 1930s* (New York: Knopf, 1984).
Huerkamp, Claudia. *Bildungsbürgerinnen: Frauen im Studium und in akademischen Berufen 1900–1945* (Göttingen: Vandenhoeck & Ruprecht, 1996).
Humbert, Gabriele. "Der DAB," *Die Studentin* 2, nos. 11/12 (1925): 81–3.
Hünemörder, Christian. "Biologie und Rassenbiologie 1933 bis 1945." In *Hochschulalltag im "Dritten Reich." Die Hamburger Universität 1933–1945*, edited by Eckart Krause (Berlin: Reimer, 1991), 1155–1187.
Hunt, Felicity, and Carol Barker. *Women at Cambridge: A Brief History* (Cambridge: University of Cambridge Publications Office, 1998).
IFUW. *International Glossary of Academic Terms* (Paris: The Federation, 1939).
IFUW. *A List of International Fellowships* (London: The Federation, 1928).
IFUW. *A List of International Fellowships for Research* (London: The Federation, 1930).
IFUW. *A List of International Fellowships for Research* (London: The Federation, 1934).
IFUW. *Report of the First Conference, London 1920* (London: The Federation, 1920).
IFUW. *Report of the Second Conference, Paris 1922* (London: The Federation, 1922).
IFUW. *Report of the Fourth Conference, Amsterdam 1926* (London: The Federation, 1926).
IFUW. *Report of the Fifth Conference, Geneva 1929* (London: The Federation, 1929).
IFUW. *Report of the Sixth Conference, Edinburgh 1932* (London: The Federation, 1932).
IFUW. *Report of the Seventh Conference, Cracow 1936* (London: The Federation, 1936).
IFUW. *Report of the Eighth Conference, Stockholm 1939* (London: The Federation, 1939).
IFUW. *Report of the Ninth Council Meeting, Brussels 1925* (London: The Federation, 1925).
IFUW. *Report of the Eleventh Council Meeting, Vienna 1927* (London: The Federation, 1927).
IFUW. *Report of the Twelfth Council Meeting, Madrid 1928* (London: The Federation, 1928).
IFUW. *Report of the Fifteenth Council Meeting, Prague 1930* (London: The Federation, 1930).
IFUW. *Report of the Nineteenth Council Meeting, Budapest 1934* (London: The Federation, 1934).
IFUW. *Report of the Twenty-Third Council Meeting, London 1938* (London: The Federation, 1938).
IFUW. *Report of the 31st Council Meeting, Oosterbeek (Holland) 1951* (London: The Federation, 1951).
IFUW. *Report of the 52nd and 53rd Councils, Karlsruhe 1968* (London: The Federation, 1968).
Iriye, Akira. *Cultural Internationalism and World Order* (Baltimore, MD: Johns Hopkins University Press, 2000).
Iriye, Akira. *Global Community: The Role of International Organizations in the Making of the Contemporary World* (Berkeley: University of California Press, 2002).
Iriye, Akira, and Pierre-Yves Saunier. *The Palgrave Dictionary of Transnational History: From the Mid-19th Century to the Present Day* (New York: Palgrave Macmillan, 2009).
James, Henry. "Refugees in Chelsea." In *Within the Rim, and Other Essays 1914–1915* (London: W. Collins, 1918), 39–62.
Jarausch, Konrad H. *After Hitler. Recivilizing Germans, 1945–1999* (Oxford: Oxford University Press, 2006).
Jensen, Kimberly. *Mobilizing Minerva: American Women in the First World War* (Urbana: University of Illinois Press, 2008).

Jensen, Kimberly. *Oregon's Doctor to the World: Esther Pohl Lovejoy and a Life in Activism* (Seattle: University of Washington Press, 2010).

Jensen, Kimberly, and Erika Kuhlman, eds. *Women and Transnational Activism in Historical Perspective* (Dordrecht: Republic of Letters Publishing, 2010).

Julian, Maureen M. "Women in Crystallography." In *Women of Science: Righting the Record*, edited by Gabriele Kass-Simon, Deborah Nash, and Patricia Farnes (Bloomington: Indiana University Press, 1990), 335–83.

Jung, Gertrud. "Dies academicus der Frau," *Die Frau* 37 (1929/30): 611–14.

Kahn, Albert. *1860–1940. Réalités d'une Utopie [Exposition, Musée Départemental Albert-Kahn, Boulogne, 28 novembre 1995–18 septembre 1996]* (Boulogne: Musée Albert-Kahn, 1995).

Kansteiner, Wulf. *In Pursuit of German Memory: History, Television, and Politics after Auschwitz* (Athens: Ohio University Press, 2006).

Kater, Michael. "Medizin und Mediziner im Dritten Reich. Eine Bestandsaufnahme," *Historische Zeitschrift* 244 (1987): 299–352.

Kater, Michael. *The Nazi Party: A Social Profile of Members and Leaders, 1919–1945* (Cambridge, MA: Harvard University Press, 1983).

Kauffmann, C. M. "Adelheid Heimann (1903–93)," *Burlington Magazine* (October 1993): 694.

Keith, Jeanette. *Rich Man's War, Poor Man's Fight: Race, Class, and Power in the Rural South during the First World War* (Chapel Hill: University of North Carolina Press, 2004).

Kempf, Rosa. "Aussprache. Lokale oder fakultätsweise Zusammenfassung der Akademikerinnen?" *Die Frau* 32 (1924/25): 120.

Kennedy, David M. *Over Here: The First World War and American Society*, 25th Anniversary ed. (Oxford: Oxford University Press, 2004).

Kennedy, Kathleen. *Disloyal Mothers and Scurrilous Citizens: Women and Subversion during World War I* (Bloomington: Indiana University Press, 1999).

Klieneberger-Nobel, Emmy. *Memoirs* (London: Academic Press, 1980).

Kluger, Richard. *The Paper: The Life and Death of the New York Herald Tribune* (New York: Knopf, 1986).

Knaggs, Isabel Ellie, and Berta Karlik. *Tables of Cubic Crystal Structures of Elements and Compounds* (London: Hilger, 1932).

Koch, Sabine. "Leben und Werk der Zellforscherin Rhoda Erdmann (1870–1935)." PhD diss., University of Marburg, 1985.

Kocka, Jürgen. "Zivilgesellschaft als historisches Problem und Versprechen." In *Europäische Zivilgesellschaft in Ost und West: Begriff, Geschichte, Chancen*, edited by Manfred Hildermeier, Jürgen Kocka, and Christoph Conrad (Frankfurt/Main: Campus, 2000), 13–39.

Kohlstedt, Sally Gregory, and Helen Longino. "The Women, Gender, and Science Question. What Do Research on Women in Science and Research on Gender and Science Have to Do with Each Other?" In *Women, Gender, and Science. New Directions*, edited by Sally Gregory Kohlstedt and Helen E. Longino (Chicago, IL: University of Chicago Press, 1997), 3–15.

Koonz, Claudia. *Mothers in the Fatherland: Women, the Family, and Nazi Politics* (London: Methuen, 1987).

Kottenhoff, Anna. "Arbeit und Ziel der Hochschulgemeinschaft Deutscher Frauen," *Die Ärztin* 17, no. 4 (1941): 157–61.

Kottenhoff, Anna. "Die Deutsche Frau in Lehre und Forschung. Dozentinnentagung in der Reichsfrauenführung," *Frauenkultur im Deutschen Frauenwerk* 4, no. 2 (1938): 2–11.

Krohn, Claus-Dieter. "Nobody Has the Right to Come to the United States," *Exilforschung* 3 (1985): 127–42.

Krohn, Claus-Dieter. "Vereinigte Staaten von Amerika." In *Handbuch der deutschsprachigen Emigration 1933–1945*, edited by Claus-Dieter Krohn, Patrick von zur Mühlen,

Gerhard Paul, and Lutz Winckler (Darmstadt: Wissenschaftliche Buchgesellschaft, 1998): 446–66.

Krumeich, Gerd, and Joachim Schröder. *Der Schatten des Weltkriegs: Die Ruhrbesetzung 1923* (Essen: Klartext, 2004).

Kuchel, Gertrud. "Die deutsche Akademikerin und der Internationale Verband Akademischer Frauen," *Die Studentin* 1, nos. 2/3 (1924): 28–30.

Kuenssberg, Katharina Freifrau von. *Vom langsamen Aufstieg der Frau: Dr. Katharina Freifrau von Kuenssberg erzählt aus ihrem Leben*. Edited by Katharina Holger (Heidelberg: DAB, 1973).

Kuhlo, Ursula. "Das Referat Ärztinnen," *Die Ärztin* 16, no. 5 (1940): 113–5.

Kushner, Tony. "An Alien Occupation—Jewish Refugees and Domestic Service in Britain." In *Second Chance: Two Centuries of German-Speaking Jews in the United Kingdom*, edited by Werner E. Mosse (Tübingen: J. C. B. Mohr/Paul Siebeck, 1991), 553–78.

Lange, Helene. "Internationale Tagungen von Akademikerinnen," *Die Frau* 31 (1923/24): 365–6.

Lange, Helene. *Lebenserinnerungen* (Berlin: Herbig, 1921/30).

Lange, Helene. "Nachtrag zur Erklärung," *Die Frau* 26 (1918): 38.

Lange, Helene. "Zur Bedeutung des Frauenstudiums (1915)," reproduced in *Die Frau* 44 (1937): 589.

Laqua, Daniel, ed. *Internationalism Reconfigured: Transnational Ideas and Movements between the World Wars* (London: Tauris, 2011).

Lässig, Simone. *Jüdische Wege ins Bürgertum: Kulturelles Kapital und sozialer Aufstieg im 19. Jahrhundert* (Göttingen: Vandenhoeck & Ruprecht, 2004).

Latzin, Ellen. *Lernen von Amerika? Das US-Kulturaustauschprogramm für Bayern und seine Absolventen* (Stuttgart: Steiner, 2005).

Laville, Helen. *Cold War Women: The International Activities of American Women's Organizations* (Manchester: Manchester University Press, 2002).

Le Cinquantenaire de l'École Normale Supérieure de Sèvres, 1881–1931 (Paris: Printory, 1932).

Lemcke, Ruth, and L. H. Collier. "Obituary Notice, Emmy Klieneberger-Nobel, 1892–1985," *Journal of Medical Microbiology* 22 (1986): 183–95.

Leubuscher, Charlotte. "Die Berufslage der deutschen Hochschuldozentinnen," *Die Frau* 33 (1925/26): 669–73.

Levin, Miriam R. *Defining Women's Scientific Enterprise: Mount Holyoke Faculty and the Rise of American Science* (Hanover, NH: University Press of New England, 2004).

Levine, David O. *The American College and the Culture of Aspiration, 1915–1940*, 3rd ed. (Ithaca, NY: Cornell University Press, 1986).

Levine, Robert M. *Tropical Diaspora: The Jewish Experience in Cuba* (Gainesville: University Press of Florida, 1993).

Levine, Susan. *Degrees of Equality: The American Association of University Women and the Challenge of Twentieth-Century Feminism* (Philadelphia, PA: Temple University Press, 1995).

Liefmann, Martha, and Else Liefmann. *Helle Lichter auf dunklem Grund: Erinnerungen von Martha Liefmann und Else Liefmann* (Bern: Christliches Verlagshaus, 1966).

Linnemann, Kai Arne. "Die Wiederkehr des akademischen Bürgers. Herrmann Nohl und die Pädagogik der Sittlichkeit." In *Akademische Vergangenheitspolitik: Beiträge zur Wissenschaftskultur der Nachkriegszeit*, edited by Bernd Weisbrod (Göttingen: Wallstein, 2002), 167–89.

Lintner, Karl. "Nachruf auf Berta Karlik," *Almanach der Österreichischen Akademie der Wissenschaften* 140 (1989/90): 309.

Little, Charles E. "The Italians and Their Schools," *Peabody Journal of Education* 10, no. 2 (1932): 77–86.

Livingstone, David N. *Putting Science in Its Place: Geographies of Scientific Knowledge* (Chicago, IL: University of Chicago Press, 2003).

Lohschelder, Britta. *"Die Knäbin mit dem Doktortitel": Akademikerinnen in der Weimarer Republik* (Paffenweiler: Centaurus, 1994).

London, Louise. "British Immigration Control Procedures and Jewish Refugees, 1933–1939." In *Second Chance: Two Centuries of German-Speaking Jews in the United Kingdom*, edited by Werner E. Mosse (Tübingen: J. C. B. Mohr/Paul Siebeck, 1991), 485–518.

London, Louise. *Whitehall and the Jews, 1933–1948: British Immigration Policy, Jewish Refugees, and the Holocaust* (New York: Cambridge University Press, 2000).

Lorenz, Charlotte. "Wieviel Frauen studieren in Deutschland?" *Die Frau* 34 (1926/27): 239–41.

Lüders, Marie-Elisabeth. *Fürchte Dich nicht: Persönliches und Politisches aus mehr als 80 Jahren, 1878–1962* (Cologne: Westdeutscher Verlag, 1963).

Lüders, Marie-Elisabeth. "Die geistige Internationale der Frauen," *Vossische Zeitung*, August 12, 1926.

Lüders, Marie-Elisabeth. *Das unbekannte Heer: Frauen kämpfen für Deutschland, 1914–1918* (Berlin: Mittler, 1936).

Lühe, Irmela von der, and Claus-Dieter Krohn, eds. *Fremdes Heimatland: Remigration und literarisches Leben nach 1945* (Göttingen: Wallstein, 2005).

Lühe, Irmela von der, Axel Schildt, and Stefanie Schüler-Springorum, eds. *"Auch in Deutschland waren wir nicht wirklich zu Hause": Jüdische Remigration nach 1945* (Göttingen: Wallstein, 2008).

Lürssen, Johanna. "Emmy Beckmann im Deutschen Akademikerinnenbund. Emmy Beckmann zum 75. Geburtstag," *Mädchenbildung und Frauenschaffen* 5, no. 4 (1955): 156–9.

Lykknes, Annette, Lise Kvittengen, and Anne Kristine Børresen. "Ellen Gleditsch: Duty and Responsibility in a Research and Teaching Career, 1916–1946," *Historical Studies in the Physical Sciences* 36, no. 1 (2005): 131–88.

Mallinson, Vernon. "Some Sources for the History of Education in Belgium," *British Journal of Educational Studies* 4, no. 1 (1955): 62–70.

Maltby, Margaret E. *History of the Fellowships Awarded by the American Association of University Women, 1888–1929, with the Vitas of the Fellows* (Washington, DC: American Association of University Women, 1929).

Marggraf, Stefanie. "Sonderkonditionen. Habilitationen von Frauen in der Weimarer Republik und im Nationalsozialismus an den Universitäten Berlin und Jena," *Feministische Studien. Zeitschrift für interdisziplinäre Frauen- und Geschlechterforschung* 20, no. 2 (2002): 40–56.

Maul, Bärbel. *Akademikerinnen der Nachkriegszeit: Ein Vergleich zwischen der Bundesrepublik Deutschland und der DDR* (Frankfurt/Main: Campus, 2002).

Maurer, Trude, ed. *Der Weg an die Universität: Höhere Frauenstudien vom Mittelalter bis zum 20. Jahrhundert* (Göttingen: Wallstein, 2010).

Maushart, Marie-Ann. *"Um mich nicht zu vergessen": Hertha Sponer—Ein Frauenleben für die Physik im 20. Jahrhundert* (Bassum: Verlag für Geschichte der Naturwissenschaften und der Technik, 1997).

Mayeur, Françoise. *L'Enseignement secondaire des jeunes filles sous la Troisième République* (Paris: Presses de la Fondation nationale des sciences politiques, 1977).

Mazon, Patricia M. *Gender and the Modern Research University: The Admission of Women to German Higher Education, 1865–1914* (Stanford, CA: Stanford University Press, 2003).

McCarthy, Kathleen. *American Creed: Philanthropy and the Rise of Civil Society 1700–1865* (Chicago, IL: University of Chicago Press, 2003).

McFadden, Margaret H., *Golden Cables of Sympathy: The Transatlantic Sources of Nineteenth-Century Feminism* (Lexington: University of Kentucky Press, 1999).

Mecenseffy, Margarethe. "Philipp IV. von Spanien und seine Heirat mit Maria Anna von Österreich." In *Historische Studien: A.F. Pribram zum 70. Geburtstag dargebracht* (Vienna: Steyermühl, 1929), 49–70.

Meier-Scherling, Anne-Gudrun. "Die Benachteiligung der Juristin zwischen 1933 und 1945," *Deutsche Richterzeitung* (January 1975): 10–13.
Meitner, Lise. "Über einige Probleme in der Ausnützung der Kernenergie," *Mädchenbildung und Frauenschaffen* 8, no. 12 (1958): 533–40.
Melchior, Margot. "Deutschlands Studentinnenheime," *Die Frau* 36 (1928/29): 586–92.
Metzler, Gabriele. *Internationale Wissenschaft und nationale Kultur: Deutsche Physiker in der internationalen Community, 1900–1960* (Göttingen: Vandenhoeck & Ruprecht, 2000).
Metzler, Gabriele. "Nationalismus und Internationalismus in der Physik des 20. Jahrhunderts." In *Wissenschaft und Nation in der europäischen Geschichte*, edited by Ralph Jessen and Jakob Vogel (Frankfurt/Main: Campus, 2002), 285–309.
Meyer, Alfred G. *The Feminism and Socialism of Lily Braun, 1865–1916* (Urbana: Indiana University Press, 1986).
Michels, Eckard. "Deutsch als Weltsprache? Franz Thierfelder, the Deutsche Akademie in Munich and the Promotion of Language Abroad, 1923–1945," *German History* 22, no. 2 (2004): 206–28.
Miller, Carol. "'Geneva—the Key to Equality': Inter-War Feminists and the League of Nations," *Women's History Review* 3, no. 2 (1994): 219–45.
Mock, James R., and Cedric Larson. *Words That Won the War: The Story of the Committee on Public Information, 1917–1919* (Princeton, NJ: Princeton University Press, 1939).
Mommsen, Hans. *Germans against Hitler: The Stauffenberg Plot and Resistance under the Third Reich* (London: Tauris, 2009).
Monheim, Maria. "Bericht über die Stimmungslage im Deutschen Ärztinnenbund nach der Gleichschaltung," *Die Ärztin* 9, no. 6 (1933): 117–18.
Nagler, Jörg. *Nationale Minoritäten im Krieg: "Feindliche Ausländer" und die amerikanische Heimatfront während des Ersten Weltkriegs* (Hamburg: Hamburger Edition, 2000).
Nazarska, Georgeta. "The Bulgarian Association of University Women, 1924–1950," *Aspasia* 1 (2007): 153–75.
Nolan, Mary. *The Transatlantic Century: Europe and America, 1890–2010* (Cambridge: Cambridge University Press, 2012).
Oberkofler, Gerhard. *Käthe Spiegel: Aus dem Leben einer altösterreichischen Historikerin und Frauenrechtlerin in Prag* (Innsbruck: Studienverlag, 2005).
Oberkofler, Gerhard, and Eduard Rabofsky. *Studien zur Geschichte der österreichischen Wissenschaft zwischen Krieg und Frieden* (Vienna: Edition Fortschrittliche Wissenschaft, 1987).
Oertzen, Christine von. "Fräuleingeschichten." In *Geschichte in Geschichten: Ein historisches Lesebuch*, edited by Barbara Duden et al. (Frankfurt/Main: Campus, 2003), 70–7.
Oertzen, Christine von. "Luise Berthold. Hochschulleben und Hochschulpolitik zwischen 1909 und 1957," *Feministische Studien. Zeitschrift für interdisziplinäre Frauen- und Geschlechterforschung* 20, no. 1 (2002): 8–22.
Oertzen, Christine von. "Rückblick aus der Emigration. Die Akademikerinnen Erna Barschak (1888–1958), Susanne Engelmann (1885–1963) und Lucie Adelsberger (1895–1971)." In *Erinnerungskartelle: Zur Konstruktion von Autobiografien nach 1945*, edited by Angelika Schaser (Bochum: Winkler, 2003), 169–95.
Oertzen, Christine von. *Strategie Verständigung: Zur transnationalen Vernetzung von Akademikerinnen, 1917–1955* (Göttingen: Wallstein, 2012).
Oertzen, Christine von. "Was ist Diskriminierung? Westdeutsche Professorinnen ringen um ein hochschulpolitisches Konzept, 1949–1989." In *Zeitgeschichte als Geschlechtergeschichte. Perspektiven auf die Bundesrepublik (1945–1980)*, edited by Julia Paulus et al. (Frankfurt/Main: Campus, 2012), 101–17.
Offen, Karen. *European Feminisms, 1700–1950: A Political History* (Stanford, CA: Stanford University Press, 2000).

Offen, Karen."The Second Sex and the Baccalauréat in Republican France," *French Historical Studies* 13, no. 2 (1983): 252–86.
Omerod, Margaret. "An International Problem," *BFUW News Sheet* no. 11 (October 1933): 5–6.
Omerod, Margaret. "An International Problem (Continued)," *BFUW News Sheet* no. 12 (January 1934): 6–7.
Opitz, Claudia, and Elke Kleinau, eds. *Geschichte der Mädchen- und Frauenbildung*, vol. 2 (Frankfurt/Main: Campus, 1996).
Pauwels, Jacques. *Women, Nazis, and Universities: Female University Students in the Third Reich, 1933–1945* (Westport, CT: Greenwood Press, 1984).
Pinar, Susana. "La introducción de la genética en España durante el primer tercio del siglo XX," *Llull* 22 (1999): 453–73.
Pitavy-Souques, Daniele. "Reid Hall. Petite Histoire d'une maison d'excellence," *Diplômées* 223 (December 2007).
Quack, Sibylle, ed. *Between Sorrow and Strength: Women Refugees of the Nazi Period* (Cambridge, MA: Cambridge University Press, 1995).
Quack, Sibylle. *Zuflucht Amerika: Zur Sozialgeschichte der Emigration deutsch-jüdischer Frauen in die USA, 1933–1945* (Bonn: Dietz, 1995).
Rathgen, Irmgard. "Die Kopenhagener Tagung des Internationalen Frauenbundes," *Die Frau* 31 (1923/24): 311–13.
Rathgen, Irmgard. "Meinungsaustausch. Zur Frage des Internationalen Bundes Akademischer Frauen," *Die Studentin* 1, no. 12 (1924): 44–5.
Rathgen, Irmgard. "Praktische Aufgaben des deutschen Akademikerinnenbundes," *Die Studentin* 2, nos. 7/8 (1925): 57–8.
Rathgen, Irmgard. "Vom gemeinsamen Aufbau." In *Dritte Generation: Für Gertrud Bäumer*, edited by Hilde Lion, Irmgard Rathgen, and Else Ulig-Beil (Berlin: F. A. Herbig, 1923), 86–8.
Rausch, Helke. "US-Amerikanische 'Scientific Philanthropy' in Frankreich, Deutschland und Großbritannien zwischen den Weltkriegen," *Geschichte und Gesellschaft* 33 (2007): 73–98.
Reid, Helen Dwight. *The Postwar Treatment of Germany: A Study Guide* (Washington, DC: American Association of University Women, 1945).
Reining, Elke. "Rosa Kempf (1874–1948). Ihr Kampf für die Rechte der Frauen," *Zeitschrift für bayerische Landesgeschichte* 64 (2001): 149–68.
Rentetzi, Maria. "Gender and Radioactivity Research in Interwar Vienna. The Case of the Institute for Radium Research." In *Women Scholars and Institutions: Proceedings of the International Conference (Prague, June 8–11, 2003)*, edited by Soňa Štrbáňová, Ida H. Stamhuis, and Katarina Mojsejová (Prague: Archiv Akad. Věd České Republiky v Nakl. Arenga, 2004), 611–38.
Rentetzi, Maria. *Trafficking Materials and Gendered Experimental Practices* (New York: Columbia University Press, 2007).
Report of the British Educational Mission (Manchester: Morris & Yeaman, 1919).
Richter, Elise. *Summe des Lebens, hg. vom Verband der Akademikerinnen Österreichs* (Vienna: WUV Universitätsverlag, 1997).
Robinson, Annabel. *The Life and Work of Jane Ellen Harrison* (Oxford: Oxford University Press, 2002).
Röder, Werner, and Herbert A. Strauss, eds. *Biographisches Handbuch der deutschsprachigen Emigration nach 1933 = International Biographical Dictionary of Central European Emigrés 1933–1945* (Munich: Saur, 1983).
Rogers, Annie Mary Anne Henley. *Degrees by Degrees: The Story of the Admission of Oxford Women Students to Membership of the University* (London: Oxford University Press, 1938).

Rosenberg, Emily S. *Spreading the American Dream: American Economic and Cultural Expansion, 1890–1945* (New York: Hill and Wang, 1982).
Rossiter, Margaret W. *Women Scientists in America*. Vol. 1, *Struggles and Strategies, 1870–1940* (Baltimore, MD: Johns Hopkins University Press, 1982).
Rossiter, Margaret W. *Women Scientists in America*. Vol. 2, *Before Affirmative Action, 1940–1972* (Baltimore, MD: Johns Hopkins University Press, 1995).
Rossiter, Margaret W. *Women Scientists in America*. Vol. 3, *Forging a New World Since 1972* (Baltimore, MD: Johns Hopkins University Press, 2012).
Rothbarth, Margarete. "Die deutschen Gelehrten und die internationalen Wissenschaftsorganisationen." In *Volkstum und Kulturpolitik: Eine Sammlung von Aufsätzen, gewidmet Georg Schreiber zum 50. Geburtstage*, edited by Heinrich Konen and Johann Peter Steffes (Cologne: Gilde-Verlag, 1932), 143–57.
Rothbarth, Margarete. "Internationale geistige Zusammenarbeit." In *Wörterbuch des Völkerrechts und der Diplomatie*, vol. 3 (Berlin: Vereinigung Wissenschaftlicher Verleger, 1928), 1–12.
Röwekamp, Marion. *Die ersten deutschen Juristinnen: Eine Geschichte ihrer Professionalisierung und Emanzipation (1900–1945)* (Cologne: Böhlau, 2011).
Rupp, Leila J. "Women's History in the New Millennium: Carroll Smith-Rosenberg's 'The Female World of Love and Ritual' after 25 Years," *Journal of Women's History* 12, no. 3 (2000): 8–9.
Rupp, Leila J. *Worlds of Women: The Making of an International Women's Movement* (Princeton, NJ: Princeton University Press, 1997).
Russell, Alys. "Colloquial English for the Foreigner," *University Women's Review*, no. 40 (July 1950): 5–6.
Russell, Bertrand, and Alys Russell. *German Social Democracy: Six Lectures; With an App. on Social Democracy and the Woman Question in Germany* (London: Longmans, Green, 1896).
Ryberg, Ewa. "Hanna Rydh—Förmedlare av Förhistorien," *Fornvännen* 85 (1990): 303–9.
Sach, Louisa. "'Gedenke, dass du eine deutsche Frau bist.' Die Ärztin und Bevölkerungspolitikerin Ilse Szagunn (1887–1971)." PhD diss., University of Berlin (Charité Universitätsmedizin), 2006.
Sachse, Carola. "Persilscheinkultur. Zum Umgang mit der NS-Vergangenheit in der Kaiser-Wilhelm-Gesellschaft." In *Akademische Vergangenheitspolitik: Beiträge zur Wissenschaftskultur der Nachkriegszeit*, edited by Bernd Weisbrod (Göttingen: Wallstein, 2002), 217–46.
Salomon, Alice. "Elisabeth Altmann-Gottheiner zum Gedächtnis," *Die Österreicherin* 9 (1930): 4.
Sandell, Marie. "'Truly International'? The International Federation of University Women's Quest for Expansion in the Interwar Period," *History of Education Researcher* 82 (2008): 74–83.
Satzinger, Helga. *Differenz und Vererbung: Geschlechterordnungen in der Genetik und Hormonforschung* (Vienna: Böhlau, 2009).
Satzinger, Helga. "Women's Places in the New Laboratories of Genetic Research in Early 20th Century: Gender, Work, and the Dynamics of Science." In *Women Scholars and Institutions: Proceedings of the International Conference (Prague, June 8–11, 2003)*, edited by Soňa Štrbáňová, Ida H. Stamhuis, and Katarina Mojsejová (Prague: Archiv Akad. Věd České Republiky v Nakl. Arenga, 2004), 265–94.
Saurer, Edith. "Frauenbewegung und soziale Netzwerke. Kommentar." In *Das Jahrhundert des Feminismus: Streifzüge durch nationale und internationale Bewegungen und Theorien*, edited by Anja Weckwert and Ulla Wischermann (Königstein: Helmer, 2006), 77–94.
Sawahn, Anke. "Sie half, wo sie konnte: Doktorin Else Wex (1884–1961). Fünf Jahrzehnte Fraueninteressen, Sozialengagement und Politik," *Cellesche Zeitung*, January 30, 1999, 42.

Schaser, Angelika. *Frauenbewegung in Deutschland 1848–1933* (Darmstadt: WBC Wissenschaftliche Buchgesellschaft, 2006).

Schaser, Angelika. *Helene Lange und Gertrud Bäumer: Eine politische Lebensgemeinschaft* (Cologne: Böhlau, 2000).

Schaser, Angelika. "Die 'undankbaren' Studentinnen. Studierende Frauen in der Weimarer Republik." In *Frauen auf dem Weg zur Elite*, edited by Günther Schulz (Munich: Boldt, 2000), 97–116.

Scheich, Elvira. "Elisabeth Schliemann (1881–1972). Patriotin im Zwiespalt." In *Autarkie und Ostexpansion: Pflanzenzucht und Agrarforschung im Nationalsozialismus*, edited by Susanne Heim (Göttingen: Wallstein, 2002), 250–79.

Scheich, Elvira. "Science, Politics, and Morality. The Relationship of Lise Meitner and Elisabeth Schiemann." In *Women, Gender, and Science. New Directions*, edited by Sally Gregory Kohlstedt and Helen E. Longino (Chicago, IL: University of Chicago Press, 1997), 143–68.

Scherer, Paul. *Lord John Russell: A Biography* (Selinsgrove/London: Susquehanna University Press/Associated University Presses, 1999).

Schmidt-Waldherr, Hiltraud. *Emanzipation durch Professionalisierung? Politische Strategien und Konflikte innerhalb der bürgerlichen Frauenbewegung während der Weimarer Republik und des Nationalsozialismus* (Frankfurt/Main: Materialis, 1987).

Schneider, Hans-Peter. "Im Namen des Menschen. Über Leben und Wirken einer großen Richterin." In *Verfassung und Verantwortung: Gesammelte Schriften und Sondervoten; Wiltraut Rupp-von Brünneck*, edited by Hans-Peter Schneider (Baden-Baden: Nomos, 1983), 1–46.

Scholtz-Klink, Gertrud. *Einsatz der Frau in der Nation: Frauenkundgebung, Reichsparteitag der Arbeit 1937*. Edited by Deutsches Frauenwerk, Nachrichtendienst der Reichsfrauenführerin (Berlin: NSDAP, 1937), 9.

Schorn, Maria. "Frau und Hochschule," *Frauenkultur im Deutschen Frauenwerk* 1, no. 11 (1935): 1–2.

Schrecker, Ellen. *Many Are the Crimes: McCarthyism in America* (Princeton, NJ: Princeton University Press, 1999).

Schröder, Iris. "Soziale Frauenarbeit als bürgerliches Projekt. Differenz, Gleichheit und weiblicher Bürgersinn in der Frauenbewegung um 1900." In *Wege zur Geschichte des Bürgertums: Vierzehn Beiträge*, edited by Klaus Tenfelde and Ulrich Wehler (Göttingen: Vandenhoeck & Ruprecht, 1994), 209–30.

Schröder, Iris. *Arbeiten für eine bessere Welt: Frauenbewegung und Sozialreform 1890–1914* (Frankfurt/Main: Campus, 2001).

Seidler, Eduard. "Lucie Adelsberger, 1945–1971," in *Auschwitz: Ein Tatsachenbericht. Das Vermächtnis der Opfer für uns Juden und für alle Menschen*. Edited by Eduard Seidler (Bonn: Bouvier, 2001), 107–35.

Sime, Ruth Lewin. *Lise Meitner: A Life in Physics* (Berkeley: University of California Press, 1996).

Simon, Helene. "Geistige Arbeiter," *Die Frau* 28 (1920/21): 44–9.

Simpson, Renate. *How the PhD Came to Britain: A Century of Struggle for Postgraduate Education* (Guildford: Society for Research into Higher Education, 1983).

Smith, Susan Margot. "Ada Comstock Notestein. Educator." In *Women of Minnesota: Selected Biographical Essays*, edited by Barbara Stuhler and Gretchen Kreuter (St. Paul: Minnesota Historical Society Press, 1998), 208–25.

Smith-Rosenberg, Carroll. "The Female World of Love and Ritual: Relations between Women in Nineteenth-Century America," *Signs* 1 (1975): 1–29.

Snider, Christy Jo. "Creating a Transnational Identity: The International Federation of University Women Confronts Racial and Religious Membership Restrictions in the 1930s." In *Women and Transnational Activism in Historical Perspective*, edited by Kimberly Jensen and Erika Kuhlman (Dordrecht: Republic of Letters Publishing, 2010), 193–218.

Solomon, Barbara Miller. *In the Company of Educated Women: A History of Higher Education in America* (New Haven, CT: Yale University Press, 1985).
Sondheimer, J. H. *History of the British Federation of University Women, 1907–1957* (London: British Federation of University Women, 1957).
Spencer, Stephanie. "Just a Book in the Library? The Sybill Campbell Collection Fostering International Friendship amongst Graduate Women," *History of Education: Journal of the History of Education Society* 42, no. 2 (2013): 257–74.
Spiegel, Käthe. "Charakterzüge der amerikanischen Geschichte," *Historische Vierteljahreszeitschrift* 28 (1934): 119–41.
Spiegel, Käthe. *Kulturgeschichtliche Grundlagen der amerikanischen Revolution* (Munich: Oldenbourg, 1931).
Spiegel, Käthe. "Die Prager Juden zur Zeit des dreißigjährigen Krieges." In *Die Juden in Prag. Bilder aus ihrer tausendjährigen Geschichte. Festausgabe der Loge Praga des Ordens B'nai B'rith zum Gedenktage ihres 25jährigen Bestandes*, edited by Samuel Steinherz (Prague: Die Bücherstube, 1927), 107–86.
Spiegel, Käthe. *Wilhelm Egon von Fürstenbergs Gefangenschaft und ihre Bedeutung für die Friedensfrage 1674–1679* (Bonn: Röhrscheid, 1936).
Stage, Sarah, and Virginia Bramble Vincenti, eds. *Rethinking Home Economics: Women and the History of a Profession* (Ithaca, NY: Cornell University Press, 1997).
Stamhuis, Ida, and Arne Monsen. "Kristine Bonnevie, Tina Tammes and Elisabeth Schiemann in Early Genetics: Emerging Chances for a University Career for Women," *Journal of the History of Biology* 40, no. 3 (2007): 427–66.
Stephenson, Jill. *Women in Nazi Germany* (Harlow: Longman, 2001).
Stoehr, Irene. "Bedingt beglückend: Die internationale Kooperation der deutschen Frauenbewegung vor 1933," In *Über Grenzen hinweg: Zur Geschichte der Frauenstimmrechtsbewegung und zur Problematik der transnationalen Beziehungen in der deutschen Frauenbewegung*, edited by Irina Hundt (Berlin: Deutscher Staatsbürgerinnenverband e.V., 2007), 128–41.
Stoehr, Irene. "Frauenerwerbsarbeit als Kriegsfall. Marie Elisabeth Lüders: Variationen eines Lebensthemas." In *Frauen arbeiten: Weibliche Erwerbstätigkeit in Ost- und Westdeutschland nach 1945*, edited by Gunilla-Friederike Budde (Göttingen: Vandenhoeck & Ruprecht, 1997), 62–77.
Stoehr, Irene. "Traditionsbewußter Neuanfang. Zur Organisation der Frauenbewegung in Berlin, 1945–1950." In *Frauenpolitik und politisches Wirken von Frauen im Berlin der Nachkriegszeit, 1945–1949*, edited by Renate Genth et al. (Berlin: Trafo, 1996), 193–223.
Stolleis, Michael. *The Law under the Swastika: Studies on Legal History in Nazi Germany* (Chicago, IL: University of Chicago Press, 1998).
Stoß, Irma. "Tagung des Deutschen Akademikerinnenbundes e.V. in Stein bei Nürnberg v. 28.–30.7.1951," *Mädchenbildung und Frauenschaffen* 1, no. 9 (1951): 1–2.
Strauss, Herbert A. "The Migration of Academic Intellectuals." In *Biographisches Handbuch der deutschsprachigen Emigration nach 1933 = International Biographical Dictionary of Central European Emigrés 1933–1945*, edited by Werner Röder and Herbert A. Strauss (Munich: Saur, 1980), vol. 1, lxvii–lxxxv.
Strupp, Christoph. "Perceptions of German Science and Research in the USA during World War I." In *Science, Technology, and Society in Russia and Germany during the First World War*, edited by Eduard Kolchinsky, Dietrich Beyrau, and Julia Lajus (St. Petersburg: Nestor Istorija, 2007), 114–35.
Studentenwerk Berlin, ed. *Ein Jahr Studentinnentagesheim (Helene-Lange-Heim), 1928–1929* (Berlin: Studentenwerk, 1929).
Sveistrup, Hans, and Agnes von Zahn-Harnack, eds. *Die Frauenfrage in Deutschland: Strömungen u. Gegenströmungen 1790–1930. Sachl. geordn. u. erl. Quellenkunde* (Burg: Hopfer, 1934).

Szagunn, Ilse. "Crosby Hall, Ein internationales Clubhaus für Akademikerinnen," *Die Frau* 34 (1926/27): 720–1.
Talbot, Marion, and Lois Kimball Mathews Rosenberry. *The History of the American Association of University Women, 1881–1931* (Boston, MA: Houghton & Mifflin, 1931).
Tanzmeister, Robert. *Elise Richter—Frau und Wissenschaftlerin*. Edited by Verband der Akademikerinnen Österreichs (Vienna: Wiener Universitätsverlag, 1998).
Thalheim, Rita. "Jewish Women Exiled in France after 1933." In *Between Sorrow and Strength: Women Refugees of the Nazi Period*, edited by Sibylle Quack (Cambridge, MA: Cambridge University Press, 1995), 51–62.
Thane, Pat, and Gisela Bock, eds. *Women and the Rise of the European Welfare State, 1880s–1950s* (London: Routledge, 1991).
Thelin, John R. *A History of American Higher Education* (Baltimore, MD: Johns Hopkins University Press, 2004).
Timm, Angelika. "Zur Biographie jüdischer Hochschullehrerinnen in Berlin bis 1933. Nach Materialien des Archivs der Humboldt-Universität zu Berlin," *Tel Aviver Jahrbuch für deutsche Geschichte* 21 (1992): 243–58.
Touaillon, Christine. *Der deutsche Frauenroman des 18. Jahrhunderts* (Vienna: Braumüller, 1919).
Tröger, Annemarie. "Die Frau im wesensgemäßen Einsatz." In *Mutterkreuz und Arbeitsbuch. Zur Geschichte der Frauen in der Weimarer Republik und im Nationalsozialismus*, edited by Frauengruppe Faschismusforschung (Frankfurt/Main: Fischer, 1981), 246–72.
Tröndle-Weintritt, Isolde. "Die Freiburger Gruppe des Deutschen Akademikerinnenbundes (DAB) und die Freiburger Akademikerinnen (FA)." In *"Nun gehen Sie hin und heiraten Sie!" Die Töchter der Alma Mater im 20. Jahrhundert*, edited by Isolde Tröndle-Weintritt and Petra Herkert (Freiburg: Kore, 1997), 374–85.
Tröndle-Weintritt, Isolde. "Dr. phil. Magda Staudinger, geb. 1902." In *"Nun gehen Sie hin und heiraten Sie!" Die Töchter der Alma Mater im 20. Jahrhundert*, edited by Isolde Tröndle-Weintritt and Petra Herkert (Freiburg: Kore, 1997), 78–94.
Tscharntke, Denise. *Re-educating German Women: The Work of the Women's Affairs Section of British Military Government 1946–1951* (Frankfurt/Main: Lang, 2003).
Turner, Ian D., ed. *Reconstruction in Post-War Germany: British Occupation Policy and the Western Zones, 1945–1955* (Oxford: Berg, 1989).
Unfried, Berthold, Jürgen Mittag, and Marcel van der Linden, eds. *Transnationale Netzwerke im 20. Jahrhundert historische Erkundungen zu Ideen und Praktiken, Individuen und Organisationen* (Leipzig: Akademische Verlags-Anstalt, 2008).
Unger, Martha. "Die Frau im Nationalsozialistischen Deutschland," *NS-Frauenkorrespondenz* 17 (April 1933).
Unterberger, Betty Miller. "Esther Delia Caukin Brunauer." In *Notable American Women: The Modern Period; A Biographical Dictionary*, edited by Barbara Sicherman and Carol Hurd Green (Cambridge, MA: Harvard University Press, 1980), 114–16.
Vaughn, Stephen L. *Holding Fast the Inner Lines. Democracy, Nationalism, and the Committee on Public Information* (Chapel Hill: University of North Carolina Press, 1980).
Velsen, Dorothee von. "Deutsche Interessen auf internationalen Frauentagungen," *Die Frau* 30 (1922/23): 43–6.
Velsen, Dorothee von. *Im Alter die Fülle: Erinnerungen* (Tübingen: Wunderlich, 1956).
Velsen, Dorothee von. "Voraussetzungen und Möglichkeiten internationaler Arbeit," *Die Frau* 29 (1922): 353–9.
Vickery, Margaret Birney. *Buildings for Bluestockings: The Architecture and Social History of Women's Colleges in Late Victorian England* (Newark: University of Delaware Press, 1999).
Vincenz, Bettina. *Biederfrauen oder Vorkämpferinnen? Der Schweizerische Verband der Akademikerinnen (SVA) in der Zwischenkriegszeit* (Baden: hier + jetzt, 2011).

Vogt, Annette. *Elsa Neumann: Berlins erstes Fräulein Doktor* (Berlin: Verlag für Wissenschafts- und Regionalgeschichte, 1999).
Vogt, Annette. *Vom Hintereingang zum Hauptportal? Lise Meitner und ihre Kolleginnen an der Berliner Universität und in der Kaiser-Wilhelm-Gesellschaft* (Stuttgart: Steiner, 2007).
Vogt, Annette. "Von Prag in die 'neue Welt': Die Wege der Chemikerin Gertrud Kornfeld." In *1933 Circuli 2003*, edited by Jana Nekvasilová (Prague: National Technical Museum, 2003), 281–97.
Vogt, Annette. *Wissenschaftlerinnen in Kaiser-Wilhelm-Instituten, A-Z* (Berlin: Archiv der Max-Planck-Gesellschaft, 2008).
Voris, Jacqueline van. *Carrie Chapman Catt: A Public Life* (New York: Feminist Press, 1996).
Walker, Linda. *The Women's Movement in Britain, 1790–1945* (London: Routledge, 2006).
Wallace, Stuart. *War and the Image of Germany: British Academics, 1914–1918* (Edinburgh: Donald, 1988).
Wassermann, Ursula. *I Was an American* (London: Bodley Head, 1955).
Weber, Marianne. "Vom Typenwandel der studierten Frau". In *Frauenfragen und Frauengedanken: Gesammelte Aufsätze* (Tübingen: Mohr, 1919 [1917]), 179–200.
Weisbrod, Bernd. "Dem wandelbaren Geist. Akademisches Ideal und wissenschaftliche Transformation in der Nachkriegszeit." In *Akademische Vergangenheitspolitik*, edited by Bernd Weisbrod (Göttingen: Wallstein, 2002), 11–35.
Wendland, Ulrike, ed. *Biographisches Handbuch deutschsprachiger Kunsthistoriker im Exil: Leben und Werk der unter dem Nationalsozialismus verfolgten und vertriebenen Wissenschaftler* (Munich: Saur, 1999).
Werner, Anja. *The Transatlantic World of Higher Education: Americans at German Universities, 1776–1914* (London: Berghahn, 2013).
Werner, Josef. *Hakenkreuz und Judenstern. Das Schicksal der Karlsruher Juden im Dritten Reich* (Karlsruhe: Badenia-Verlag, 1988).
Westerdyk, Johanna. "Help the German University Women!" *BFUW News Sheet* no. 11 (October 1933): 4–5.
Wetterer, Angelika. "Ausschließende Einschließung—marginalisierende Integration. Gechlechterkonstruktionen in Professionalisierungsprozessen." In *Vielfältige Verschiedenheiten: Geschlechterverhältnisse in Studium, Hochschule und Beruf*, edited by Ayla Neusel and Angelika Wetterer (Frankfurt/Main: Campus, 1999), 223–54.
Wierling, Dorothee. *Eine Familie im Krieg: Leben, Sterben und Schreiben 1914–1918* (Göttingen: Wallstein, 2013).
Wierling, Dorothee. "Studentinnen in der Weimarer Republik." In *Bürgerliche Gesellschaft in Deutschland: Historische Einblicke, Fragen, Perspektiven*, edited by Lutz Niethammer (Frankfurt/Main: Fischer, 1990), 364–82.
Wigmore, John Henry. *Science and Learning in France, with a Survey of Opportunities for American Students in French Universities: An Appreciation by American Students* (Chicago, IL: Society for American Fellowships in French Universities, 1917).
Willich, Johanna. "Das internationale Akademikerinnen-Heim. Crosby Hall," *Die Studentin* 4, no. 4 (1927/28): 54.
Winnewisser, Brenda P. "The Emigration of Hedwig Kohn, Physicist, 1940," *Mitteilungen der Österreichischen Gesellschaft für Wissenschaftsgeschichte* 18 (1998): 41–58.
Wittke, Carl Frederick. *The Anti-German Hysteria of World War One* (Munich: Saur, 1994).
Wöbse, Anna-Katharina. "'To Cultivate the International Mind.' Der Völkerbund und die Förderung der globalen Zivilgesellschaft," *Zeitschrift für Geschichtswissenschaft* 54, no. 10 (2006): 837–40.
"Women of Attainment: Dr. Gertrude Kornfeld," *Museum Service*. Edited by Rochester Museum of Arts and Sciences, March/April 1939: 65

Wood, Molly M. "'Commanding Beauty' and 'Gentle Charm': American Women and Gender in the Early Twentieth-Century Foreign Service," *Diplomatic History* 31, no. 3 (2007): 505–30.

Wyman, David. *Paper Walls: America and the Refugee Crisis 1938–1941* (Amherst: University of Massachusetts Press, 1968).

Yasutake, Rumi. *Transnational Women's Activism: The United States, Japan, and Japanese Immigrant Communities in California, 1859–1920* (New York: New York University Press, 2004).

Zahn-Harnack, Agnes von. "Ein Jahrzehnt nach Kriegsausbruch." In *Agnes von Zahn-Harnack: Schriften und Reden, 1914–1950*, edited by Marga Anders and Ilse Reicke (Tübingen: Hopfer, 1964), 80–5.

Zahn-Harnack, Agnes von. "From a Berlin Hospital," *AAUW Journal* 42, no. 2 (1948): 124–5.

Zahn-Harnack, Agnes von. "Geschichte des Deutschen Akademikerinnenbundes, 1926–1933." In *Agnes von Zahn-Harnack: Schriften und Reden, 1914–1950*, edited by Marga Anders and Ilse Reicke (Tübingen: Hopfer, 1964), 1–8.

Zahn-Harnack, Agnes von. "Letter from Germany," *AAUW Journal* 42, no. 4 (1948/49): 205–8.

Zahn-Harnack, Agnes von. "Margarete Fürstin Andronikow-Wrangell," *Die Frau* 39 (1931/32): 571–3.

Zahn-Harnack, Agnes von. "Die vierte Tagung des Internationalen Akademikerinnenbundes," *Die Frau* 33 (1925/26): 732–4.

Zepp, Marianne. *Redefining Germany: Reeducation, Staatsbürgerschaft und Frauenpolitik im US-amerikanisch besetzten Nachkriegsdeutschland* (Göttingen: V&R Unipress, 2007).

Zilius-Falkenberg, Waltraut. "Die Frau in der wissenschaftlichen Forschung." In *Das weite Wirkungsfeld. Frauenschaffen in Deutschland*, edited by Reichsfrauenführung, Hauptabteilung Presse/Propaganda (Berlin: Reichsfrauenführung, 1941), 112–26.

Zimmermann, Susan. "The Challenge of Multinational Empire for the International Women's Movement: The Habsburg Monarchy and the Development of Feminist Inter/National Politics," *Journal of Women's History* 17, no. 2 (2005): 87–117.

Index

GPSR Compliance
The European Union's (EU) General Product Safety Regulation (GPSR) is a set of rules that requires consumer products to be safe and our obligations to ensure this.

If you have any concerns about our products, you can contact us on

ProductSafety@springernature.com

In case Publisher is established outside the EU, the EU authorized representative is:

Springer Nature Customer Service Center GmbH
Europaplatz 3
69115 Heidelberg, Germany

www.ingramcontent.com/pod-product-compliance
Ingram Content Group UK Ltd.
Pitfield, Milton Keynes, MK11 3LW, UK
UKHW020226160926
475607UK00004B/108
* 9 7 8 1 1 3 7 4 3 8 8 8 1 *